Viscoelasticity of Engineering Materials

Viscoelasticity of Engineering Materials

Y. M. Haddad

University of Ottawa
Ottawa, Canada

 **SPRINGER-SCIENCE+BUSINESS MEDIA, B.V.**

First edition 1995

© 1995 Springer Science+Business Media Dordrecht
Originally published by Chapman & Hall in 1995
Softcover reprint of the hardcover 1st edition 1995

Typeset in 10/12 Palatino by Techset Composition Ltd., Salisbury, Wilts.

ISBN 978-94-010-4555-1   ISBN 978-94-011-1272-7 (eBook)
DOI 10.1007/978-94-011-1272-7

A catalogue record for this book is available from the British Library

Library of Congress Catalog Card Number: 94–68787

ⓐ Printed on permanent acid-free text paper, manufactured in accordance with
   ANSI/NISO Z39.48-1992 and ANSI/NISO Z39.48-1984 (Permanence of Paper).

To the loving memory of my parents

Contents

Contents

Preface

I express my full indebtedness to all researchers whose work is referenced in this book. Without their outstanding contributions to knowledge, this book would not have been written.

I convey my thanks to Professor D. R. Axelrad (McGill University), who was the first person to introduce the fascinating subject of rheology to me and to Professor J. T. Pindera (University of Waterloo) for his kind encouragement and stimulating discussions on the subject matter. I am indebted to Dr J. H. Gittus, Editor- in-Chief of *Res Mechanica*, for originally inviting me to write a book on viscoelasticity.

Permission granted to the author for the reproduction of figures and/or data by the following scientific societies, journals and publishers is gratefully acknowledged: Academic Press, American Chemical Society, American Institute of Physics, British Textile Technology Group, Elsevier Applied Science Publishers, Gebrüder Borntraeger, *Helvetica Chimica Acta*, Hermann, International Union of Crystallography, John Wiley & Sons, Pergamon Press, Springer-Verlag Heidelberg, Steinkopff Verlag, *Tappi Journal*, Taylor and Francis Ltd., and the Institute of Physics. In the same context, the author wishes to express his sincere thanks and gratitude to Professors M. F. Ashby (University of Cambridge, United Kingdom), N. Davis (The Pennsylvania State University), H. F. Frost (Thayer School of Engineering), F. A. Leckie (University of Illinois at Urbana-Champaigne), E. H. Lee (Stanford University), J. M. Morrison (AT & T Bell Laboratories), A. K. Mukherjee (University of California, Davis) and Dr H. J. Sutherland (Sandia National Laboratories).

I wish to thank Mrs Denise Champion-Demers for her excellent efforts in the efficient execution of the word processing of the manuscript. I would also like to extend my thanks to past and present graduate students, Mrs Ping Yu and Messrs S. Tanary, W. Zhao, S. Iyer and G. Molina for their conscientious assistance during the preparation of the text.

I also wish to express sincere thanks and deep appreciation to Dr Philip Hastings (Senior Editor), the sub editorial staff at Chapman & Hall, to David Norris (Copy Editor) and to Chapman & Hall for the reviewing, editing and the efficient production and distribution of the text.

I am grateful to my wife Dawn and daughter Leila for their understanding, patience and support.

I hope that the work presented in this book will provide guidance to science and engineering students, educators and researchers who are working in the field. Also, it is hoped that the book will be of value to scientists and engineers who are involved in the production and processing of viscoelastic materials and the study of their properties.

Y. M. Haddad
University of Ottawa, Ottawa, Canada

List of symbols

$a(t)$	Amplitude of wave front (time dependent)
$a_T, a_G(T)$	Temperature shift factor
A	Energy per unit mass
A_0	Mean free energy
$\mathbf{A}$	Displacement vector of a particle along the plane of the wave
$^{\alpha\beta}A$	Junction area between two overlapping fibres α and β
c_1, c_2	Magnitude of wave velocity (dilatational, rotational)
$\mathbf{d}(t)$	Interfibre bond displacement
$H(\cdot)$	Heaviside step function
dH	Heat per unit mass
dS	Line element
$D(\cdot)/Dt$	Material derivative
e	Specific internal energy (per unit mass)
$\mathbf{e}_k \ (k = 1, 2, 3)$	Unit vectors associated with an external Cartesian frame of reference
E	Elastic modulus
$E_1(\omega)$	Storage modulus (frequency dependent)
$E_2(\omega)$	Loss modulus (frequency dependent)
$E^*(i\omega)$	Dynamic complex modulus
$\mathbf{E}(t)$	Nonlinear strain measure
$E_E(\varepsilon)$	Equilibrium tangent modulus (strain dependent)
$\check{E}_E(\varepsilon)$	Equilibrium second-order modulus (strain dependent)
$E_I(\varepsilon)$	Instantaneous tangent modulus (strain dependent)
$\check{E}_I(\varepsilon)$	Instantaneous second-order modulus (strain dependent)
$\mathbf{f}(\cdot)$	Constitutive functional
$f(\lambda)$	Retardation spectrum
$\mathbf{F}$	Deformation measure
$F(t)$	Creep (compliance) function
$F_{ijkl}(t)$	Creep tensorial function

$\mathbf{F}_d(t)$	Delayed compliance function
$\mathbf{F}_e(0^+)$	Elastic compliance function
$\mathbf{F}_v(t)$	Viscous compliance function
$F_1(t)$, $F_2(t)$	Creep function (shear, dilatation)
$\mathbf{g}$	Deformation gradient
$\mathbf{g}(T)$	Temperature gradient
$\mathbf{h}$	Heat flux vector
$\mathbf{I}$	Identity matrix
I_1, I_2, I_3	Stress invariants
II_1, II_2, II_3	Strain invariants
$\mathbf{J}(t)$	Shear creep compliance (time dependent)
$J_1(\omega)$	Storage compliance (frequency dependent)
$J_2(\omega)$	Loss compliance (frequency dependent)
$J^*(i\omega)$	Dynamic creep compliance
$k = E/E'$	Ratio of elastic moduli E and E'
K	Bulk modulus
$K(t)$	Geometric function (time dependent)
$\mathbf{L}(\cdot)$	Constitutive function
$\mathbf{L}$	Velocity gradient
$\mathbf{n}$	Unit normal vector
$\overset{1}{\mathbf{n}}, \overset{2}{\mathbf{n}}, \overset{3}{\mathbf{n}}$	Unit vectors associated with the stress tensor principal axes
$N(s)$, $N'(s)$	Frequency distribution (creep, relaxation)
$p = d/dt$	Time derivative operator
p, q	Probabilities
$\boldsymbol{q}$	Heat flux vector
P, Q	Linear differential operators
$\mathbf{P}, \mathbf{Q}$	Nonlinear measures of input and output
$\mathbf{r}(t), \mathbf{R}$	Position vector (deformed, undeformed state)
$\mathbf{R}$	Orthogonal motion indicating rigid rotation
$\mathbf{R}(t)$, $R_{ijkl}(t)$	Stress relaxation function
$R_1(t)$, $R_2(t)$	Stress-relaxation function (shear, dilatation)
R_∞	Equilibrium (relaxed) modulus
$R'(t)$	Time derivative of the relaxation function $R(t)$
$s = 1/\lambda$	Frequency
s	Laplace transform parameter
S	Specific entropy
$S\text{–}S, S_1\text{–}S_1, S_2\text{–}S_2$	Scanning lines
t, τ	Time parameter
T	Absolute temperature
$\mathbf{T}$	Stress traction vector
$\mathbf{u}$	Displacement vector
$\mathbf{u}_I$	Irrotational displacement field

$\mathbf{u}_R$	Rotational displacement field
$\mathbf{U}$	Positive definite symmetric matrix indicating pure stretch
$\mathbf{v}$	Velocity
v_{ij}	Rate of deformation tensor
$V(t)$	Velocity magnitude of a nonlinear wave
$\mathbf{V}$	Positive definite symmetric matrix indicating pure rotation
W	Elastic energy per stress cycle
$\mathbf{x}, \mathbf{X}$	Position vector (deformed, undeformed state)
$\hat{\mathbf{x}}(t)$	Spatial position of a propagating wave at time t
Z_i	Cartesian components of a heat flux vector (per unit area per unit time)
B.	Refers to an intermolecular bond
j.	Refers to a junction area between two microelements
$\cdot_R$	Denotes recovery
f.	Refers to a single fibre or a fibre-segment
$(\cdot)^T$	Designates a transpose
$[\cdot]$	Indicates a jump in a function across the trajectory of a wave at a particular instant of time
$\bar{\cdot}$	Indicates a transform
$\|\cdot\|$	Denotes a norm
α	Angle of incidence of a dilatational wave; coefficient of linear thermal expansion; propagation constant
α, β	Normalization factors
α, β	Structural microelements
γ	Specific rate (per unit mass) of entropy production
$\gamma_m \ (m = 1, 2, \cdots)$	Exponent factors
Γ	Time rate of entropy production
Γ_1^2, Γ_2^2	Wave operator (dilatational, rotational)
$\mathbf{\Gamma}(t)$	Material operator (time dependent)
$\Gamma(\lambda)$	Relaxation spectrum
δ	Phase lag
$\delta(t)$	Depth of penetration (time dependent)
$\Delta = \varepsilon_{kk} = \varepsilon_{11} + \varepsilon_{22} + \varepsilon_{33}$	Dilatation
δ, Δ	Interfibre bond length (deformed, undeformed state)
∇^2	Laplace's operator
$\delta_{ij}, \delta_{\alpha\beta}$	Kronecker delta
$\varepsilon_{ijk}, \varepsilon_{\alpha\beta}$	Alternating tensor
ε_{ij}	Infinitesimal strain tensor
ε'_{ij}	Deviatoric strain tensor
η	Viscosity modulus
θ	Empirical temperature
κ	Morse constant
λ	Relaxation time

List of symbols

$\lambda = \eta/\eta'$	Ratio of viscosity coefficients η and η'
$\hat{\lambda}$	Critical strain gradient
Λ	Rate of dissipation of energy
ν	Poisson's ratio
ξ	Microstress
ξ, ξ'	Reduced time parameters
Ξ	Total energy
Π	Constitutive functional
ρ, ρ_0	Mass density (current, reference configuration)
$\boldsymbol{\sigma}, \sigma_{ij}$	Cauchy stress tensor
σ'_{ij}	Stress deviator
$\bar{\sigma}$	Mean stress
$\sigma_1, \sigma_2, \sigma_3$	Principal components of a stress
σ_1	Instantaneous value of the stress
Σ	Piola–Kirchhoff stress tensor
$\phi_1(\cdot), \phi_2(\cdot), \cdots$	Material functions
χ	Body force
$\psi_1(\cdot), \psi_2(\cdot), \cdots$	Material functions
ψ	Energy per unit mass
$\boldsymbol{\psi}, \boldsymbol{\psi}', \boldsymbol{\psi}''$	Nonlinear material functions
$\psi_{ij}(t)$	Relaxation function (time dependent)
$\varnothing$	Null set

Introduction

With the recent advances in material science and the parallel extensive industrial demands on advanced industrial materials such as high polymers and polymeric base composite systems, the subject of viscoelasticity has gained recently a strong momentum in the realms of engineering techniques and applications.

High polymeric materials are organic substances of high molecular weight, the technical importance of which depends on their particular microstructure (e.g. Leaderman, 1943; Bernal, 1958). This class of materials may include, for example, rubber in its various forms, synthetic rubber-like materials, commercial plastics and natural and synthetic textile fibres. Other examples of a viscoelastic material would include a wide range of inorganic polymeric systems such as silicones and glass resins, constituents of polymeric base systems, natural fibres such as wood and the byproducts of such fibres as, for instance, paper and board, building materials such as concrete, and a large class of biomaterials, among others. As will be demonstrated in this book, these materials are time dependent in response and possess a 'time memory'.

In the mechanics of deformable media, the response behaviour of an elastic solid is dealt with within the realm of the classical theory of elasticity. The most direct description of such response is in accordance with the well-known Hooke's law (Robert Hooke, *De Potentia Restitutiva*, London, 1678). This law forms the basis of the mathematical theory of elasticity (the reader is referred to Love (1944) for an introductory review of the history of the mathematical theory of elasticity). That is, provided that the occurring deformations are small, the stress is considered to be directly proportional to the strain and it is independent of the strain rate. Such a response is termed consequently as 'perfectly elastic' or 'Hookean'.

In a simple uniaxial test, the load–deformation curve of the perfectly elastic solid will follow the same path for both increasing and decreasing load. Thus, the material test specimen will regain its original dimensions instantaneously on removal of the load. Under a constant level of loading, the occurring deformation is constant, i.e. time independent. Further, when such a solid is subjected to a sinusoidally oscillating loading, the deformation will also be found to be sinusoidal and practically in phase with the load. All the energy is stored and recovered in each cycle.

On the other hand, the mechanical response of a viscous fluid is dealt with within the domain of the classical theory of fluid dynamics. In this case, the most direct description of the response is in accordance with Newton's law whereby the stress is considered to be proportional to the occurring rate of strain but independent of the strain itself. This is provided that the rate of strain is small. When a 'Newtonian viscous fluid' is subjected to a sinusoidally oscillating load, the deformation will be found to be $90°$ out of phase with the load.

The classical theories of linear elasticity and Newtonian fluids, though impressively well structured, do not adequately describe the response behaviour and flow of most real materials. That is, between the above two described responses of the elastic solid and the viscous fluid, a real material may exhibit. even if both strain and strain rate are infinitesimal, the combined response characteristics of these two media. Attempts to characterize the behaviour of such real materials under the action of external loading, consequently, gave rise to the science of rheology within which the phenomenon now labelled 'viscoelasticity' is well defined and intended to convey mechanical behaviour combining response characteristics of both an elastic solid and a viscous fluid. A viscoelastic material is, thus, characterized by a certain level of rigidity of an elastic solid body, but, at the same time, it flows and dissipates energy by frictional losses as a viscous fluid. A few characteristics of a viscoelastic material may be cited as follows.

- When a viscoelastic material is subjected to a constant stress, it does not hold a constant deformation (as it would be the case for a solid material), but it continues to flow with time, i.e. it creeps. Immediately, on removal of the load, the specimen is found to have taken an amount of 'residual' strain the magnitude of which depends on the length of time for which the load is applied and on the level of loading. Following removal of the load, a noticeable reduction in the amount of residual strain gradually takes place with the passage of time. This residual strain may even disappear entirely in the course of time. The latter phenomenon which occurs following the removal of the load is referred to as 'creep recovery'. A specimen of viscoelastic material, tested as mentioned above, eventually regains its original dimensions. Consequently, the creep of such material under load cannot be regarded as a phenomenon of plasticity, as in the case of polycrystalline solids, but rather as a 'delayed elasticity' (Leaderman, 1943). Boltzmann (1874) denoted this property by the term 'elastic aftereffect' (*Nachwirkung* in German). In a simple uniaxial test of a viscoelastic material, a load–deformation loop (hysteresis) is obtained, i.e. the descending load curve corresponds to a larger amount of strain than the ascending load curve. Neither of the two curves is completely linear. The shape of the resulting hysteresis looop is dependent on the magnitude of load, rate of application and removal of the load and temperature.
- Further, when a viscoelastic material is subjected to sinusoidally oscillating stress, the resulting strain, through sinusoidal, is neither exactly in phase with the stress (as it would be the case for a perfectly elastic solid) nor $90°$ out of phase (as it would be for a perfectly viscous fluid); it is somewhere between. The magnitude

of the strain and the phase angle between the stress and strain are generally frequency and temperature dependent. On loading and unloading a viscoelastic material specimen, some of the energy input is stored and recovered in each cycle and some of it is dissipated as heat.

The particular nature of viscoelastic response considered in the foregoing proves the existence of a property of 'passive resistance' in such materials. This is in contrast to the instantaneous response and reversibility that usually characterize pure elastic behaviour. This passive resistance is of viscous nature and reflects what is usually called the property of 'hereditary response' of the material. That is, the present state of response depends not only on the present state of loading input but also on previous states (Boltzmann, 1874, 1877, 1878). This property is revealed experimentally in different time-dependent phenomena pertaining to the viscoelastic response such as creep, stress relaxation and intrinsic attenuation and dispersion of propagating waves.

The phenomenological theory of viscoelasticity dates from the nineteenth century (Leaderman (1943) and Markovitz (1977) present interesting reviews of the history of viscoelasticity) but, unfortunately, the application of the theory to actual engineering applications is only a development of the last 50 years. This contrasts with the situation of the related field of linear elasticity whereby technological requirements have traditionally stimulated significant research over the last two centuries. Such technological stimulus was lacking for the development of a formal theory of viscoelasticity as engineering design has traditionally made use of materials whose mechanical response behaviour would be adequately described by the laws of classical elasticity. Research in the theory of viscoelasticity has been, however, recently enhanced by the introduction of engineering components that are fabricated from advanced industrial materials such as those mentioned at the beginning of this introduction. A large class of these newly developed materials exhibit, as mentioned earlier, mechanical response behaviour outside the scopes of the more conventional theories of linear elasticity and viscosity. Consequently, a development of the theory of viscoelasticity has become of parallel necessity with the gradual introduction of such new materials.

It is often considered that the response behaviour of a viscoelastic material is a fundamental property of its molecular structure. Hence, the viscoelastic response prediction of polymeric systems (references which focus on polymeric materials include, for instance, Eirich (1956), Staverman and Schwarzl (1956), Ferry (1970) and Doi and Edwards (1986), among others), for instance, has been often considered from the point of view of reaction-rate theory (Tobolosky and Eyring, 1943; Mark and Tobolosky, 1950), that is by treating flow as a bond breakage–bond formation process (e.g. Peters, 1955). Most of the formulations, in this context, however, have referred solely to the deformation of the critical weak bonds in the microstructure by the reasoning that the deformation of the much stronger bonds is likely to be negligible at low stresses. This is, however, an oversimplification of the actual flow process occurring in the real, complex microstructure of viscoelastic material systems such as those mentioned before (e.g. Takehiro and San-Ichiro, 1955; Bernal, 1958). It is well recognized that such materials behave in a manner which depends primarily on the material source, microstructure and previous history, in addition to the current state of loading and environmental conditions. In a large class of viscoelastic materials, such as natural amorphous or semiamorphous types of materials, for example, environmental effects, e.g. moisture content, could enhance the deterioration of the internal microstructure and, hence, the amount of occurring deformation. Furthermore, the energy required to produce a certain deformation may change abruptly at particular temperatures owing to internal transitions in the material (e.g. Kauman, 1966). In the case of polymeric materials, for instance, the level of order (crystallinity), the extent of alignment of the morphological units (orientation) and the degree of polymerization are among the many factors that could influence the viscoelastic response characteristics of such materials. Thus, while in the major part of this book, the continuum mechanics approach is maintained primarily for the characterization of the response behaviour of viscoelastic materials, it is emphasized that such a response is essentially dependent on the effects of a large number of significant microscopic and macroscopic parameters such as those introduced above.

The contents of this book are presented in ten chapters and four appendices. Chapter 1 contains the continuum mechanics background necessary for the presentation of the viscoelasticity theory and pertaining subjects in the subsequent Chapters 2–10. In Chapter 2, the basic formalism of the mechanical response of the linear viscoelastic material is presented. Consequently, in Chapter 2, the formulations are considered entirely within the context of the infinitesimal linearized deformation theory. Here, the ideas set down by L. Boltzmann (1844–1906) and V. Volterra (1860–1940) are taken as fundamental within the context of linear superposition of input histories (Boltzmann, 1874, 1877, 1878; Volterra, 1913; Volterra and Peres, 1936; Leaderman, 1943, 1958). In this chapter, the author confines his attention to the one-dimensional linear theory of isothermal viscoelasticity in the time domain under variable levels of stress or strain inputs. The transition to dynamic viscoelasticity is discussed in Chapter 3 whereby the formulation of Chapter 2 is extended to include the possibility of characterizing the linear viscoelastic behaviour of materials in the frequency domain. In this context, the relationships between the material functions characterizing the viscoelastic response in both the time and frequency domains are considered. In Chapter 4, the formulation of the three-dimensional viscoelastic response behaviour is dealt with whereby the one-dimensional constitutive relations of Chapter 2 are replaced by their corresponding tensorial equivalents. Remarks concerning the thermodynamic restrictions on isothermal linear viscoelasticity are also presented in Chapter 4. Chapter 5 presents the basic formulations of the constitutive relations in thermoviscoelasticity whereby the dependence of the performance of the viscoelastic material on both fixed reference and transient-temperature fields are considered. The fundamental areas considered in Chapter 5 include the thermodynamics of the deformation process, the rheological equations of state and the thermodynamical derivation of the constitutive relations. The treatment of the thermoviscoelastic response of the so-called 'thermorheologically simple materials' and 'thermorheologically complex materials' are dealt with. Various viewpoints concerning constitutive formalism in thermoviscoelasticity as expressed by various authors are presented. In Chapter 6, the more complex subject of nonlinear viscoelasticity is dealt with. Here, one deals primarily with the nonlinear analysis of the deformation process and its effect on the constitutive formalism of the viscoelastic response. In this chapter, illustrations are first given concerning the nonlinear viscoelastic behaviour; then, the characterization of the nonlinear response is dealt with, with the inclusion of the concept of coordinate invariance (the objectivity principle). In this context, the various approaches to the formulation of the governing constitutive equations, as proposed by researchers in the field, are presented. The problem of experimental determination of the pertaining nonlinear viscoelastic material functions in both the one- and the three-dimensional situations is discussed. Chapter 7 deals with a few aspects of the broad subject of numerical analysis in viscoelasticity. In this, the constitutive formalism in linear viscoelasticity is treated to characterize the nonlinear viscoelastic response of materials. This is carried out through an analytical model that includes a differential approximation of the time-dependent behaviour and an optimization procedure. The procedure is illustrated numerically for arbitrary ranges of stress, strain and temperature and a selected extent of time concerning the viscoelastic response.

In Chapter 8, the author deals with the important subject of wave propagation in viscoelastic materials. In recent years, there has been considerable interest in the subject of wave propagation in engineering materials in general, from both theoretical and experimental points of view. Such interest has been motivated primarily by the advancements in testing and measurement techniques. With the recent progress in fields such as electronics and laser optics, stress waves of high frequency can be now produced and detected easily. This has been particularly pronounced in important domains of nondestructive testing such as ultrasonics and acoustic emission. The combination of these two techniques has led further to the newly developed acousto-ultrasonic method (Tanary, 1988; Vary, 1988) as a modern nondestructive tool for the evaluation and prediction of the mechanical properties of engineering materials. The latter technique has been applied recently by Iyer (1993) and Iyer and Haddad (1993) for the characterization and prediction of the viscoelastic behaviour of a class of polymeric materials. Another equally important cause of the ensuing interest in the subject of viscoelastic wave propagation is, as mentioned earlier, the continuous emergence of newly developed industrial viscoelastic materials such as plastics and polymeric composite material systems. In this, the study of the phenomenon of wave motion has been able to identify microstructural problems and to assist in the development of homogeneous and inhomogeneous material systems. Further, any new development within the realm of smart materials and structures (e.g. Yoshiki and Shun-Ichi, 1988; Rogers, 1989; Iyer and Haddad, 1994) is expected to depend on the understanding and utilization of the wide range of mechanical performance of viscoelastic constituents and, hence, on the ability to employ wave propagation as a successful detecting mechanism for a feedback concerning the status of such materials.

In view of the time dependency of the response behaviour in viscoelasticity which is further complicated by the form of the constitutive relations and, hence, the associated boundary conditions, serious attempts to solve viscoelastic boundary value problems have lagged considerably behind those in classical elasticity. It is only in the last few decades that viscoelastic boundary value problems have been actively considered. The classification of boundary value problems in viscoelasticity, their formulations and possible methods of solutions are considered in Chapter 9. The extension of the phenomenological theory of viscoelasticity to include microscopic effects is discussed in Chapter 10. In this chapter, a case study is presented concerning the formulation of the viscoelastic response of a two-dimensional fibrous system with the inclusion of the real microstructure. In this, the material system is regarded as a two-dimensional network of randomly oriented viscoelastic fibres which are bonded together by intermolecular bonds. Thus, the mechanics of the discrete microstructure introduces the relevant field quantities as random variables or functions of such variables and their corresponding distribution functions. The model is presented in an explicit form for a cellulosic system by using available experimental data which permitted the numerical evaluation of the response behaviour of such a system. The analysis is introduced in a general representative form which can be modified for particular applications concerning other classes of structured composite material systems.

Throughout the text, generalized tensorial notations are used. For simplification, however, the presentation has been limited to Cartesian tensors only. The reader is referred to Appendix A for a brief introduction on the subject matter. Appendix B presents the definitions and a summary of the properties of delta and step functions. These functions are used in the connection of presentation of the theory of viscoelasticity in Chapter 2 and throughout the text. The important subjects of Laplace and Fourier transformations are dealt with, respectively, in Appendix C and Appendix D. The two types of transformations are utilized frequently in the book.

In the presentation, vectors and unindexed tensorial quantities in general are indicated by bold. The author has used majuscules to identify the undeformed configuration or state X_I ($I = 1, 2, 3$) and minuscules to designate the corresponding deformed state x_i ($i = 1, 2, 3$). Equations, figures and tables are numbered within the chapter; for example, Fig. 1 of Chapter 2 is identified by 'Fig. 2.1'.

REFERENCES

Bernal, J. D. (1958) Structure arrangements of macromolecules. *Discuss. Faraday Soc.*, **25**, 7–18.

Boltzmann, L. (1874) Zür Theorie der elastichen Nachwirkung, Sitzungsber, Kaiserl. *Akad. Wiss. Wien, Math. Naturwiss. Kl.*, **70**(2), 275–306.

Boltzmann, L. (1877) Zür Theorie der elastischen Nachwirkung. *Akad. Wiss. Wien, Math. Naturwiss. Kl.*, **76**, 815–42.

Boltzmann, L. (1878) Zür Theorie der elastischen, Nachwirkung. *Ann. Phys. Chem., N.F.*, **5**, 430–2.

Doi, M. and Edwards, S. F. (1986) *The Theory of Polymer Dynamics*, Clarendon, Oxford.

Eirich, F. R. (1956) *Rheology Theory and Applications*, Academic Press, New York.

Ferry, J. D. (1970) *Viscoelastic Properties of Polymers*, 2nd edn, Wiley, New York.

Iyer, S. S. (1993) On the characterization of the viscoelastic response of a class of materials using acousto-ultrasonics – a pattern recognition approach. Master's Thesis, University of Ottawa, Canada.

Iyer, S. S. and Haddad, Y. M. (1993) On the characterization of the linear viscoelastic response of a class of materials by acousto-ultrasonics – a pattern recognition approach. CANCOM '93 – 2nd Canadian International Composites Conference and Exhibition (eds W. Wallace, R. Gauvin and S. V. Hoa), Canadian Association for Composite Structures and Materials, Ottawa, pp. 479–89.

Iyer, S. S. and Haddad, Y. M. (1994) Intelligent materials – an overview. *Int. J. Pressure Vessels Piping*, **58**, 335–44.

Kauman, W. G. (1966) On the deformation and setting of the wood cell wall. *Holz Roh- Werkst.*, **24**(11), 551–6.

Leaderman, H. (1943) *Elastic and Creep Properties of Filamentous Materials and Other High Polymers*, Textile Foundation, Washington, DC.

Leaderman, H. (1958) In *Viscoelasticity Phenomena in Amorphous High Polymeric Systems*, Vol. II (ed. F. Eirich), Academic Press, New York, pp. 1–6.

Love, A. E. H. (1944) *A Treatise on the Mathematical Theory of Elasticity*, 4th edn, Dover Publications, New York, pp. 1–31.

Mark, H. and Tobolosky, A. V. (1950) *Physical Chemistry of High Polymeric Materials*, Interscience Publishers, New York.

Markovitz, H. (1977) Boltzmann and the beginnings of linear viscoelasticity. *Trans. Soc. Rheol.*, **21**(3), 381–98.

Peters, L. (1955) A note on nonlinear viscoelasticity. *Textile Res. J.*, **29** (March), 262–5.

Rogers, C. A. (ed.) (1989) *Smart Materials, Structures, and Mathematical Issues*, Selected Papers presented at the US Army Research Office Workshop, Virginia Polytechnic Institute and State University, Blacksburg, VA, September 15–16, 1988, Technomic, Lancaster, PA.

Staverman, A. J. and Schwarzl, F. (1956) Linear Deformation Behaviour of High Polymers, in *Die Physik der Hochpolymeren*, Vol. IV (ed. H. A. Stuart), Springer, Berlin.

Takehiro, S. and San-Ichiro, M. (1955) On the helical configuration of a polymer chain. *J. Chem. Phys.*, **23**(4), 707–11.

Tanary, S. (1988) Characterization of adhesively bonded joints using acousto-ultrasonics. Master's Thesis, University of Ottawa, Ottawa.

Tobolosky, A. and Eyring, H. J. (1943) Mechanical properties of polymeric materials. *J. Chem. Phys.*, **11**, 125–34.

Vary, A. (1988) The acousto-ultrasonic approach, in *Acousto- Ultrasonics: Theory and Applications* (ed. J. C. Duke, Jr), Plenum, New York, pp. 1–21.

Volterra, V. (1913) *Fonctions de Lignes*, Gauthier-Villard, Paris.

Volterra, V. and Peres, J. (1936) *Thérie Générale des Fonctionnelles*, Gauthier-Villard, Paris.

Yoshiki, S. and Shun-Ichi, H. (1988) Development of polymeric shape memory material. *Mitsubishi Tech. Bull.*, no. 184, Mitsubishi Heavy Industries Ltd., New York.

1

Continuum mechanics background

1.1 INTRODUCTION

Continuum mechanics is a branch of general mechanics that deals with the evolution of the mechanical response process in solids and fluids when such media are considered as idealized continua. The emphasis here is on the concept 'continuous medium', whereby the actual microstructure of the medium is disregarded and the medium is pictured as a continuum without gaps or empty spaces. Hence, the configuration of the assumed continuous medium would be described by a continuous mathematical model whose geometrical points are identified with material particles of the actual physical medium. Further, when such a continuum changes its configuration under some boundary conditions, such change is assumed to be continuous, i.e. neighbourhoods evolve into neighbourhoods. Thus, the mathematical functions entering the analysis of, for instance, a deformation process, are assumed to be continuous functions with continuous derivatives. Any creation of new boundary surfaces, such as those developed by internal fracture, would then be seen as extraordinary events that might require alternative formulations outside the realm of continuum mechanics.

A basic concept in continuum mechanics is that of the definition of stress at a point (Cauchy, 1827, 1828). Reference is being made here to a geometric point that has no volume and may be associated with a mathematical limit of an elemental region of the continuum when the volume of such a region shrinks down to zero. This is, in essence, similar to the definition of the derivative in differential calculus and follows immediately from the postulate of continuity of the medium. Through its connection with the definition of the derivative, the concept of the stress at a point makes the powerful methods of calculus available for the analysis of the deformation process or flow in continuous media.

Two other physical postulates are often encountered in continuum mechanics presentations. They are concerned with the following.

1. *Homogeneity*. A medium is homogeneous if it has identical properties at all points. Hence, under this assumption, the medium is uniform and its properties are independent of the position.

2. *Isotropy.* A medium is isotropic with respect to a certain material property if such property has the same value in all directions, i.e. independent of orientation of any reference coordinate system that may be chosen to measure the property. Hence, such a material property remains constant in any plane that passes through a point in the material.

The above-mentioned postulates, together with other basic assumptions and principles of continuum mechanics, are demonstrated during the course of development of the present chapter.

Continuum mechanics theory may be divided into the following three domains of interest.

1. Generalized assumptions and principles that could be applicable to all continuous media.
2. Specialized theories pertaining to an idealized class of media. Such specialized theories are built on the foundations of the generalized assumptions and principles referred to under (1) above.
3. Constitutive equations defining the response behaviour of a particular idealized medium under specified boundary conditions. The boundary conditions may appear in the form of forces and/or displacements and velocities which could arise from contact with other bodies, thermal effects, chemical interactions, electromagnetic fields and other environmental changes. In most cases, however, we are concerned, in continuum mechanics, with media subjected to forces of mechanical origin and thermal influences.

The aim of this chapter is to provide the reader with a concise introduction of the basic assumptions and principles of continuum mechanics with an emphasis on those specifically used in the remainder of the book. Section 1.2 deals with the introduction of a number of general principles of the continuum mechanics theory. Section 1.3 treats the analysis of stress from a continuum mechanics point of view. Section 1.4 considers the kinematics and the measures of strain in a continuous medium. A reader who is not familiar with Cartesian tensor operations is advised to consult Appendix A concurrently with this chapter. The reader is referred to other text-books and references for a comprehensive study of the continuum mechanics theory. Some chosen references are provided at the end of this chapter.

1.2 GENERAL PRINCIPLES

1.2.1 Conservation of mass

From a classical mechanics point of view, mass is assumed to be conserved. Hence, the mass of a material body is considered as a time-independent property. In continuum mechanics, it is further postulated that mass is an absolutely continuous function of volume. Hence, it is assumed that a positive quantity ρ, referred to as density, can be defined at every point in the body by (e.g. Fung, 1965)

$$\rho(x) = \max_{l \to x} \frac{\text{mass of } \Omega_l}{\text{volume of } \Omega_l} \tag{1.1}$$

where Ω_l is a suitably chosen infinite sequence of particle sets shrinking down upon the point of the medium. This point is identified by the current position vector x referred to a particular coordinate frame of reference. The counterpoint of the latter, in the undeformed configuration, is given the majuscule symbol X. The mapping between the two positions x and X is dealt with in section 1.4 where we discuss the deformation kinematics and the measures of strain in a continuous medium.

The conservation of mass is expressed by

$$\int \rho(\mathbf{x}) \, dx_i = \int \rho(\mathbf{X}) \, dX_I \tag{1.2}$$

where the integrals extend over the same sets of particles. Hence,

$$\rho(\mathbf{x}) = \rho(\mathbf{x}) \left| \frac{\partial x_i}{\partial X_I} \right| \quad \text{and} \quad \rho(\mathbf{x}) = \rho(\mathbf{X}) \left| \frac{\partial X_I}{\partial x_i} \right|. \tag{1.3}$$

Equation (1.2) or (1.3) relates, then, the densities of the body for different configurations of the deformation process.

1.2.2 Material derivative of field functions

The derivative of a field function $\phi(x_i, t)$ that is attributable to the motion of a point in a continuum is expressed by the so-called 'mathematical derivative'. The latter is denoted by a dot or, alternatively, by the symbol D/Dt. It can be shown that the material derivative of a field function $\phi(\mathbf{x}, t)$ takes the form

$$\dot{\phi}(\mathbf{x}, t) = \frac{D\phi(x, t)}{Dt}$$

$$= \left(\frac{\partial \phi}{\partial t} \right)_{x = \text{constant}} + v_i \frac{\partial \phi}{\partial x_i} \tag{1.4}$$

where v_i is the velocity associated with the current position x_i. The first term on the right-hand side of (1.4) is due to the time dependence of the function ϕ ($\cdot$) and is interpreted as the 'local part of the field function'. The second term is contributed by the motion of the particle in the current field of the function ϕ ($\cdot$) and is referred to as the 'convective' part of $\dot{\phi}(\mathbf{x}, t)$. The field function $\phi(\mathbf{x}, t)$ may take the form of a tensor field of any order.

1.2.3 Continuity of mass (Continuity Equation)

In the derivation of the continuity equations, one of the following two approaches may be taken.

1. One considers a constant mass in a volume that varies with the time.
2. Alternatively, one may consider a constant volume while taking into account the variation in mass between entry and exit.

Consider the second approach. Thus, the rate of mass increase of an arbitrary fixed volume V is equal to the influx of matter through its surface S, i.e.

$$\int_V \frac{\partial \rho}{\partial t}\, dV = - \int_S \rho v_i \mathbf{n}_i\, dS. \tag{1.5}$$

In the above equation, the negative sign accounts for the influx being opposite to the direction of the outward unit normal $\mathbf{n}_i$ and v_i is the velocity of the material entering the control volume. Recalling the three-dimensional form of the divergence theorem of Gauss (e.g. Flügge, 1972), then equation (1.5) is written as

$$\int_V \left[\frac{\partial \rho}{\partial t} + (\rho v_i)_{,i} \right] dV = 0$$

whereby the **continuity of mass equation** is expressed as

$$\frac{\partial \rho}{\partial t} + (\rho v_i)_{,i} = 0. \tag{1.6}$$

An alternative expression may be found by carrying through the indicated partial differentiation in (1.6) to obtain

$$\frac{\partial \rho}{\partial t} + \rho_{,i} v_i + \rho v_{i,i} = 0. \tag{1.7}$$

Employing the derivative operator $D(\cdot)/Dt$ introduced earlier by equation (1.4), then, in terms of this operator, the continuity of mass equation (1.7) is expressed as

$$\frac{D\rho}{Dt} + \rho v_{i,i} = 0. \tag{1.8}$$

1.2.4 Continuity of momentum

Following the approach adopted above in deriving the continuity of mass equation, we consider an arbitrary fixed volume V whereby the total rate of change of linear momentum has two components; one associated with the change of mass within the volume V and the other associated with the influx of the mass through the bounding surface S. Thus,

$$\frac{d}{dt} \int_V \rho v_i\, dV = \int_V \frac{\partial(\rho v_i)}{\partial t}\, dV + \int_S (\rho v_i) v_j\, \mathbf{n}_j\, dS \tag{1.9}$$

which can be written by using the divergence theorem of Gauss as

$$\frac{\mathrm{d}}{\mathrm{d}t} \int_V \rho v_i \, \mathrm{d}V = \int_V \left[\frac{\partial \rho}{\partial t} v_i + \rho \frac{\partial v_i}{\partial t} + (\rho v_i v_j)_{,j} \right] \mathrm{d}V$$

$$= \int_V \left[\left(\frac{\partial \rho}{\partial t} + \rho_{,j} v_j + \rho v_{j,j} \right) v_i + \rho \frac{\partial v_i}{\partial t} + \rho v_{i,j} v_j \right] \mathrm{d}V. \quad (1.10)$$

In view of the continuity of mass equation (1.7), expression (1.10) reduces to

$$\frac{\mathrm{d}}{\mathrm{d}t} \int_V \rho v_i \, \mathrm{d}V = \int_V \rho \left(\frac{\partial v_i}{\partial t} + v_{i,j} v_j \right) \mathrm{d}V \qquad (1.11)$$

Expression (1.11) is referred to as the **continuity of momentum equation**.

1.3 ANALYSIS OF STRESS

1.3.1 Body and surface forces

The external forces acting at any time on a free body are classified, from a continuum mechanics point of view, in two categories, namely **body** forces and **surface** forces.

Body forces act on the elements of mass or volume inside the body. Hence, in continuum mechanics, they are expressed as forces per unit mass or forces per unit volume. Examples of body forces are those due to gravity, magnetic effects and inertia. In this book, unless otherwise stated, we regard the body forces acting on a free body as expressed per unit volume. Hence, the term 'volumetric forces' may be used as a reference to these body forces throughout the text.

We shall denote the body force per unit volume acting on an infinitesimal volume element $\mathrm{d}V$ of the body by the vector χ. If the resultant of body forces acting on an elemental volume ΔV is designated by $\Delta \mathbf{B}$, then the body force is defined by

$$\chi = \lim_{\Delta V \to 0} \frac{\Delta \mathbf{B}}{\Delta V}. \qquad (1.12)$$

In general, the vector χ varies from point to point in the free body at any given instant of time and may also vary with time at any particular point of the body. It can be also dependent on other state variables such as the temperature. The vector sum of all body forces acting on a free body of a finite volume V, at any particular instant of time, is then given by the space integral over this volume. However, in many applications, the body forces are likely to be uniform, e.g. gravity forces or, otherwise, they may be small enough so that they could be assumed to be negligible.

Surface forces are exerted on the bounding surface of the free body. Hence, they are usually expressed per unit area of the surface on which they are acting.

The limit of the surface force per unit area, as the unit area tends to zero, is often referred to as the **stress vector**, or, alternatively, traction vector. It is denoted by the symbol **T**. Accordingly, the force acting on an elemental area $\mathrm{d}S$ of the bounding surface is **T** d*S* and the vector sum of all surface forces acting on a finite region *S* of this surface is given by the corresponding vector surface integral, that is $\int_S$ **T** d*S*.

Generally, when applied to a given solid body, the definitions of body and surface forces are taken with reference to the current deformed configuration of the body. However, in many applications of the theory of elasticity, the occurring deformations might be so small that the definitions of such forces could be expressed with reference to the undeformed configuration of the body without a significant error. Alternatively, in applications concerning fluid flow, one may use as a reference a given fixed volume of space, through which different substance or fluid passes at different times. In this context, the concept of the imaginary control surface in a fluid is well recognized in the studies of fluid mechanics (e.g. Malvern, 1969).

1.3.2 Stress vector

We consider, in Fig. 1.1, a continuous free body being acted upon by an externally applied equilibrium force system $\mathbf{F}_1$, $\mathbf{F}_2$, ..., $\mathbf{F}_N$. Suppose, now, that the body be divided, through a bounding surface *S*, into two regions R_1 and R_2. If R_1, for example, is to be in equilibrium, forces must be exerted on *S* by R_2. These forces would be in equilibrium with the externally applied forces on R_1. We consider a point p on *S* with unit normal **n** surrounded by an elemental surface area ΔS of *S*. The forces acting on ΔS are statistically equivalent to a force and a couple. At the limit, as ΔS

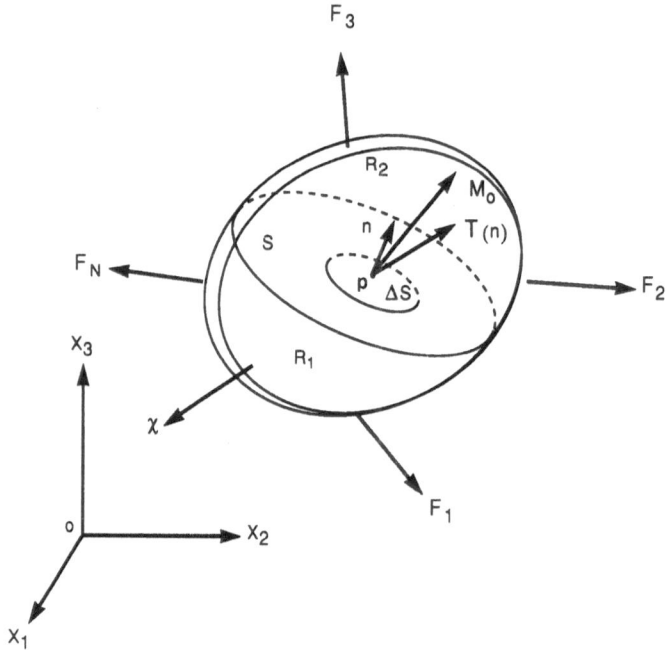

Fig. 1.1 Body and surface forces.

shrinks down to a point, the couple (per unit area) produced, at this point, by the continuous distribution of the internal force may be taken to be zero. This does not, of course, exclude the possibility of the existence of a 'couple stress' whose value may be different from zero. Such couple stress has been proposed in 'higher order' continuum mechanics theories (Malvern, 1969). However, for our purpose, we shall assume, following classical continuum mechanics, that there is no couple stress acting on ΔS and that the action of one body on another across an infinitesimal surface area can be presented solely by the **stress vector**. The latter is defined, at the point p of the elemental surface area ΔS with normal **n** (Fig. 1.1), by

$$\mathbf{T}(\mathbf{n}) = \lim_{\Delta S \to 0} \frac{\Delta \mathbf{F}}{\Delta S}. \tag{1.13}$$

That is, $\mathbf{T}(\mathbf{n})$ is a stress vector function defined on the elemental surface area ΔS and is dependent on the unit normal **n**.

The stress state at any point in a continuum can be determined in terms of the three stress vectors acting on three mutually perpendicular planes intersecting at this point. Hence, the stress vectors acting on planes perpendicular to a rectangular coordinate system embedded at a point in the continuum are considered to be of particular interest. Figure 1.2 shows three such stress vectors acting on the centre points of three faces of a cube surrounding a chosen particular point of interest O. The point O represents, in turn, the centre of the cube and, at the same time, the

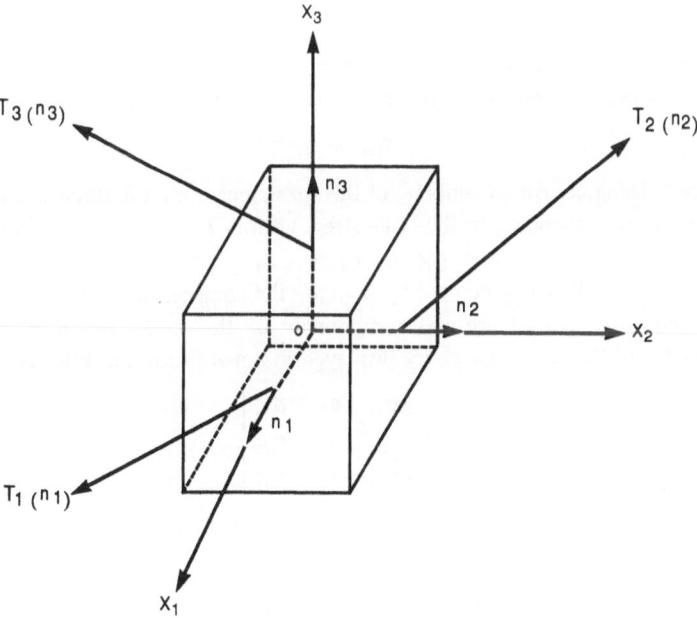

Fig. 1.2 Average traction vectors acting on three planes perpendicular to the rectangular coordinate axes.

origin of the rectangular coordinate system. With reference to Fig. 1.2, let

- $T_1(n_1)$ be the stress vector acting on a plane whose normal n_1, is pointing in the positive x_1 direction,
- $T_2(n_2)$ be the stress vector acting on a plane with normal, n_2, directed in the positive x_2 direction, and
- $T_3(n_3)$ be the stress vector acting on a plane with normal, n_3, in the direction of the positive x_3 axis.

The three stress vectors mentioned above are considered to be the average traction vectors on the corresponding faces of the cube. As the cube shrinks down towards its centre point O, the limit approached by the stress vector on each face is taken, from continuum mechanics point of view, as the stress vector at the point O on a plane perpendicular to one of the coordinate axes. Hence, the three stress vectors may be seen to represent the local stress state at the point O. This approach illustrates a basic concept in continuum mechanics, i.e. the concept of stress at a point, mentioned earlier in the introduction to this chapter. This concept refers to a geometric point in space visualized as a mathematical limit in a manner similar to the definition of the derivative in differential calculus. This approach has been, in fact, a key to the development of continuum mechanics as it immediately makes the powerful methods of calculus available for the study of the deformation and flow in a physical continuum.

1.3.3 Cauchy's stress tensor

As mentioned in the foregoing, the three stress vectors $T_1(n_1)$, $T_2(n_2)$ and $T_3(n_3)$ are taken as a representation of the stress tensor at the particular point under consideration. The stress tensor is denoted by $\boldsymbol{\sigma}$. It is the linear vector function which associates with each argument unit vector n the traction vector $T(n)$. That is,

$$T(n) = n\boldsymbol{\sigma}. \tag{1.14}$$

The nine rectangular components σ_{ij} of the stress tensor are the three sets of stress components corresponding to the three stress vectors $T_1(n_1)$, $T_2(n_2)$ and $T_3(n_3)$. This is illustrated in Fig. 1.3 where the set $(\sigma_{11}, \sigma_{12}, \sigma_{13})$ constitutes the components of the stress vector $T_1(n_1)$ while $(\sigma_{21}, \sigma_{22}, \sigma_{23})$ are the components of the stress vector $T_2(n_2)$ and $(\sigma_{31}, \sigma_{32}, \sigma_{33})$ are the components of the stress vector $T_3(n_3)$. The components of the stress tensor are displayed in a matrix form as follows:

$$\sigma_{ij} = \begin{bmatrix} \sigma_{11} & \sigma_{12} & \sigma_{13} \\ \sigma_{21} & \sigma_{22} & \sigma_{23} \\ \sigma_{31} & \sigma_{32} & \sigma_{33} \end{bmatrix}.$$

With reference to Fig. 1.3, the following sign convention of the stress components σ_{ij} is adopted.

1. σ_{ij} $(i \neq j)$ is the shear component of the stress tensor acting on a face of the cube whose normal is in the ith direction and the component itself is acting in the jth direction.

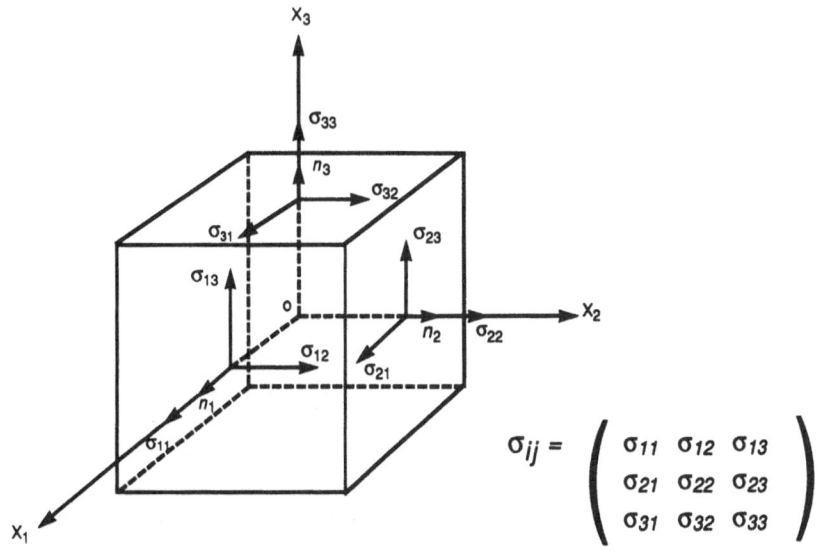

Fig. 1.3 Stress tensor components.

2. $\sigma_{\bar{\imath}\bar{\imath}}$ (no sum over i) is the normal component of the stress tensor. It is acting on a face whose normal is in the ith direction and the component is acting in the same direction. The normal component of stress is positive if drawn outward from the face upon which it acts.
3. If the outward normal **n** to a surface is in the positive direction of the coordinate axis, then the shear component of stress acting on this surface is positive if it is in the positive direction of the associated-with axis. Alternatively, if the outward normal to a surface is in the negative direction of the coordinate axis, then the shear component is positive if it is in the negative direction of the associated-with axis. With reference to Fig. 1.3, positive σ_{23}, for instance, represents an upward-acting stress component on the right-hand side and down-acting component on the left-hand side of the cube. Negative σ_{23}, on the other hand, acts downward on the right-hand side and upward on the left-hand side of the cube. Further, the components on the negative sides of the cube will have senses opposite to those on the positive sides.

When the normal component of stress, $\sigma_{\bar{\imath}\bar{\imath}}$ (no sum over i), is positive it represents a tensile stress but, if it is negative, it is compressive. The algebraic sign of a tangential shear component does not, however, have an intrinsic physical significance. Hence, positive and negative shear components represent the same kind of loading but in different directions.

The state of stress at a particular point in a continuous medium can be specified fully by the second-order tensor σ_{ij} with nine components. However, as will be dealt with later, because of the symmetry property of the stress tensor, only six of the nine components are independent. This is under the assumption that there are no distributed body or surface couples acting on the free body. In general, the components of the stress tensor are functions of the coordinates and time and they may be also dependent on other state variables such as the temperature. Accordingly, they form a second-order tensor field (Appendix A). In continuum mechanics, these functions and their partial derivatives are assumed to be continuous.

1.3.4 Stress boundary conditions: Traction vector on an arbitrary plane, Cauchy's tetrahedron

With reference to Fig. 1.4, we consider the free continuous body to be the tetrahedron or triangular pyramid OABC. The latter is enclosed by the three rectangular coordinate planes through the point O (the origin) and a fourth plane ABC not passing through O.

Let

- $T(n)$ be the stress vector at a point of the oblique surface ABC whose normal is **n**,
- ΔS be the area ABC,
- h be the perpendicular distance from the origin O to ΔS,
- $\Delta V = (1/3)h\,\Delta S$ be the elemental volume of the tetrahedron OABC,
- χ_i be the components of the body force vector per unit volume and
- σ_{ij} be the components of the stress tensor.

We apply Newton's second law of motion, that is the sum of forces acting on the tetrahedron is equal to the rate of change of linear momentum. Let v_i denote the

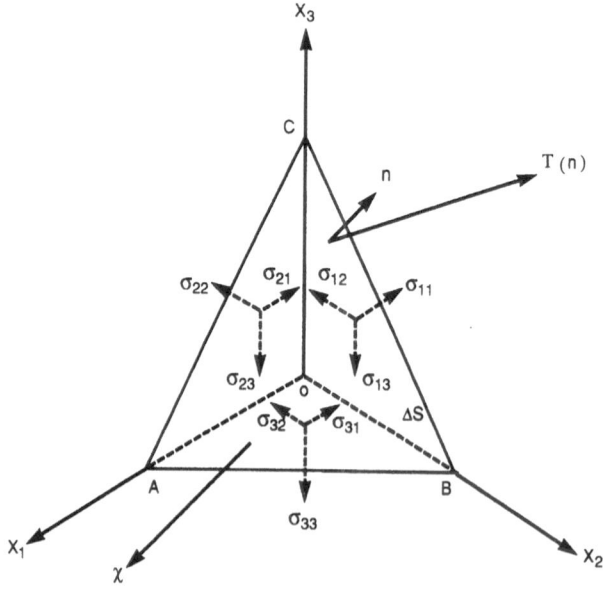

Fig. 1.4 Cauchy's tetrahedron.

components of the velocity vector, then, with reference to Fig. 1.4,

$$\int_{ABC} T_i \, dS - \int_{OAB} \sigma_{3i} \, dS - \int_{OCA} \sigma_{2i} \, dS$$

$$- \int_{OBC} \sigma_{1i} \, dS + \int_{\Delta V} \chi_i \, dV = \frac{d}{dt} \int_{\Delta V} \rho v_i \, dV \quad (1.15)$$

where ρ is the density.

Equation (1.15) can be written with the inclusion of the components of the unit vector **n** as

$$\int_{\Delta S} [T_i - (\sigma_{3i} n_3 + \sigma_{2i} n_2 + \sigma_{1i} n_1)] \, dS + \int_{\Delta V} \chi_i \, dV = \frac{d}{dt} \int_{\Delta V} \rho v_i \, dV$$

or

$$\int_{\Delta S} (T_i - \sigma_{ji} n_j) + \int_{\Delta V} \chi_i \, dV = \frac{d}{dt} \int_{\Delta V} \rho v_i \, dV. \quad (1.16)$$

From both the 'mean value theorem of calculus' and the concept of 'continuity of momentum' (1.11), equation (1.16) can be written in the form

$$(\langle T_i \rangle - \langle \sigma_{ji} n_i \rangle) \, \Delta S = (\langle \rho \dot{v}_i \rangle - \langle \chi_i \rangle) \, \Delta V \quad (1.17)$$

where $\langle \cdot \rangle$ indicates the mean value. Taking the limits as both ΔS and ΔV tend to zero, i.e. h also tends to zero, then equation (1.17) is approximated by

$$T_i = \sigma_{ji} n_j \quad (1.18)$$

where σ_{ji} is a second-order Cartesian tensor since both T_i and n_j are components of vectors.

Equation (1.18) establishes the '**stress boundary conditions**' on the free body (tetrahedron) considered and states that for every second-order (symmetric) tensor σ_{ji}, defined at some point in the continuum, there is, associated with each direction (specified by the unit normal **n** at that point), a traction vector **T** given by the form of this equation.

1.3.5 Symmetry of the stress tensor

As dealt with in the foregoing, we consider an arbitrary region of a continuum. This region, Fig. 1.5, is of volume V which is enclosed by a surface S with an outward unit normal **n**. The forces acting on an elemental portion of S are **T** dS while the body forces on an elemental portion of V are $\chi \, dV$.

Applying the principle of conservation of linear momentum to this arbitrary region, then

$$\int_S T_i \, dS + \int_V \chi_i \, dV = \frac{d}{dt} \int_V \rho v_i \, dV. \quad (1.19)$$

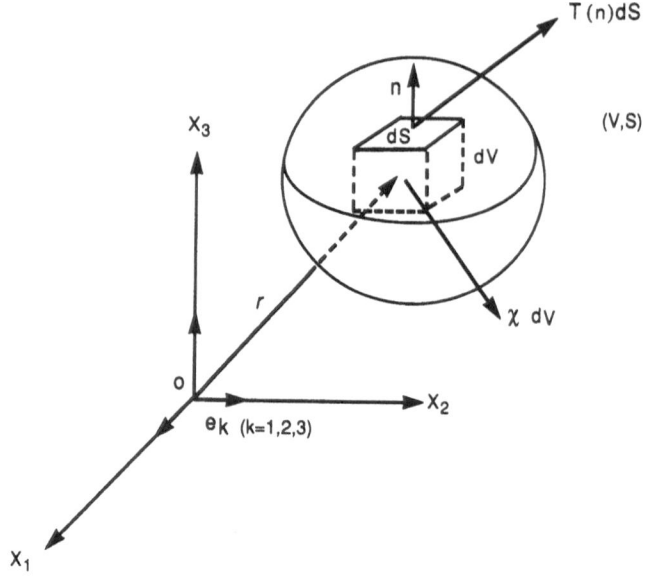

Fig. 1.5 An elemental volume (dV, dS) of a continuous region (V, S).

Thus, in view of (1.18) and the conservation of momentum relation (1.11), equation (1.19) can be written as

$$\int_S \sigma_{ji} n_j \; \mathrm{d}S + \int_V \chi_i \; \mathrm{d}V = \int_V \rho \dot{v}_i \; \mathrm{d}V. \tag{1.20}$$

If we now apply the divergence theorem to the first integral of (1.20), then

$$\int_V (\sigma_{ji,j} + \chi_i - \rho \dot{v}_i) \; \mathrm{d}V = 0 \tag{1.21}$$

or

$$\sigma_{ji,j} + \chi_i = \rho \dot{v}_i \tag{1.22}$$

which is known as 'Cauchy's first equation of motion'.

With reference to Fig. 1.5, one considers the sum of moments of forces about the origin O to be equal to the rate of change of angular momentum;

$$\int_S \mathbf{r} \times \mathbf{T} \; \mathrm{d}S + \int_V \mathbf{r} \times \boldsymbol{\chi} \; \mathrm{d}V = \frac{\mathrm{d}}{\mathrm{d}t} \int_V \mathbf{r} \times (\rho \mathbf{v}) \; \mathrm{d}V$$

or

$$\int_S \varepsilon_{ijk} x_i T_j \mathbf{e}_k \; \mathrm{d}S + \int_V \varepsilon_{ijk} x_i \chi_j \mathbf{e}_k \; \mathrm{d}V = \frac{\mathrm{d}}{\mathrm{d}t} \int_V \rho \varepsilon_{ijk} x_i \dot{x}_j \mathbf{e}_k \; \mathrm{d}V \tag{1.23}$$

where ε_{ijk} is the alternating tensor and e_k are the components of the unit vector associated with the coordinate system x_i as shown in Fig. 1.5.

Applying the three-dimensional divergence theorem of Gauss and the boundary conditions (1.18) to the left-hand side of (1.23), the latter equation is expressed as

$$\int_V e_k\{\varepsilon_{ijk}[(x_i\sigma_{lj})_{,l} + x_i\chi_j]\}\, dV = e_k \int_V \frac{\partial}{\partial t}(\rho\varepsilon_{ijk}x_i\dot{x}_j)\, dV$$

$$+ e_k \int_S \rho\varepsilon_{ijk}x_i v_j v_l n_k\, dS. \tag{1.24}$$

Equation (1.24) can be written with reference to equations (1.7) and (1.11) as

$$\int_V e_k\{\varepsilon_{ijk}[(x_i\sigma_{lj})_{,l} + x_i\chi_j]\}\, dV = e_k\varepsilon_{ijk}\int_V \rho x_i\dot{v}_j\, dV$$

or

$$e_k\varepsilon_{ijk}\int_V [\sigma_{ij} + x_i(\sigma_{lj,l} + \chi_j - \rho\dot{v}_j)]\, dV = 0. \tag{1.25}$$

In the above equation, since the expression in parentheses is equal to zero, from Cauchy's first equation of motion (1.22), equation (1.25) becomes

$$e_k \int_V \varepsilon_{ijk}\sigma_{jk}\, dV = \mathbf{0} \tag{1.26}$$

which implies that

$$\varepsilon_{ijk}\sigma_{jk} = 0 \tag{1.27}$$

or, alternatively, in view of the properties of the skew-symmetric tensor ε_{ijk} (Appendix A),

$$\sigma_{ij} = \sigma_{ji}. \tag{1.28}$$

It is apparent that equation (1.28) expresses the symmetry of the stress tensor. This equation is referred to as '**Cauchy's second law of motion**' and it implies the 'conservation of moment of momentum'.

1.3.6 Principal axes of stress, principal planes and principal stresses

Regardless of the state of stress at a given point in a continuum, it is always possible to choose a special set of rectangular axes through this point so that the shear stress components vanish when the stress components are referred to this set of axes. This set of rectangular axes are referred to as **principal axes** or **principal directions**. Thus, on a plane perpendicular to a principal axis, the traction vector is entirely normal.

The principal planes through the point perpendicular to the three principal axes of stress are referred to as the **principal planes of stress**.

The normal stress components on the three principal planes are known as **principal stresses**. The principal stresses are physical quantities whose magnitudes do not depend on the particular coordinate system to which the stress components are referred. Accordingly, they are invariants pertaining to the stress state at the particular point under consideration.

Let **n** define the direction of a principal axis and σ designate the corresponding principal stress; then the stress vector acting on the surface defined by **n** can be expressed, with reference to (1.18), by

$$T_i = \sigma n_i = \sigma_{ji} n_j \qquad (1.29)$$

This is with the understanding that

$$n_i = \delta_{ji} n_j \qquad (1.30)$$

where δ_{ji} is the Kronecker delta;

$$\delta_{ij} = \begin{cases} 1 & \text{if } i = j, \\ 0 & \text{if } i \neq j. \end{cases}$$

Combining equations (1.29) and (1.30), it follows that

$$(\sigma_{ji} - \sigma\delta_{ji})n_j = 0. \qquad (1.31)$$

Expression (1.31) represents three equations to be solved for the components of the unit normal **n**, i.e. n_1, n_2, n_3. In this context, we search a set of nontrivial solutions for which

$$n_1^2 + n_2^2 + n_3^2 = 1. \qquad (1.32)$$

Hence, equation (1.31), subject to (1.32), poses an eigenvalue problem. Since the matrix of σ_{ij} is real and symmetric, then there exist three real-valued principal stresses and a set of orthogonal principal axes. The three principal stresses are denoted by σ_1, σ_2, σ_3. The algebraically greatest of the three principal stresses is the algebraically greatest normal stress component acting on any plane through the point. At the same time, the algebraically smallest of the principal stresses is the algebraically smallest normal stress component on any plane through the point. Referring to equation (1.31), this equation has a set of nonvanishing solutions n_1, n_2, n_3 if, and only if, the determinant of its coefficients vanishes, i.e.

$$|\sigma_{ij} - \sigma\delta_{ij}| = 0. \qquad (1.33)$$

Equation (1.33) represents a cubic equation in σ. The roots of this equation are the principal stresses σ_1, σ_2 and σ_3. For each value of the principal stress, a unit normal **n** is involved.

Expansion of (1.33) leads to

$$|\sigma_{ij} - \sigma\delta_{ij}| = \begin{vmatrix} \sigma_{11} - \sigma & \sigma_{12} & \sigma_{13} \\ \sigma_{21} & \sigma_{22} - \sigma & \sigma_{23} \\ \sigma_{31} & \sigma_{32} & \sigma_{33} - \sigma \end{vmatrix} \qquad (1.34)$$

$$= -\sigma^3 + I_1\sigma^2 - I_2\sigma + I_3$$

where the coefficients I_1, I_2, I_3 denote, respectively, the following scalar expressions of the stress components:

$$I_1 = \sigma_{11} + \sigma_{22} + \sigma_{33}$$

$$= \sigma_{kk}; \qquad (1.35a)$$

$$I_2 = \begin{vmatrix} \sigma_{22} & \sigma_{23} \\ \sigma_{32} & \sigma_{33} \end{vmatrix} + \begin{vmatrix} \sigma_{11} & \sigma_{13} \\ \sigma_{31} & \sigma_{33} \end{vmatrix} + \begin{vmatrix} \sigma_{11} & \sigma_{12} \\ \sigma_{21} & \sigma_{22} \end{vmatrix}$$

$$= \tfrac{1}{2}(\sigma_{ij}\sigma_{ij} - \sigma_{ii}\sigma_{jj})$$

$$= -(\sigma_{11}\sigma_{22} + \sigma_{22}\sigma_{33} + \sigma_{33}\sigma_{11}) + \sigma_{23}^2 + \sigma_{31}^2 + \sigma_{12}^2; \quad (1.35b)$$

$$I_3 = \begin{vmatrix} \sigma_{11} & \sigma_{12} & \sigma_{13} \\ \sigma_{21} & \sigma_{22} & \sigma_{23} \\ \sigma_{31} & \sigma_{31} & \sigma_{33} \end{vmatrix} = \tfrac{1}{6}\varepsilon_{ijk}\varepsilon_{pqr}\sigma_{ip}\sigma_{jq}\sigma_{kr}. \qquad (1.35c)$$

It is recognized that both equation (1.34) and the coefficients I_1, I_2, I_3 represented by (1.35) do not depend on the choice of the coordinate axes; hence, they are invariants of coordinate transformation. For this reason, the coefficients I_1, I_2, and I_3 of equation (1.34) are conventionally referred to as invariants of the stress tensor.

With reference to equations (1.34) and (1.35), the coefficient I_1 is the first invariant of the stress tensor. It is also referred to as the trace of the stress matrix. That is the sum of the elements on the main diagonal in the matrix of rectangular Cartesian components of σ_{ij}, the three normal stresses; equation (1.35a). The coefficient I_2, the second invariant, is a homogeneous quadratic expression in the stress components. It is the sum of the three minor determinants of the three diagonal elements in the determinant of the stress matrix; equation (1.35b). The third invariant, I_3, is the determinant of the stress matrix and it is a homogeneous cubic expression in the stress components; equation (1.35c).

Further, the invariants I_1, I_2 and I_3 of σ_{ij} can be expressed in terms of the roots of the cubic equation (1.34), i.e. the three principal stresses σ_1, σ_2 and σ_3:

$$I_1 = \sigma_1 + \sigma_2 + \sigma_3;$$

$$I_2 = \sigma_1\sigma_2 + \sigma_2\sigma_3 + \sigma_3\sigma_1; \qquad (1.36)$$

$$I_3 = \sigma_1\sigma_2\sigma_3.$$

We state now an important property of a symmetric second-order tensor:

For a symmetric second-order tensor, the three principal stresses are all real and the three principal planes are mutually orthogonal.

To illustrate the above property, let $\overset{1}{\mathbf{n}}, \overset{2}{\mathbf{n}}, \overset{3}{\mathbf{n}}$ be unit vectors in the directions of the principal axes, with components $\overset{1}{n_j}, \overset{2}{n_j}, \overset{3}{n_j}$ ($j = 1, 2, 3$) which are the solutions of equation (1.31) corresponding, respectively, to the principal stresses $\sigma_1, \sigma_2, \sigma_3$, i.e.

$$(\sigma_{ij} - \sigma_1 \delta_{ij})\overset{1}{n_j} = 0,$$

$$(\sigma_{ij} - \sigma_2 \delta_{ij})\overset{2}{n_j} = 0,$$

$$(\sigma_{ij} - \sigma_3 \delta_{ij})\overset{3}{n_j} = 0,$$

(1.37)

Multiplying the first equation of (1.37) by $\overset{2}{n_i}$ and the second equation of (1.37) by $\overset{1}{n_i}$, summing over i and subtracting the resulting equations, it follows that

$$(\sigma_2 - \sigma_1)\overset{1}{n_i}\overset{2}{n_i} = 0.$$

(1.38)

Assuming tentatively that (1.34) has a complex root and recalling that the coefficients of this equation are real valued, then a complex conjugate root must also exist and the set of roots may be written as

$$\sigma_1 = a + ib, \ \sigma_2 = a - ib \text{ and } \sigma_3$$

where a and b are real numbers and i stands for the imaginary number $\sqrt{-1}$. In this case, equations (1.37) would show that $\overset{1}{n_j}$ and $\overset{2}{n_j}$ are complex conjugate to each other and may be written as

$$\overset{1}{n_i} = \alpha_j + i\beta_j, \ \overset{2}{n_i} = \alpha_j - i\beta_j$$

where α and β are real numbers. Thus,

$$\overset{1}{n_i}\overset{2}{n_j} = (\alpha_j + i\beta_j)(\alpha_j - i\beta_j)$$
$$= \alpha_1^2 + \alpha_2^2 + \alpha_3^2 + \beta_1^2 + \beta_2^2 + \beta_3^2 \neq 0.$$

Hence, with reference to (1.38),

$$\sigma_1 - \sigma_2 = 2ib, \text{ or } b = 0,$$

which means that the original assumption of the existence of complex roots is incorrect and $\sigma_1, \sigma_2, \sigma_3$ must be all real.

Further, if $\sigma_1 \neq \sigma_2 \neq \sigma_3$, then, recalling equation (1.38),

$$\overset{1}{n_i}\overset{2}{n_i} = 0, \ \overset{2}{n_i}\overset{3}{n_i} = 0 \text{ and } \overset{3}{n_i}\overset{1}{n_i} = 0,$$

(1.39)

i.e. the principal stresses are orthogonal to each other.

On the other hand, if $\sigma_1 = \sigma_2 \neq \sigma_3$, one may determine an infinite number of pairs of unit normals $\overset{1}{n_i}$ and $\overset{2}{n_i}$ with $\overset{3}{n_i}$ a vector orthogonal to $\overset{1}{n_i}$ and $\overset{2}{n_i}$.

However, if $\sigma_1 = \sigma_2 = \sigma_3$, then any set of orthogonal axes may be considered as principal axes.

It is evident that, if the reference axes x_1, x_2, x_3 were selected to coincide with the principal axes, then the stress tensor would be expressed as

$$\sigma_{ij} = \begin{bmatrix} \sigma_1 & 0 & 0 \\ 0 & \sigma_2 & 0 \\ 0 & 0 & \sigma_3 \end{bmatrix}. \tag{1.40}$$

1.3.7 Spherical and deviatoric components

The stress tensor σ_{ij} may be expressed as the sum of two parts.

● One tensorial part represents a spherical or hydrostatic state of stress in which each normal stress is equal to $-p$, where p is the mean normal pressure, and all shear stresses are zero:

$$-p = \bar{\sigma} = \tfrac{1}{3}(\sigma_{11} + \sigma_{22} + \sigma_{33}) = \tfrac{1}{3}\sigma_{kk} = \tfrac{1}{3}I_1 \tag{1.41}$$

where $\bar{\sigma}$ is the mean stress and I_1 is the first invariant of σ_{ij} expressed earlier by (1.35a).

● The second part, denoted below by σ'_{ij}, defines the deviator as

$$\sigma'_{ij} = \sigma_{ij} - \bar{\sigma}\delta_{ij}. \tag{1.42}$$

Thus,

$$\sigma_{ij} = \bar{\sigma}\delta_{ij} + \sigma'_{ij}. \tag{1.43}$$

To determine the principal values of the stress deviator, the procedure illustrated in section 1.3.6 for the determination of the principal stresses of the stress tensor may be followed by replacing the determinant equation (1.33) by

$$|\sigma'_{ij} - \sigma'\delta_{ij}| = 0. \tag{1.44}$$

1.3.8 Piola–Kirchhoff stress tensor

As presented earlier, the Cauchy equations of motion (1.22) and (1.28) apply to the current deformed configuration. Here, the Cauchy stress tensor field is a function of the spatial coordinate **x** and it was concluded by (1.28) to be a symmetric tensor for the case where there is no couple stress or assigned couples. In the applications of the theory of elasticity, however, it is often assumed that there exists a natural state to which the body would return when it is unloaded. In this case, it is generally preferred that both the stress and the equations of motion be expressed as functions of the material point **X** (the reference state) and, hence, to derive the equations of motion in this state. The first and second Piola–Kirchhoff stress tensors (due to Piola in 1833 and Kirchhoff in 1853) are two alternatives for the definition of the stress in the reference state (Truesdell and Toupin, 1960; Malvern, 1969). The two Piola–Kirchhoff tensors are expressed in terms of the force per unit undeformed area.

The first Piola–Kirchhoff stress tensor is the simpler one. However, it has the disadvantage of being antisymmetric. The second Piola–Kirchhoff stress tensor, on the other hand, is symmetric; thus it has often been used in the formulations of the theory of elasticity. In this context, we present below the definition of the second Piola–Kirchhoff tensor.

Second Piola–Kirchhoff stress tensor

The second Piola–Kirchhoff stress tensor Σ gives a force $d\mathbf{P}_0$ on a unit area dS in the undeformed configuration in relation to the actual force $d\mathbf{P}$ on the corresponding elemental area $d\mathbf{s}$ in the deformed configuration according to the following interpretations

$$d\mathbf{P}_0 = \mathbf{g}^{-1} \cdot d\mathbf{P} \tag{1.45}$$

where $\mathbf{g} = X\nabla\mathbf{x}$ is the inverse of the spatial deformation gradient.

The following relation can be proven (Malvern, 1969) between the second Piola–Kirchhoff stress and the Cauchy stress:

$$\Sigma = \frac{\rho_0}{\rho} \mathbf{g}^{-1} \cdot \sigma \cdot (\mathbf{g}^{-1})^{\mathsf{T}} \tag{1.46a}$$

or, in indicial notations,

$$\Sigma_{Jl} = \frac{\rho_0}{\rho} X_{J,l} \sigma_{,li} \partial_i X_l \tag{1.46b}$$

which shows that the second Piola–Kirchhoff stress tensor Σ is symmetric.

The inverse relation to (1.46) is

$$\sigma = \frac{\rho}{\rho_0} \mathbf{g} \cdot \Sigma \cdot \mathbf{g}^{\mathsf{T}} \tag{1.47a}$$

or,

$$\sigma_{ji} = \frac{\rho}{\rho_0} x_{j,J} \Sigma_{Jl} \partial_l X_i. \tag{1.47b}$$

The symmetry of σ imposes the following condition on Σ so that the latter can be symmetric

$$\Sigma \, \mathbf{g}^{\mathsf{T}} = \mathbf{g} \, \Sigma^{\mathsf{T}} \tag{1.48}$$

where the superscript T indicates the transpose of the associated matrix. Meantime, the equation of motion corresponding to (1.22) in the reference state is expressed by

$$\nabla \cdot (\Sigma \cdot \mathbf{g}^{\mathsf{T}}) + \rho_0 \chi_0 = \rho_0 \frac{\partial^2 \mathbf{x}}{\partial t^2} \tag{1.49a}$$

or

$$\partial_j(\Sigma_{jl}\,\partial_l x_i) + \rho_0(\chi_0)_i = \rho_0\,\frac{\partial^2 x_i}{\partial t^2}. \qquad (1.49b)$$

PROBLEMS

1.1 Briefly discuss the basic postulates and concepts of continuum mechanics leading to the definitions of the stress vector, body force and stress tensor.

1.2 Label the stress tensor components shown.

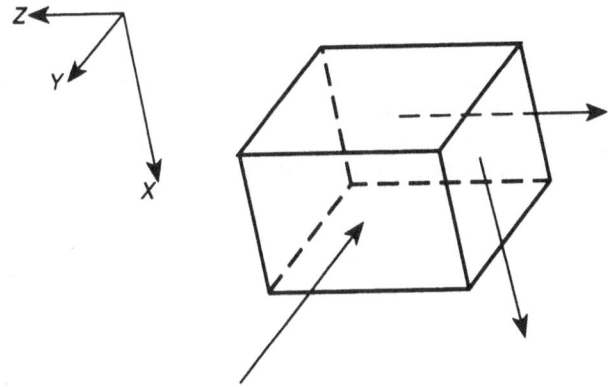

1.3

1. Evaluate the invariants for the stress tensor

$$\sigma_{ij} = \begin{bmatrix} 6 & -3 & 0 \\ -3 & 6 & 0 \\ 0 & 0 & 8 \end{bmatrix}.$$

Determine the principal stress values for this state of stress and show that the diagonal form of the stress tensor yields the same values for the stress invariants.

2. Decompose the stress tensor

$$\sigma_{ij} = \begin{bmatrix} 12 & 4 & 0 \\ 4 & 9 & -2 \\ 0 & -2 & 3 \end{bmatrix}.$$

into its spherical and deviatoric parts and show that the first invariant of the deviator is zero.

3. Determine the principal deviator stress values for the stress tensor

$$\sigma_{ij} = \begin{bmatrix} 10 & -6 & 0 \\ -6 & 10 & 0 \\ 0 & 0 & 1 \end{bmatrix}.$$

1.4 At a point P of a continuous medium, the stress tensor referred to the axes x_i
($i = 1, 2, 3$) is given by

$$\sigma_{ij} = \begin{bmatrix} 15 & -10 & 0 \\ -10 & 5 & 0 \\ 0 & 0 & 20 \end{bmatrix}$$

If the new axes x_i' are chosen by a rotation about the origin for which the
transformation matrix is

$$a_{ij} = \begin{bmatrix} 3/5 & 0 & -4/5 \\ 0 & 1 & 0 \\ 4/5 & 0 & 3/5 \end{bmatrix}$$

determine the traction vectors on each of the primed coordinate planes by
projecting the traction vectors of the original axes onto the primed directions.
Determine the stress deviator components σ_{ij}'. Verify your result using the
transformation relation (consult Appendix A).

1.4 DEFORMATION AND STRAIN

All engineering materials when subjected to external loading may undergo deformation and/or motion. While section 1.3 considers the nature of loads applied to a
body, the present section treats the kinematics of deformation and the various
measures of strain. Thus, in this section, the relationships between the initial positions
of the material points of the continuum and their subsequent positions are considered
without taking into consideration the type of material that we are dealing with or
the imposed boundary conditions.

1.4.1 Lagrangian and Eulerian descriptions

In studying the motion of a continuous medium, we fix our attention on a single
material point with which we associate the geometry of a mathematical Euclidean
point and study its path (trajectory). Such trajectory can be established by determining
the position vector **p** of the point at time t that was initially at position characterized
by a position vector **P** at time $t = 0$ (Fig. 1.6).

This can be expressed with reference to the Cartesian coordinate system shown
in Fig. 1.6 as

$$\mathbf{P} = X_I \mathbf{e}_I, \ \mathbf{p} = X_i \mathbf{e}_i \tag{1.50}$$

where X_I are the values of the rectangular Cartesian components of the position
vector **P**, i.e. at time $t = 0$ (the initial position) corresponding to a current position
coordinates x_i at time t. In (1.50), $\mathbf{e}_I$ and $\mathbf{e}_i$ are the components of the unit base
vectors associated with the rectangular Cartesian frames of reference in the un-
deformed and deformed states, respectively. We have used majuscules to identify

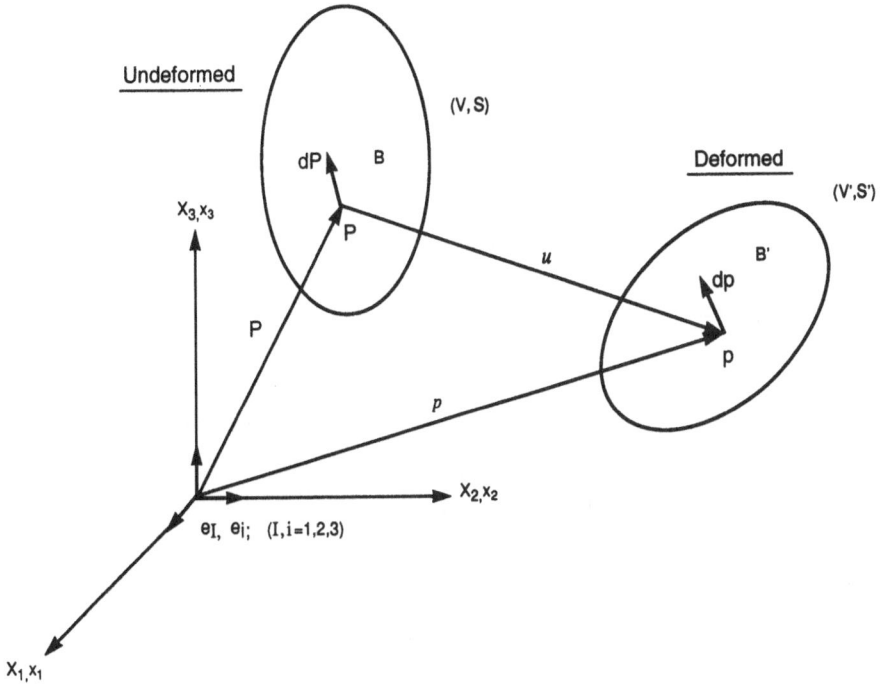

Fig. 1.6 Deformation kinematics of a continuous body.

the undeformed configuration X_I ($I = 1, 2, 3$) and minuscules to designate the deformed configuration x_i ($i = 1, 2, 3$). Hence, the identity of the material point is X_I and its subsequent motion is described by

$$x_i = x_i (X_I, t), \quad X_I = X_I (x_i, t). \tag{1.51}$$

The X_I coordinates are known in continuum mechanics by the **material** or **Lagrangian** description while the x_i coordinates are called the **spatial** or **Eulerian** description. Equation (1.51) expresses the evolution of the deformation process in the body as a function of time. From a continuum mechanics point of view, the functions $x_i (X_I, t)$ are single-valued continuous functions whose Jacobian of transformation does not vanish, namely

$$J = |x_{i,I}| = \tfrac{1}{6} \varepsilon_{ILM} \varepsilon_{ilm} x_{i,I} x_{l,L} x_{m,M} \neq 0 \tag{1.52}$$

where ε_{ILM}, ε_{ilm} are the alternating tensors. Equation (1.52) expresses the form of the implicit function theorem of calculus. It is a basic relation for securing the axiom of continuity in continuum mechanics. It points out the fact that the matter is indestructible, i.e. no region of positive, finite volume of matter may be deformed into a zero or an infinite volume. Another axiom secured by this equation is that the matter is impenetrable, that is the motion carries every region into a region, every surface into a surface and every curve into a curve.

With reference to Fig. 1.6, one may also consider a line element d**P** deformed into d**p**. The two line elements are expressed, respectively, by

$$d\mathbf{P} = \mathbf{e}_I \, dX_I = \mathbf{e}_I X_{I,i} \, dx_i,$$
$$d\mathbf{p} = \mathbf{e}_i \, dx_i = \mathbf{e}_i x_{i,I} \, dX_I. \tag{1.53}$$

In equations (1.53), $X_{I,i}$ and $x_{i,I}$ are the deformation gradients referred, respectively, to the spatial and material coordinates. These deformation gradients are the most 'basic or primitive measures of strain'.

Alternative measures of strain can be expressed from the definition of the square of the line element. Thus, in the undeformed configuration (Fig. 1.6),

$$dS^2 = d\mathbf{P} \cdot d\mathbf{P} = dX_K \, dX_K.$$

The above equation can be written, using the chain rule of partial differentiation, as

$$dS^2 = X_{K,j} X_{K,i} \, dx_j \, dx_i$$
$$= c_{ji} \, dx_j \, dx_i \tag{1.54}$$

where

$$c_{ji} = X_{K,j} X_{K,i}$$

is known as the 'Cauchy deformation tensor'.

Similarly, one may express, with reference to Fig. 1.6, the square of the line element in the deformed configuration as

$$ds^2 = d\mathbf{p} \cdot d\mathbf{p} = dx_k \, dx_k = x_{k,J} x_{k,I} \, dX_J \, dX_I$$
$$= C_{JI} \, dX_J \, dX_I$$

where

$$C_{JI} = x_{k,J} x_{k,I} \tag{1.55}$$

is referred to as the 'Green deformation tensor'.

Both c_{ji} and C_{JI} are symmetric and positive definite. They can be shown to be tensors.

Alternatively, one may consider that the difference between ds^2 and dS^2 to be a measure of strain. Thus,

$$ds^2 - dS^2 = dx_i \, dx_i - dX_L \, dX_L$$
$$= C_{KL} \, dX_K \, dX_L - \delta_{KL} \, dX_K \, dX_L$$
$$= (C_{KL} - \delta_{KL}) \, dX_K \, dX_L$$
$$= 2E_{KL} \, dX_K \, dX_L$$

where relation (1.55) is used and

$$E_{KL} = \tfrac{1}{2}(C_{KL} - \delta_{KL}) \tag{1.56}$$

is the 'material or Lagrangian strain tensor'.

Similarly,

$$ds^2 - dS^2 = dx_i \, dx_i - dX_L \, dX_L$$
$$= \delta_{ij} \, dx_i \, dx_j - c_{ij} \, dx_i \, dx_j$$
$$= (\delta_{ij} - c_{ij}) \, dx_i \, dx_j$$
$$= 2\varepsilon_{ij} \, dx_i \, dx_j$$

where relation (1.54) is used and

$$\varepsilon_{ij} = \tfrac{1}{2}(\delta_{ij} - c_{ij}) \tag{1.57}$$

is the '**spatial or Eulerian strain tensor**'. It can be shown that E_{KL} and ε_{kl} are related

$$E_{KL} = \varepsilon_{kl} \, x_{k,K} \, x_{l,L}$$

and

$$\varepsilon_{kl} = E_{KL} \, X_{K,k} \, X_{L,l}. \tag{1.58}$$

The reader may find it of interest to prove the validity of the two expressions (1.58).
With reference to Fig. 1.6, one may write the following relation:

$$\mathbf{p} = \mathbf{P} + \mathbf{u}$$

or

$$x_i \mathbf{e}_i = X_K \mathbf{e}_K + u_K \mathbf{e}_K$$

and

$$dx_i \, \mathbf{e}_i = dX_K \, \mathbf{e}_K + u_{K,L} \, dX_L \, \mathbf{e}_K$$
$$= (\delta_{KL} + u_{K,L}) \, dX_L \, \mathbf{e}_K.$$

Hence

$$dS^2 = dx_i \, dx_i = (\delta_{KL} + u_{K,L})(\delta_{KM} + u_{K,M}) \, dX_L \, dX_M.$$

Now, if we choose $ds^2 - dS^2$ as a measure of strain, it can be shown that

$$ds^2 - dS^2 = (u_{M,L} + u_{L,M} + u_{K,L} u_{K,M}) \, dX_L \, dX_M$$
$$= 2E_{LM} \, dX_L \, dX_M$$

where

$$E_{LM} = \tfrac{1}{2}(u_{M,L} + u_{L,M} + u_{K,L} u_{K,M}) \tag{1.59}$$

is an alternative expression for the material or Lagrangian strain tensor expressed previously by (1.56).
One may also choose, with reference to Fig. 1.6, the formulation

$$\mathbf{P} = \mathbf{p} - \mathbf{u},$$

i.e.

$$X_l \mathbf{e}_l = x_k \mathbf{e}_k - u_k \mathbf{e}_k$$

and

$$dX_l \, \mathbf{e}_l = dx_k \, \mathbf{e_k} - u_{k,l} \, dx_l \, \mathbf{e}_k$$
$$= (\delta_{kl} - u_{k,l}) \, dx_l \, \mathbf{e}_k.$$

Thus

$$dS^2 = dX_l \, dX_l = (\delta_{kl} - u_{k,l})(\delta_{km} - u_{k,m}) \, dx_l \, dx_m$$
$$= (\delta_{lm} - u_{m,l} - u_{l,m} + u_{k,l} u_{k,m}) \, dx_l \, dx_m$$

and one can show that

$$ds^2 - dS^2 = (u_{m,l} + u_{l,m} - u_{k,l} u_{k,m}) \, dx_l \, dx_m$$
$$= 2\varepsilon_{lm} \, dx_l \, dx_m$$

where

$$\varepsilon_{lm} = \tfrac{1}{2}(u_{m,l} + u_{l,m} - u_{k,l} u_{k,m}) \qquad (1.60)$$

is another expression for the spatial or Eulerian strain tensor given earlier by equation (1.57). Similar to the stress tensor, both the Lagrangian strain tensor and the Eulerian strain tensor are symmetric tensors; hence, each has only six independent components in a three-dimensional space.

1.4.2 Infinitesimal strain and rotation

We considered in the foregoing the formulations of the three-dimensional nonlinear measures of strain as based only on the implications of assumed continuity. In this section, we introduce some of the simplifying assumptions in the theory of continuum mechanics with the aim of reducing the complexity of mathematics which otherwise would be involved. The definitions of infinitesimal strains and rotations depend upon the following two assumptions:

1. that the occurring deformations $u_i \, (x_k, t)$ or $u_I \, (X_K, t)$ are much smaller than the least dimension of the free body under consideration;
2. that the deformation gradient $u_{i,j}$ or $u_{I,J} \ll 1$.

Recall the expression of nonlinear measure of strain given by equation (1.59), that is

$$E_{IJ} = \tfrac{1}{2}(u_{I,J} + u_{J,I} + u_{K,I} u_{K,J}) \qquad (1.61)$$

This expression becomes the infinitesimal or linear Lagrangian (material) strain tensor $\hat{E}_{IJ}$ on requiring the nonlinear term in it to be zero, i.e.

$$u_{K,I} u_{K,J} = 0.$$

Accordingly

$$\hat{E}_{IJ} = \tfrac{1}{2}(u_{I,J} + u_{J,I}) = u_{(I,J)} \tag{1.62}$$

is the infinitesimal Lagrangian strain tensor. It is apparent from the above expression that $\hat{E}_{IJ}$ represents the symmetric portion of the tensor $u_{I,J}$. This is indicated in (1.62) by $u_{(I,J)}$ following our notations in Appendix A.

Introducing the Lagrangian infinitesimal rotation tensor

$$\hat{\Omega}_{IJ} = \tfrac{1}{2}(u_{I,J} - u_{J,I}) = u_{[I,J]} \tag{1.63}$$

where $u_{[I,J]}$ is the skew-symmetric portion of $u_{I,J}$. Thus,

$$u_{I,J} = \hat{E}_{IJ} + \hat{\Omega}_{IJ} = u_{(I,J)} + u_{[I,J]}. \tag{1.64}$$

Repeating the same procedure for the infinitesimal or linear Eulerian strain tensor, then, with reference to (1.60),

$$\hat{\varepsilon}_{lm} = \tfrac{1}{2}(u_{l,m} + u_{m,l}) = u_{(l,m)} \tag{1.65}$$

where

$$u_{l,m} = \hat{\varepsilon}_{lm} + \hat{\omega}_{lm} \tag{1.66}$$

and

$$\hat{\omega}_{lm} = \tfrac{1}{2}(u_{l,m} - u_{m,l}) = u_{[l,m]} \tag{1.67}$$

is the infinitesimal Eulerian rotation tensor, which is skew-symmetric (Appendix A).

In three dimensions, it is possible to express a dual vector $\boldsymbol{\omega}: \omega_k$ in terms of the skew-symmetric tensor $\hat{\omega}_{ij}$. That is

$$\omega_k = \tfrac{1}{2}\varepsilon_{kij}\hat{\omega}_{ij}$$

or

$$\boldsymbol{\omega} = \tfrac{1}{2} \operatorname{curl} \mathbf{u} \tag{1.68}$$

where ε_{kij} is the alternating tensor.

At the same time, since $\hat{\omega}_{ij}$ is antisymmetric, it can be shown that (1.68) has a unique inverse, i.e.

$$\hat{\omega}_{ij} = \varepsilon_{ijk}\omega_k. \tag{1.69}$$

Hence, $\hat{\omega}_{ij}$ may be called the dual tensor of a vector ω_k. The latter vector is referred to as the rotation vector of the displacement field u_i.

1.4.3 Equivalence between infinitesimal Lagrangian strain and infinitesimal Eulerian strain

With reference to Fig. 1.6, the deformation vector $\mathbf{u}$ can be expressed as

$$\mathbf{u} = \mathbf{p} - \mathbf{P}$$

which results in the expression

$$u_{i,j} = \delta_{ij} - X_{I,J} \tag{1.70}$$

or, alternatively,

$$u_{I,j} = \delta_{ij} - X_{I,j}. \tag{1.71}$$

Thus, with reference to (1.70) and (1.71), one can write

$$u_{i,j} = u_{I,j} = u_{I,K}X_{K,j}$$

$$= u_{I,K}(\delta_{KJ} - u_{k,j})$$

$$= u_{I,J} - u_{I,J}u_{k,j}$$

which can be approximated, by neglecting the nonlinear term, as

$$u_{i,j} = u_{I,j} = u_{I,J}.$$

That is, within the framework of linear strains expressed by equations (1.62) and (1.65),

$$\hat{\varepsilon}_{kl} = \hat{E}_{KL}. \tag{1.72}$$

At this point, it should be emphasized that both $\hat{\varepsilon}_{kl}$ and $\hat{E}_{KL}$ cannot be considered as strain measures and they are, in fact, only approximations of strain measures within the context of the infinitesimal strain theory. Following (1.72), we shall use in the remaining chapters of the book, the notation ε_{ij} to denote the infinitesimal strain, i.e.

$$\varepsilon_{ij} = \tfrac{1}{2}(u_{i,j} + u_{j,i}) \tag{1.73}$$

unless otherwise mentioned.

1.4.4 Principal strains, principal directions and strain invariants

In section 1.3.6, we considered the treatment of an eigenvalue problem to determine the principal values of stress, principal directions and the three invariants appearing in equation (1.34). This procedure applies to every symmetric second-order tensor. Hence, on following the same analysis, it can be shown that the invariants of the strain tensor are expressed by

$$II_1 = \varepsilon_{11} + \varepsilon_{22} + \varepsilon_{33} = \varepsilon_{kk}, \tag{1.74a}$$

$$II_2 = \tfrac{1}{2}(\varepsilon_{ij}\varepsilon_{ij} - \varepsilon_{ii}\varepsilon_{jj})$$

$$= -(\varepsilon_{11}\varepsilon_{22} + \varepsilon_{22}\varepsilon_{33} + \varepsilon_{33}\varepsilon_{11}) + \varepsilon_{23}^2 + \varepsilon_{31}^2 + \varepsilon_{12}^2 \tag{1.74b}$$

and

$$II_3 = \begin{vmatrix} \varepsilon_{11} & \varepsilon_{12} & \varepsilon_{13} \\ \varepsilon_{21} & \varepsilon_{22} & \varepsilon_{23} \\ \varepsilon_{31} & \varepsilon_{32} & \varepsilon_{33} \end{vmatrix} = \det \varepsilon_{ij}$$

$$= \tfrac{1}{6}\varepsilon_{ijk}\varepsilon_{pqr}\varepsilon_{ip}\varepsilon_{jq}\varepsilon_{kr}.$$

The first strain invariant II_1 has a simple geometrical meaning in the case of infinitesimal strain. It represents the change in volume per unit volume, i.e.

$$\mathrm{II}_1 = \Delta V/V = \varepsilon_{kk}. \tag{1.75}$$

For this reason, ε_{kk} is called the cubical dilatation. If a two-dimensional state (plane strain) is considered, the first invariant represents the change of area per unit area of the surface under strain. In the definition of finite strain, the sum of principal strains does not have such a simple interpretation.

Whereas in our analysis of the stress, section 1.3, we have only dealt with one definition of the stress, we have encountered, here, five different measures of strain. The invariants of the latter are interrelated in terms of algebraic relations of each other. The reader is referred, in this context, to relevant texts in continuum mechanics (as in the list of further reading at the end of this chapter).

1.4.5 Compatibility conditions

We deal now with the problem of determining the displacement components u_i when the strain components are known. Consider, for the simplification of presentation, the expression of linear strain as presented by (1.73), i.e.

$$\varepsilon_{ij} = \tfrac{1}{2}(u_{i,j} + u_{j,i}). \tag{1.76}$$

The problem is how one would integrate the differential equation (1.76) to determine the components u_i.

Since we have six equations corresponding to the six independent components of ε_{ij} for three unknown functions u_i, the system of equation (1.76) will not have a single-valued solution in general if the functions ε_{ij} were arbitrarily assigned. However, we would expect that a solution for this equation may exist only if the functions ε_{ij} satisfy certain conditions. The conditions of integrability of (1.76) are referred to as the compatibility conditions. The latter are to be satisfied by the strain components and may be determined by eliminating the components u_i from (1.76).

Differentiating (1.76) twice with respect to the coordinates, it follows that

$$\varepsilon_{ij,kl} = \tfrac{1}{2}(u_{i,jkl} + u_{j,ikl}). \tag{1.77}$$

We interchange subscripts in the above relation, i.e.

$$\begin{aligned}
\varepsilon_{kl,ij} &= \tfrac{1}{2}(u_{k,lij} + u_{l,kij}), \\
\varepsilon_{jl,ik} &= \tfrac{1}{2}(u_{j,lik} + u_{l,jik}), \\
\varepsilon_{ik,jl} &= \tfrac{1}{2}(u_{i,kjl} + u_{k,ijl}).
\end{aligned} \tag{1.78}$$

This leads to the following restriction specifying the compatibility conditions:

$$\varepsilon_{ij,kl} + \varepsilon_{kl,ii} - \varepsilon_{ik,jl} - \varepsilon_{jl,ik} = 0. \tag{1.79}$$

Equation (1.79) was first obtained by St. Venant in 1860 and is named after him. This equation represents 81 equations of which six only are essential. The remaining equations are repetitions owing to the symmetry of the strain tensor. The six compatibility equations are independent and may be expressed in an uncondensed notation as

$$\frac{\partial^2 \varepsilon_{11}}{\partial x_2^2} + \frac{\partial^2 \varepsilon_{22}}{\partial x_1^2} - 2\frac{\partial^2 \varepsilon_{12}}{\partial x_1 \partial x_2} = 0,$$

$$\frac{\partial^2 \varepsilon_{22}}{\partial x_3^2} + \frac{\partial^2 \varepsilon_{33}}{\partial x_2^2} - 2\frac{\partial^2 \varepsilon_{23}}{\partial x_2 \partial x_3} = 0, \qquad (1.80)$$

$$\frac{\partial^2 \varepsilon_{33}}{\partial x_1^2} + \frac{\partial^2 \varepsilon_{11}}{\partial x_3^2} - 2\frac{\partial^2 \varepsilon_{31}}{\partial x_3 \partial x_1} = 0,$$

and

$$\frac{-\partial^2 \varepsilon_{11}}{\partial x_2 \partial x_3} + \frac{\partial}{\partial x_1}\left(-\frac{\partial \varepsilon_{23}}{\partial x_1} + \frac{\partial \varepsilon_{31}}{\partial x_2} + \frac{\partial \varepsilon_{12}}{\partial x_3}\right) = 0,$$

$$\frac{-\partial^2 \varepsilon_{22}}{\partial x_3 \partial x_1} + \frac{\partial}{\partial x_2}\left(\frac{\partial \varepsilon_{23}}{\partial x_1} - \frac{\partial \varepsilon_{31}}{\partial x_2} + \frac{\partial \varepsilon_{12}}{\partial x_3}\right) = 0,$$

$$\frac{-\partial^2 \varepsilon_{33}}{\partial x_1 \partial x_2} + \frac{\partial}{\partial x_3}\left(\frac{\partial \varepsilon_{23}}{\partial x_1} + \frac{\partial \varepsilon_{31}}{\partial x_2} - \frac{\partial \varepsilon_{12}}{\partial x_3}\right) = 0.$$

In the solution of a boundary value problem, within the realm of continuum mechanics, if the displacement field is unknown and the displacement components u_i are required to be continuous and single-valued functions of the coordinates, the compatibility requirement would then be fulfilled. On the other hand, if the displacements are not explicitly retained as unknowns, the compatibility conditions must then be imposed on the strain field to ensure that there exists a continuous single-valued displacement distribution corresponding to the strain distribution.

PROBLEMS

1.5 The vector $t_i = \varepsilon_{ijk} T_{jk}$ is called a 'dual vector' of the tensor T_{jk}. Show that, if the dual vector vanishes, the tensor is symmetric.

1.6 Show that the strain tensor, using the Eulerian approach, is given by

$$\varepsilon_{ik} = \tfrac{1}{2}(u_{j,k} + u_{k,j} - u_{i,j}u_{i,k}).$$

Compare with the Lagrangian strain tensor and show why no distinction between the two is necessary in the case of small displacement theory.

1.7 Show that the large deformation strain is given by

$$\varepsilon_L = \varepsilon_{ij}\, l_i\, l_j$$

where

$$\varepsilon_L = \frac{ds - dS}{dS}$$

and

$$l_i = \frac{dx_i}{dS}.$$

1.8 The most general form of a linear relationship between the stress and strain components of an elastic isotropic solid is represented by the generalized Hooke's law as follows:

$$\sigma_{ij} = A_{ijkl}\,\varepsilon_{kl}.$$

Prove that the elastic constants are the components of a Cartesian tensor of the fourth order (consult Appendix A).

1.9 Assume that the strain tensor is given as

$$\begin{bmatrix} 2 & 3 & 2 \\ 3 & 2 & 1 \\ 2 & 1 & 2.5 \end{bmatrix}.$$

Determine the principal strains and corresponding principal directions.

1.10 Let a_i be the original direction cosines of an 'undeformed' line element dS in a solid and a_j^* be the final direction cosines of the 'deformed' element ds. Show that

$$a_j^* = \frac{(\delta_{jk} + u_{j,k})a_k}{(1 + \varepsilon_L)^{1/2}}$$

where

$$\varepsilon_L = \frac{ds - dS}{dS}.$$

Show also that the rotation of a line element can be written as

$$\omega_i = \tfrac{1}{2}\,\mathrm{curl}\,u_i = \tfrac{1}{2}\varepsilon_{ijk}(\partial_j u_k - \partial_k u_j).$$

1.11 Find

1. the Green strain tensor C_{KL},
2. the Cauchy strain tensor c_{kl},
3. the Lagrangian strain tensor E_{KL},
4. the Eulerian strain tensor ε_{kl} and
5. the infinitesimal strain tensor

for the following two displacement fields:

- simple extension

$$z_1 = (1 + \varepsilon)\,Z_1, \quad z_2 = Z_2, \quad z_3 = Z_3;$$

- simple shear

$$z_1 = Z_1 + KZ_2, \quad z_2 = Z_2, \quad z_3 = Z_3.$$

1.12 The displacement field of a body is described by

$$u_i = A\,\frac{z_1 z_3}{r^3}, \quad u_2 = A\,\frac{z_2 z_3}{r^3}, \quad u_3 = A\left(\frac{z_3^2}{r^3} + \frac{\lambda + 3\mu}{\lambda + \mu}\,\frac{1}{r}\right),$$

where $r = (z_i z_i)^{1/2}$ and A, λ and μ are constants.

1. Determine the infinitesimal strains ε_{kl} and rotations ω_{kl}.
2. Sketch the deformed shape of a spherical cavity $r = r_0$.
3. Determine the principal strains.
4. Determine the principal axes.

1.5 CONSTITUTIVE RELATIONS

Different materials of the same geometry may respond differently under identical external effects. Such a difference in response is often attributed to the inherent constitution of the material. Consequently, the response behaviour of a particular material, or of a class of such materials, is described mathematically by so-called 'constitutive relations'. These constitutive equations define the response behaviour of idealized media within a specific range of external effects. Accordingly, they only approximate the response characteristics of real materials. In general terms, constitutive relations establish the connection between the stimuli acting on the material specimen and the evolution of the occurring response. In the majority of situations, the stimuli are the external forces, or the stresses caused by them, and the evolution of the response is expressed by the histories of both the deformation, or the calculated strain, and the temperature. In a continuum mechanics sense, a general form of a constitutive equation may be expressed as (Hunter, 1976)

$$\sigma_{ij} = f_{ij} \text{ (history of deformation, history of temperature)} \tag{1.81}$$

where σ_{ij} is the stress tensor and f_{ij} are the components of a second-order tensorial response function. In view of the form of the constitutive equation (1.81), continuum mechanics constitutive formulations are deterministic and the science of continuum mechanics itself is a branch of classical deterministic physics.

The constitutive relations of different classes of engineering materials are particular forms of (1.81). For elastic materials, for instance, the stress tensor σ_{ij} is a function of the current strain and temperature. In case of viscoelastic materials, as will be seen in the following chapters, constitutive equations should account for the time history of the deformation process and that of the temperature. Hence, constitutive equations for viscoelastic materials, in general, are of the form of equation (1.81). In this equation, the choice of the independent variables pertaining to f_{ij} is usually guided by the experimental results but, in most situations, this choice is restricted by a number of physical principles. In this context, a properly formulated constitutive equation must satisfy certain invariance principles (Eringen, 1962, 1967; Hunter, 1976) as follows.

1. The constitutive relation is invariant with respect to different stationary coordinate systems. This requirement is readily satisfied by expressing the constitutive law in tensorial form.
2. The constitutive relation is invariant with respect to coordinate systems in an arbitrary relative motion. This condition is usually dealt with within the context of 'material frame indifference' (Hunter, 1976). This is translated into the requirement that the transformation law relating the components of the tensorial function f_{ij} in different coordinate frames in relative motion is exactly the same as the ordinary tensor transformation law.

Conformity of the form of the constitutive relation to the invariance requirements mentioned above together with the assumption of isotropy of the continuous body impose restrictions on the form of (1.81) and lead to explicit forms of constitutive equations for particular materials under specific conditions. The material functions or parameters characterizing the explicit form of the constitutive relation would be then characteristic of the particular material under consideration.

REFERENCES

Cauchy, A. L. (1827) De la pression ou tension dans un corps solide, in *Exercices de Mathématique* (see Love (1944, pp. 8–9)).
Cauchy, A. L. (1828) Sur les équations qui expriment les conditions d'équilibre ou les lois de mouvement intérieur d'un corps solide, in *Exercices de Mathématique* (see Love (1944, pp. 8–9)).
Eringen, A. C. (1962) *Nonlinear Theory of Continuous Media*, McGraw-Hill, New York.
Eringen, A. C. (1967) *Mechanics of Continua*, Wiley, New York.
Flügge, W. (1972) *Tensor Analysis and Continuum Mechanics*, Springer, Berlin.
Fung, Y. C. (1965) *Foundations of Solid Mechanics*, Prentice-Hall, Englewood Cliffs, NJ.
Hunter, S. C. (1976) *Mechanics of Continuous Media*, Ellis Horwood, Chichester.
Malvern, L. E. (1969) *Introduction to the Mechanics of a Continuous Medium*, Prentice-Hall, Englewood Cliffs, NJ.
Truesdell, C. and Toupin, R. A. (1960) The classical field theories, in *Handbuck du Physik*, Vol. III/1 (ed. S. Flügge), Springer, Berlin.

FURTHER READING

Bowen, R. M. (1989). *Introduction to Continuum Mechanics for Engineers*, Plenum, New York.
Chung, T. J. (1988) *Continuum Mechanics*, Prentice-Hall, Englewood Cliffs, NJ.
Coleman, B. D. and Noll, W. (1960) An approximation theorem for functionals with applications in continuum mechanics. *Arch. Ration. Mech. Anal.*, **6**, 355–70.
Davis, J. L. (1987) *Introduction to Dynamics of Continuous Media*, Macmillan, New York.
Fredrick, D. and Chang, T. S. (1965) *Continuum Mechanics*, Allyn and Bacon, Boston, MA.
Green, A. E. and Adkins, J. E. (1960) *Large Elastic Deformations and Nonlinear Continuum Mechanics*, Clarendon, Oxford.

Gurtin, M. E. (1981) *An Introduction to Continuum Mechanics*, Academic Press, New York.

Gurtin, M. E. and Williams, W. O. (1967) An axiomatic foundation for continuum thermodynamics. *Arch. Ration. Mech. Anal.*, **26**, 83–117.

Jaunzemis, W. (1967) *Continuum Mechanics*, Macmillan, New York.

Jessop, H. T. (1950) The determination of the principal stress differences at a point in a three-dimensional photoelastic model. *Br. J. Appl. Phys.*, **1**, 184–9.

Lai, W. M., Rubin, D. and Krempl, E. (1978) *Introduction to Continuum Mechanics (SI/Metric Units)*, Pergamon, New York.

Leigh, D. C. (1968) *Nonlinear Continuum Mechanics*, McGraw-Hill, New York.

Love, A. E. H. (1944) *A Treatise on the Mathematical Theory of Elasticity*, 4th edn, Dover Publications, New York.

Mindlin, R. D. and Tiersten, H. F. (1962) Effect of couple-stresses in linear elasticity. *Arch. Ration. Mech. Anal.*, **11**, 415–48.

Noll, W. (1958) A mathematical theory of the mechanical behaviour of continuous media. *Arch. Ration. Mech. Anal.*, **2**, 197–226.

Prager, W. (1961) *Introduction to Mechanics of Continua*, Ginn, Boston, MA.

Sneddon, I. N. and Hill, R. (eds.) (1960–1963) *Progress in Solid Mechanics*, Vol. 1 (1960), Vol. 2 (1961), Vol. 3 (1963), Vol. 4 (1963), North-Holland, Amsterdam.

Sommerfeld, A. (1950) *Mechanics of Deformable Bodies*, Academic Press, New York.

Truesdell, C. (1965a) The nonlinear field theories of mechanics, in *Encyclopedia of Physics*, Vol. III/3 (ed. S. Flügge), Springer, Berlin.

Truesdell, C. (1965b) *The Elements of Continuum Mechanics*, Springer, New York.

Washizu, K. (1958) A note on the condition of compatability. *J. Math. Phys.*, **36**, 306–12.

Williams, W. O. (1970) Thermodynamics of rigid continua. *Arch. Ration. Mech. Anal.*, **36**, 270–84.

2

Linear viscoelasticity

2.1 INTRODUCTION

As discussed in the introduction of this book, many engineering materials such as polymeric and rubberlike materials, and metals at elevated temperatures, flow when subjected to stress or strain. Such flow is accompanied by the dissipation of energy due to some internal loss mechanism (for example, bond breakage and bond formation reaction, dislocations, formation of substructures in metals). Materials of this type are 'viscoelastic' in response. The description 'viscoelastic' is due to the fact that such materials exhibit both 'elastic' and 'viscous' properties. Figure 2.1 illustrates the differences in strain response of elastic, viscous and viscoelastic specimens when the three specimens are subjected to a constant stress of unit magnitude. The stress is applied at time $t = 0$ to undisturbed specimens and maintained constant for time duration t_1 (Fig. 2.1(a)). As shown in Fig. 2.1(b), the strain–time response of the elastic specimen has the same form as the applied stress. On application of the load, the strain reaches instantaneously a certain level ε_0 and then remains constant. For the viscous fluid (Fig. 2.1(c)) the material flows at a constant rate and the strain response is proportional to the time. For the viscoelastic specimen (Fig. 2.1(d)) there is a relatively rapid increase in the strain response for small values of t immediately after the application of the load. As t increases, the slope of the curve decreases and, as $t \to \infty$, the slope may approach zero or finite value provided that the applied stress maintained is constant.

On removal of the load at time t_1, the strains in the three specimens will recover in the manners shown in Fig. 2.1. The perfectly elastic solid will recover instantaneously on removal of the load (Fig. 2.1(b)), but the viscous fluid will not recover (Fig. 2.1(c)). Meantime, on removal of the load, the viscoelastic specimen will recover immediately its elastic deformation; however, the retarded part of the response will require time for recovery.

Under constant stress, the creep strain in a viscoelastic material may be divided, with reference to Fig. 2.2, into the following three components (e.g. Lethersich, 1950).

1. Instantaneous (immediate) elastic strain $\varepsilon_e(0^+)$. In a polymeric material, for instance, this part of the strain is attributable to bond stretching and bending including the

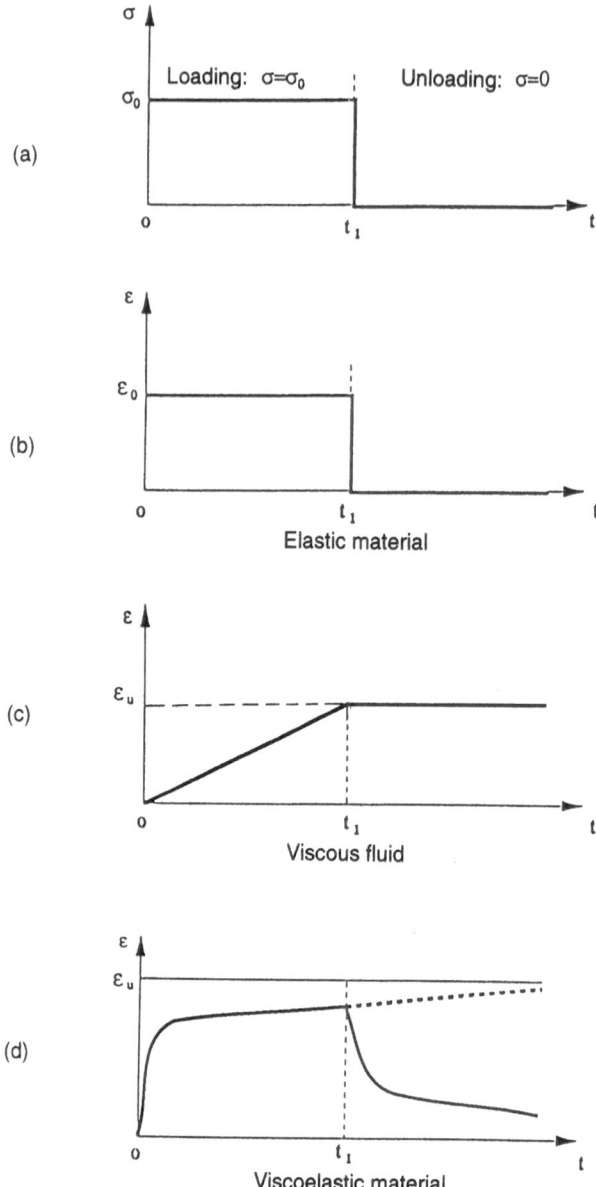

Fig. 2.1 Comparison of strain response for elastic, viscous and viscoelastic material specimens under constant stress of unit magnitude until time t_1.

deformation of weak Van der Waals bonds between the molecular chains. This strain is reversible and disappears on removal of the stress.

2. Delayed elastic strain $\varepsilon_d(t)$. The rate of increase of this part of strain decreases steadily with time. It is also elastic, but, after the removal of the load, it requires

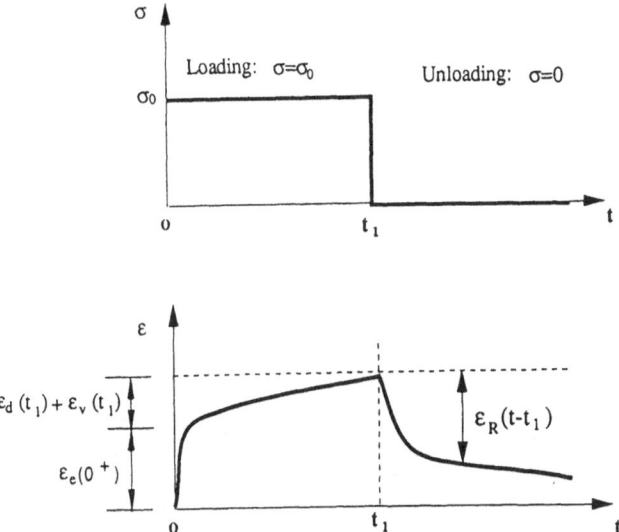

Fig. 2.2 Creep and recovery of a viscoelastic material specimen subjected to a constant stress of unit magnitude until time t_1.

time for complete recovery. It is often called 'primary creep' or 'elastic aftereffect', among other terms. In a polymeric material, the delayed elastic strain is attributable to, for instance, chain uncoiling.

3. Viscous flow $\varepsilon_v(t)$. It is an irreversible component of strain which may or may not increase linearly with time of stress application. In a polymeric substance, it is characteristic of interchain slipping. It is often referred to as 'secondary creep' or 'nonrecoverable strain'.

On unloading the viscoelastic specimen at t_1, the instantaneous elastic response recovers immediately and the delayed elastic response recovers gradually, but the viscous flow remains (e.g. Ward, 1983).

From a phenomenological point of view, two aspects of viscoelastic behaviour are dealt with, i.e. creep response under constant stress and stress-relaxation response under constant strain. As will be dealt with later in this chapter, the correlation between these two aspects of response constitutes an important characteristic of the development of the linear viscoelasticity theory.

2.1.1 Creep response (under constant stress)

In a simple creep experiment, the undisturbed material specimen is subjected initially, at time $t = 0$, to a stress $\sigma_0 = \sigma(0^+)$ which is maintained constant during the experiment; meanwhile, the time-dependent strain $\varepsilon(t)$ is observed. In the linear viscoelastic case, the creep response follows, in general, the pattern discussed above in conjunction with Figs 2.1(d) and 2.2. In this case, the total creep strain can be considered to be the sum of the three separate parts $\varepsilon_e(0^+)$, $\varepsilon_d(t)$, and $\varepsilon_v(t)$ mentioned earlier. Further, the magnitudes of these individual parts of the strain are proportional to the magnitude of the stress input. Accordingly, a creep compliance function $F(t)$, which is a function of time only, may be defined, in the linear viscoelastic case, as

$$F(t) = \frac{\varepsilon(t)}{\sigma(0^+)} = F_e(0^+) + F_d(t) + F_v(t). \tag{2.1}$$

In the above relation, the compliance function $F_v(t)$ which defines the Newtonian flow can be neglected for solid materials with large flow viscosities, e.g. rigid polymers at ordinary temperatures. Linear amorphous polymers, on the other hand, would demonstrate a finite $F_v(t)$ at temperatures above their glass transitions. However, at low temperatures, the viscoelastic behaviour of the latter polymers may be influenced only by the compliances $F_e(t)$ and $F_d(t)$. The same could be valid for the case of high linked polymers and, to a reasonable approximation, in the case of highly crystalline polymers. In general, the separation of the creep compliance $F(t)$, for a particular material at any given temperature, into the compliances $F_e(0^+)$, $F_d(t)$ and $F_v(t)$ may not be an easy task and could involve arbitrary division.

2.1.2 Creep and recovery

With reference to Fig. 2.2, consider the case where the stress σ_0 is applied to an undisturbed specimen at time $t = 0$ and removed at time $t = t_1$. Thus, on the assumption of linear viscoelastic behaviour, the total creep strain $\varepsilon(t)$ at any instant of time $t > t_1$ is given by the superposition of the two individual strains, i.e. $\varepsilon_e = \sigma_0 F(t)$ corresponding to loading the specimen at $t = 0$ and $\varepsilon_R = -\sigma_0 F(t - t_1)$ corresponding to unloading at $t = t_1$. That is,

$$\varepsilon(t) = \sigma_0 F(t) - \sigma_0 F(t - t_1). \tag{2.2}$$

The recovery strain, $\varepsilon_R(t - t_1)$, is defined as the difference between the anticipated creep under the initial stress and the actual measured creep strain. This is shown in Fig. 2.2. Examples of the creep response and creep recovery of a number of engineering materials are shown in Figs 2.3–2.5.

2.1.3 Stress relaxation (under constant strain)

In a simple stress-relaxation experiment, the material specimen is subjected initially to a constant strain $\varepsilon(t) = \varepsilon(0^+)$ and the time-dependent stress response is observed (Fig. 2.6). As shown in the latter figure for a viscoelastic material, the stress-relaxation response is monotonically decreasing with the time. On the assumption of a linear viscoelastic behaviour, a stress-relaxation modulus, which is a function of time only,

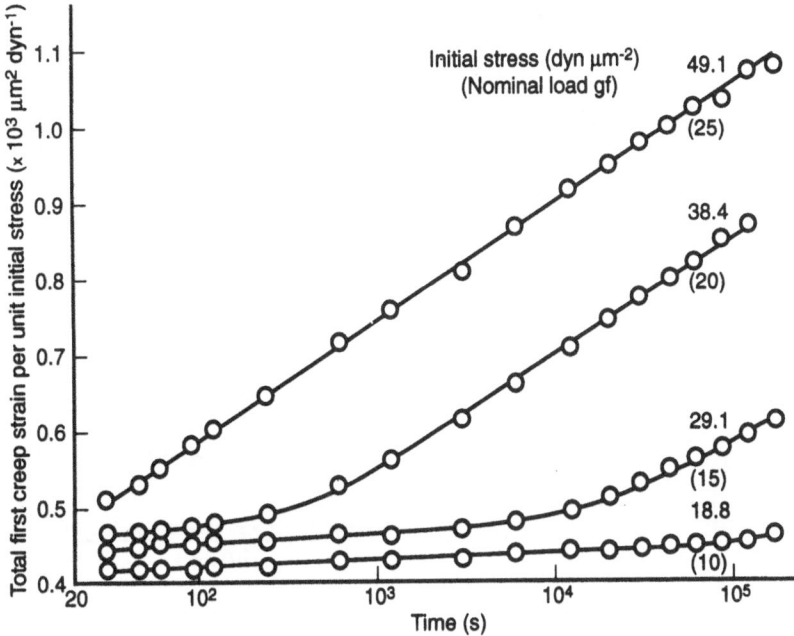

Fig. 2.3 First creep curves reduced in deformation by factors of initial stress. Individual dry summerwood fibres of a longleaf pine pulp after conditioning at 50% RH and 23°C. (Source: Hill, R. L. (1967) The creep behaviour of individual pulp fibres under tensile stress. *Tappi*, **50**(8), 432–40. Reprinted by permission of Tappi.)

is defined as

$$R(t) = \frac{\sigma(t)}{\varepsilon(0^+)}. \tag{2.3}$$

In a stress-relaxation experiment, such as described here, viscous flow affects the limiting value of stress. In the presence of viscous flow, the stress may decay to zero at sufficient long times. On the other hand, if there is no viscous flow, the stress decays to a finite value. This would result in an equilibrium or 'relaxed' modulus $R_\infty = R(\infty)$ at infinite time (e.g. Lockett, 1972; Gittus, 1975). Examples of the stress-relaxation response of a number of materials are given in Figs 2.7–2.9.

The particular nature of the class of viscoelastic materials considered in the above examples proves the existence of a property of 'passive resistance' in such materials. This is in contrast to the instantaneous response and reversibility that usually characterize pure elastic behaviour. This passive resistance is of viscous nature and reflects what is usually called the property of 'hereditary response' of the material. That is, the present state of response depends not only on the present state of loading input but also on previous states. This property is revealed experimentally, in different

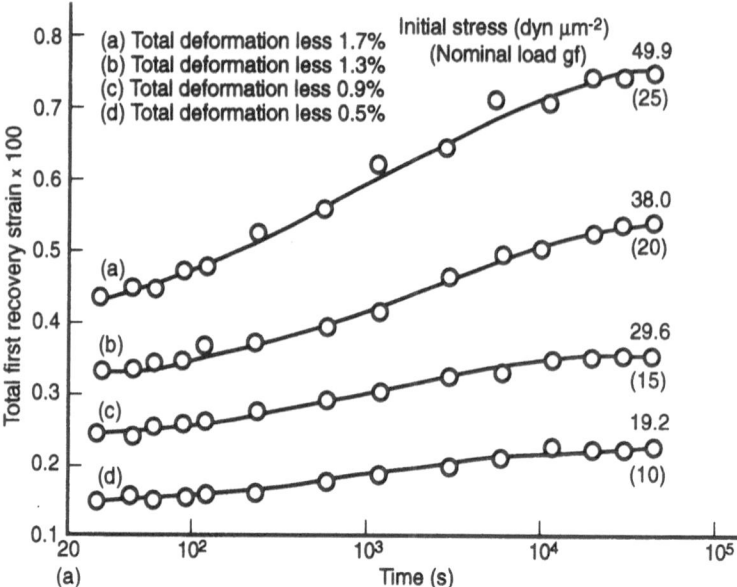

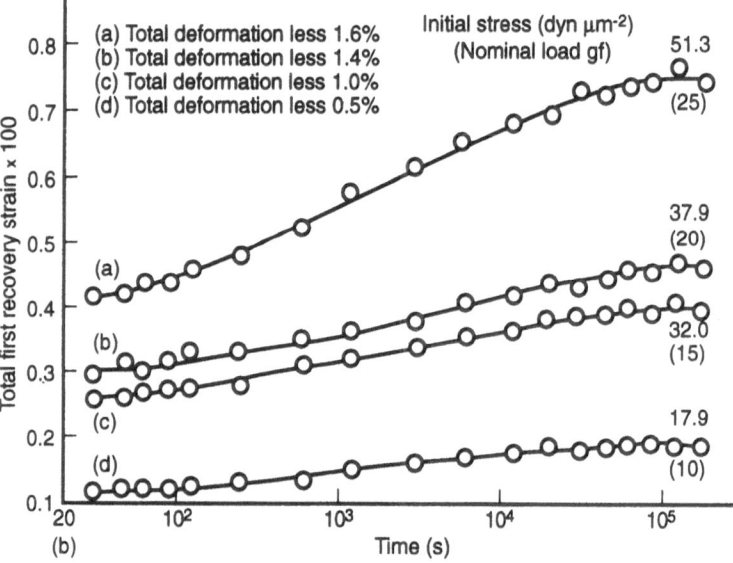

Fig. 2.4 First recovery curves after different time periods of first creep. Individual dry summerwood fibres of a longleaf pine pulp: (a) first recovery after 12 h of first creep; (b) first recovery after 48 h of first creep. (Source: Hill, R. L. (1967) The creep behaviour of individual pulp fibres under tensile stress. *Tappi*, **50**(8), 432–40. Reprinted by permission of Tappi.)

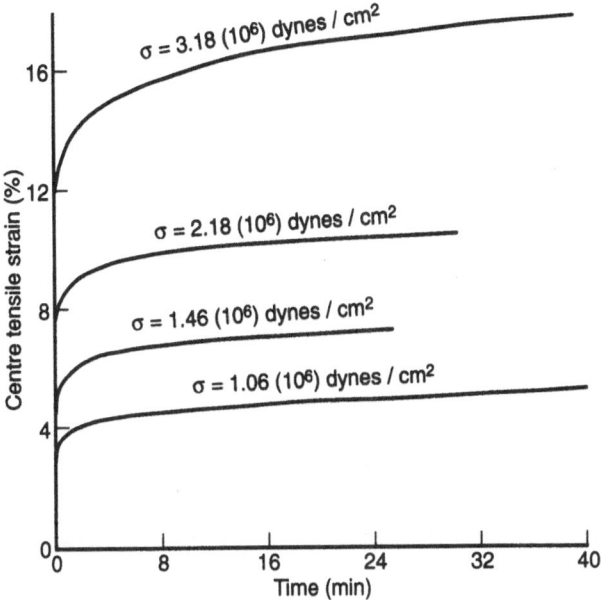

Fig. 2.5 Effect of load on creep of a composite solid propellant at 75°F. (Reprinted with permission from: Blatz, P. J. (1956) Rheology of composite solid propellants. *J. Ind. Eng. Chem.*, **48**(4), 727–9. Copyright (1956) American Chemical Society.)

time-dependent phenomena such as creep, stress relaxation and intrinsic attenuation of propagating waves.

While, in this text, the continuum mechanics approach is maintained primarily for the characterization of the viscoelastic response of materials, it is emphasized that such response is essentially dependent on the effects of a large number of significant microscopic and macroscopic parameters such as those discussed in the introduction of the book and in Chapter 10. Hence, in the present chapter, the presentation is phenomenological and aims at the introduction of the basic formulism of the mechanical response of the linear viscoelastic material. Consequently, the formulations are considered entirely within the context of the infinitesimal linearized deformation theory. Here, as discussed below, the ideas set down by L. Boltzmann (1844–1906) (discussed in Markovitz (1977)) and V. Volterra (1860–1940) are taken as fundamental within the context of linear superposition of input histories (Boltzmann, 1874; Volterra, 1913; Volterra and Peres, 1936; Leaderman, 1943, 1958).

In the present chapter, we confine ourselves to the one-dimensional linear theory of isothermal viscoelasticity under static loading. The transition to the dynamic case is discussed in Chapter 3. Meantime, the generalization to the three-dimensional formulation is dealt with in Chapter 4. The dependence of the viscoelastic performance of engineering materials on the service temperature is treated in Chapter 5. For further studies on the subject of linear viscoelasticity, the reader is referred to the books by

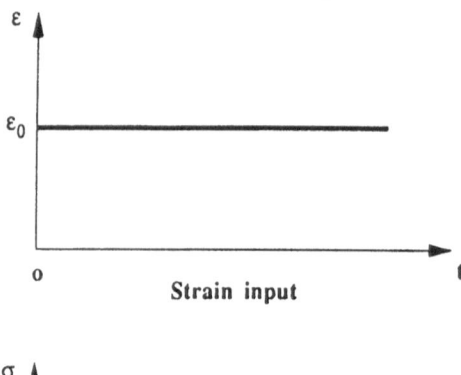

Strain input

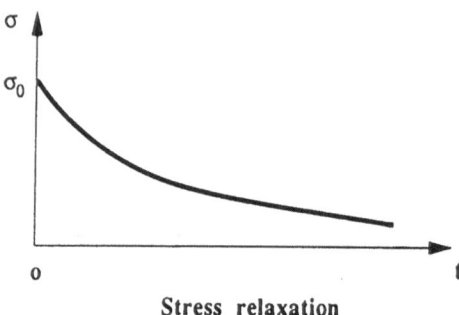

Stress relaxation

Fig. 2.6 Stress relaxation of a viscoelastic material specimen subjected to constant strain of unit magnitude.

Gross (1953), Eirich (1956, 1957), Bland (1960), Ferry (1961), Christensen (1971), Gittus (1975), Flügge (1975), Tschoegl (1989) and Ward (1983), among others. The technical reviews by Scott-Blair (1949), Lee (1960), Coleman and Noll (1961), Gurtin and Sternberg (1962), Halpin (1968), Leitman and Fisher (1973) and Schapery (1974), among others, must also be mentioned.

2.2 DIFFERENTIAL REPRESENTATION OF
LINEAR VISCOELASTIC BEHAVIOUR: MECHANICAL MODELS

The following linear differential relation is often used as a linear viscoelastic constitutive equation connecting the stress to the strain:

$$P\sigma(t) = Q\varepsilon(t) \tag{2.4}$$

where P and Q are linear differential operators with respect to the time t. In a general form, these operators are expressed as

$$P = \sum_{i=0}^{p} a_i \frac{\partial^i}{\partial t^i}$$
$$Q = \sum_{i=0}^{q} b_i \frac{\partial^i}{\partial t^i} \tag{2.5}$$

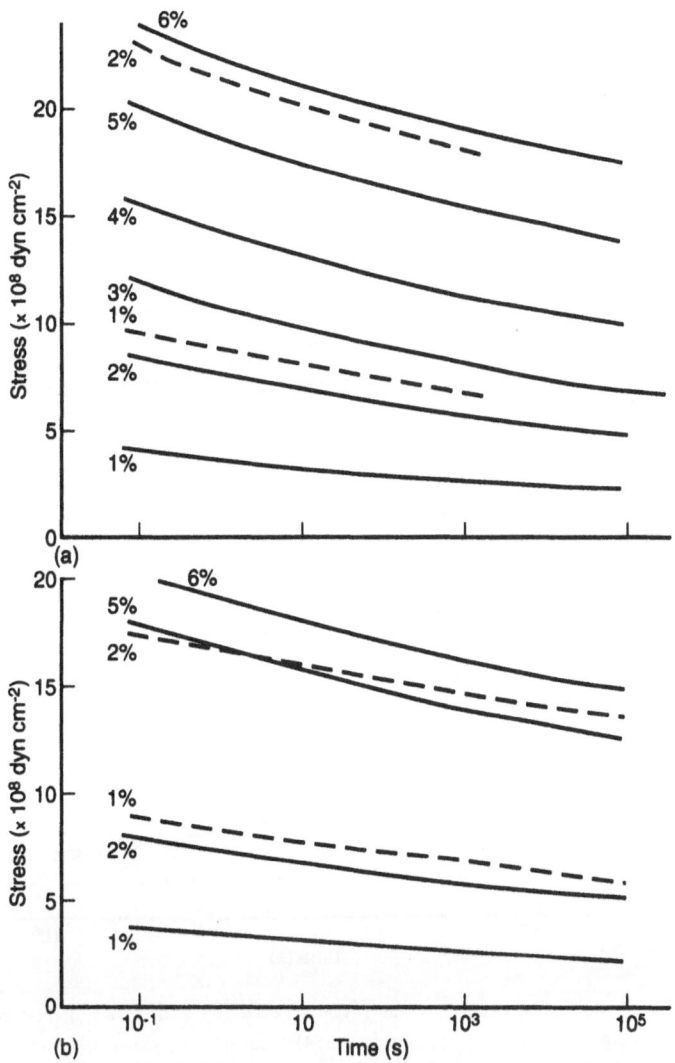

Fig. 2.7 Relaxation of stress in cotton (———) and flax (– – –) at 65% RH and 25°C: (a) first relaxation; (b) fourth relaxation. (Source: Meredith, R. (1954) Relaxation of stress in stretched cellulose fibres. *J. Textile Inst.*, **45**, T438–T460. Reprinted by permission of the British Textile Technology Group.)

where a_i and b_i are material constants. The number of the constants a_i, b_i will depend on the viscoelastic response of the particular material under consideration. By combining (2.4) and (2.5), the former equation is written as

$$a_0\sigma + a_1\frac{d\sigma}{dt} + a_2\frac{d^2\sigma}{dt^2} + \cdots = b_0\varepsilon + b_1\frac{d\varepsilon}{dt} + b_2\frac{d^2\varepsilon}{dt^2} + \cdots. \tag{2.6}$$

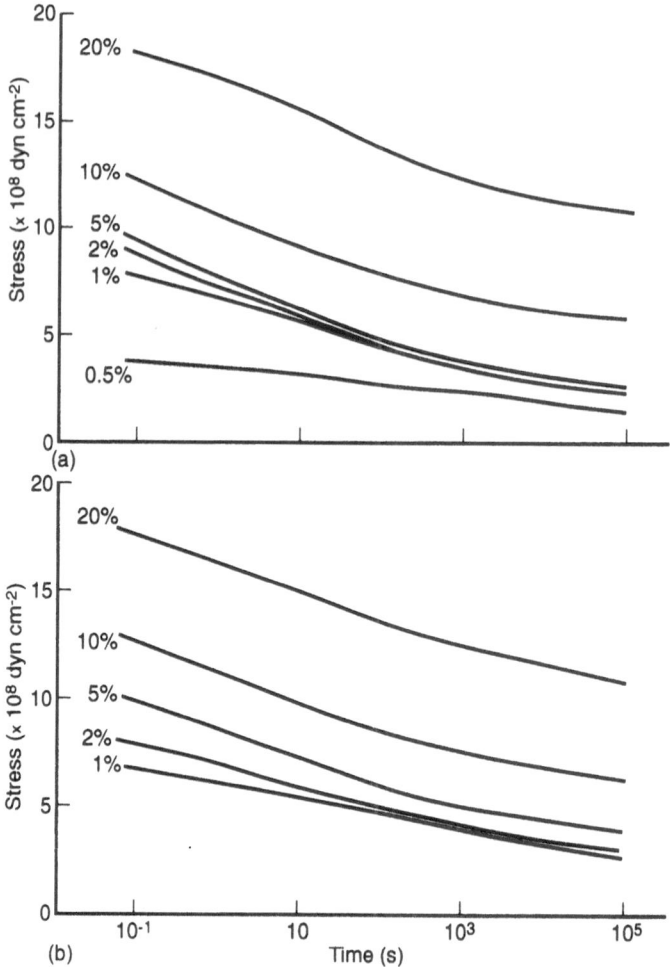

Fig. 2.8 Relaxation of stress in viscous rayon at 65% RH and 20°C: (a) first relaxation; (b) second relaxation. (Source: Meredith, R. (1954) Relaxation of stress in stretched cellulose fibres. *J. Textile Inst.*, **45**, T438–T460. Reprinted by permission of the British Textile Technology Group.)

However, it might be sufficient to represent the viscoelastic response over a limited time scale by considering only one or two terms on each side of (2.6). This would be, then, equivalent to describing the linear viscoelastic behaviour by mechanical models constructed of linear elastic elements, which obey Hooke's law, and viscous dashpots, which obey Newton's law of viscosity. Thus, the viscoelastic behaviour of material, in general, may be investigated by the use of mechanical models which consist of finite networks of springs and dashpots. There are also corresponding electrical models containing resistances and capacitances (or, instead, conductances) which may be used. The invention of mechanical models for the identification of the

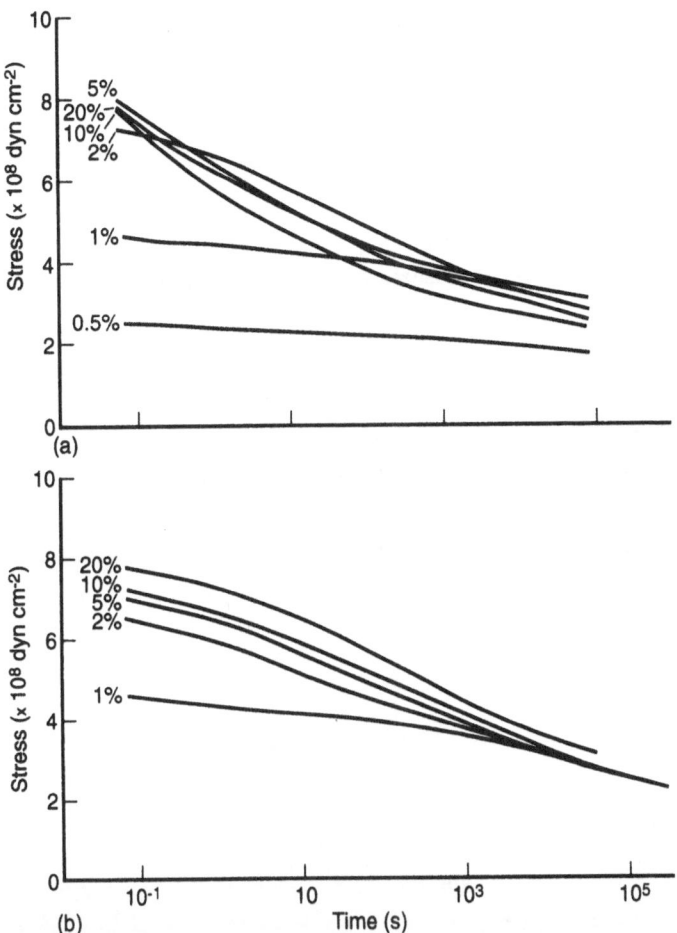

Fig. 2.9 Relaxation of stress in cellulose acetate at 65% RH and 20°C: (a) first relaxation; (b) second relaxation. (Source: Meredith, R. (1954) Relaxation of stress in stretched cellulose fibres. *J. Textile Inst.*, **45**, T438–T460. Reprinted by permission of the British Textile Technology Group.)

viscoelastic responses of materials dates back to the 19th century and coincides with the first introduction of man-made polymers. These models give an indication of the significance of internal parameters of state as represented by the response of the model elements. The presentation below is limited to a number of basic models that are often used. For more information, however, concerning the finite networks of mechanical models for viscoelastic behaviour, the reader is referred, for instance, to Alfrey (1948), Gross (1953), Stuart (1956), Eirich (1956, 1957), Bland (1960), Ferry (1961), Christensen (1971), Flügge (1975) and Gittus (1975).

2.2.1 Single one-dimensional models

Three single mechanical models are dealth with here.

(a) The Maxwell model

The Maxwell model is one idealization of the viscoelastic response of real materials. This model is a combination of a linear spring and a dashpot in series as shown in Fig. 2.10(a). The dashpot is visualized as a piston moving in a viscous fluid. Under

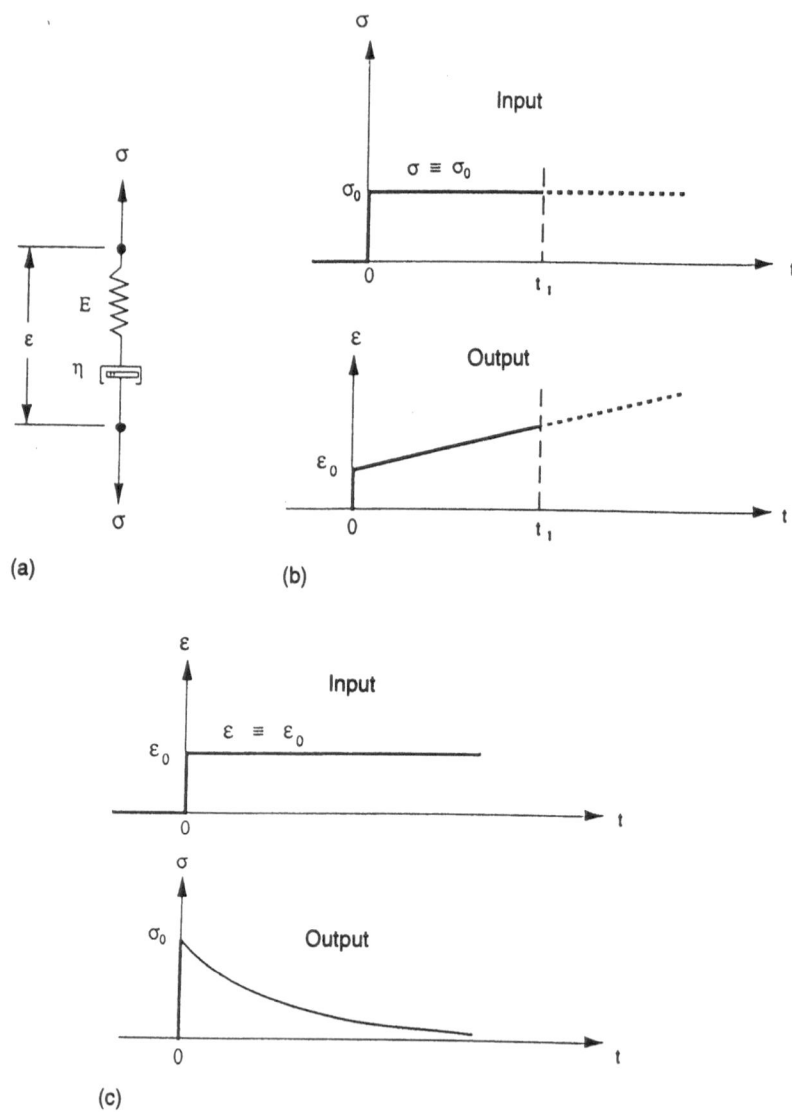

Fig. 2.10 The Maxwell model: (a) the model; (b) creep response; (c) relaxation response.

uniaxial tension, an instantaneous extension of the spring occurs. This is the elastic (Hookean) response of the model. At the same time, the fluid in the dashpot passes slowly through an orifice in the piston resulting in an extension of the overall length of the dashpot. This is a viscous flow which is referred to as the 'time-dependent' response of the Maxwell model. As the spring is in series with the dashpot, the elastic and the viscous strains are additive.

Considering the Maxwell model of Fig. 2.10(a) under uniaxial tension, the following stress–strain relations are obtained: for the spring element 's',

$$\sigma = E\varepsilon_s \tag{2.7}$$

where ε_s is the longitudinal strain displayed by the spring and E is its elastic modulus; for the dashpot

$$\sigma = \eta\dot{\varepsilon}_d \tag{2.8}$$

in which η is the viscosity modulus of the fluid in the dashpot and $\dot{\varepsilon}_d$ is the time rate of the resulting longitudinal strain.

For the spring and dashpot in series (the Maxwell model), the total strain is given by

$$\varepsilon = \varepsilon_s + \varepsilon_d. \tag{2.9}$$

Equation (2.9) can be written, in view of (2.7) and (2.8), in terms of the material properties of the model elements, as

$$\sigma + \lambda\dot{\sigma} = \eta\dot{\varepsilon} \tag{2.10}$$

in which

$$\lambda = \eta/E. \tag{2.11}$$

It is recognized that the response equation (2.10) for the Maxwell model is connected to the differential response relation (2.5) through

$$p = 1, \ a_0 = \frac{1}{\eta}, \ a_1 = \frac{1}{E},$$

$$q = 1, \ b_0 = 0, \ b_1 = 1.$$

In a creep experiment, we apply at $t = 0$ a constant stress $\sigma = \sigma_0$ and we seek the time-dependent creep strain $\varepsilon(t)$. Accordingly, equation (2.10) is a differential equation for ε and has the solution

$$\varepsilon(t) = \frac{\sigma_0}{\eta} t + c \tag{2.12}$$

where c is a constant of integration. The latter can be found subject to the initial condition at $t = 0$, i.e.

$$\varepsilon_0 = \varepsilon(0) = \frac{\sigma_0}{E}, \tag{2.13}$$

which corresponds to the instantaneous elastic response in the spring element. Thus by combining equations (2.12) and (2.13), it follows that

$$c = \varepsilon_0 = \frac{\sigma_0}{E}. \tag{2.14}$$

Utilizing equations (2.12) and (2.14), the creep constitutive equation for the Maxwell model becomes

$$\varepsilon(t) = \frac{\sigma_0}{E}\left(1 + \frac{t}{\lambda}\right) = \sigma_0 F(t), \quad \lambda = \eta/E, \tag{2.15}$$

where $F(t)$ is the creep compliance function or simply the 'creep function' which takes in view of (2.15) for the Maxwell model the form

$$F(t) = E^{-1}(1 + t/\lambda). \tag{2.16}$$

The creep response equation (2.15) is presented in Fig. 2.10(b) for $0 \le t \le t_1$. With reference to this figure, equation (2.15) shows that the instantaneous response at time $t = 0^+$ (i.e. immediately after the application of the load) of the Maxwell model is elastic with a modulus E. The latter is the elastic (or spring) constant. Further, in view of equation (2.15), one can see that the Maxwell model shows a typical properties of a fluid, i.e. its capability of unlimited deformation under finite stress. This is illustrated by the broken lines in Fig. 2.10(b). Such performance would constitute a limiting disadvantage if one attempts to employ the Maxwell model in the prediction of the creep behaviour of real viscoelastic materials. In addition, the Maxwell model cannot demonstrate the time-dependent viscoelastic contraction which occurs in a real viscoelastic material if, during creep, the external stress is removed. Removing the stress from the Maxwell model simply allows an instantaneous elastic strain recovery to occur as a result of the contraction of the spring. There will be no subsequent time-dependent strain recovery as there would be no force acting on the piston to move it back through the fluid of the dashpot when the external stress has been removed (e.g. Gittus, 1975).

On the other hand, if we apply at $t = 0$ a constant strain, i.e. $\varepsilon(t) = 0$ for $t < 0$ and $\varepsilon(t) = \varepsilon_0$ for $t > 0$ which corresponds to a stress-relaxation experiment, then, with reference to equation (2.10), it follows that

$$\sigma + \lambda\dot{\sigma} = 0. \tag{2.17}$$

Integrating this equation with respect to time, one obtains

$$\sigma(t) = \sigma_0 \exp(-t/\lambda) \tag{2.18}$$

whereby the initial condition $\sigma = \sigma_0$ at $t = 0$ has been used. Equation (2.18) indicates that, in a stress-relaxation experiment, the stress decays exponentially with a characteristic time parameter $\lambda = \eta/E$; hence this parameter is referred to as the 'relaxation time' of the Maxwell model at constant strain.

Substituting for σ_0 in (2.18) in terms of the initial strain, this equation may be written in the form

$$\sigma(t) = \varepsilon_0 R(t) \tag{2.19}$$

where $\varepsilon_0 = \varepsilon(0^+)$ and $R(t)$ is referred to as the stress-relaxation modulus or the 'relaxation function'. The latter is expressed in view of equations (2.18) and (2.19) as

$$R(t) = E \exp(-t/\lambda) \tag{2.20}$$

The relaxation response of the Maxwell model is presented in Fig. 2.10(c).

An additional shortcoming of the Maxwell model becomes apparent from examining the form of the relaxation function $R(t)$ (equation (2.20)), which contains only one exponential decay term. This may not suffice for the representation of the stress-relaxation behaviour of real viscoelastic materials. Such real behaviour, as discussed earlier, might not necessarily decay to zero at infinite time as (2.20) suggests.

An alternative procedure for obtaining the constitutive equations for the Maxwell model is by using the Laplace transform (Appendix C). In the relaxation phase, if the strain is applied at time $t = 0^+$ such that the conditions are $\varepsilon(0)$ and $\sigma(0) = 0$ for $t < 0$, then the Laplace transform of (2.10) gives

$$E \, d_t \bar{\varepsilon}(s) = \left(\frac{E}{\eta} + d_t \right) \bar{\sigma}(s) \tag{2.21}$$

where d_t designates the time derivative operator, i.e. $d_t = d/dt$, an overbar $^-$ indicates the Laplace transform of the pertaining variable and s in the Laplace parameter. An operational form corresponding to (2.21) may be written as

$$\bar{\sigma}(s) = d_t \bar{R}(s) \bar{\varepsilon}(s) \tag{2.22}$$

where

$$\bar{R}(s) = \frac{E}{d_t + E/\eta}. \tag{2.23}$$

The inversion of $\bar{R}(s)$ of (2.23) is the relaxation function $R(t)$ given earlier by equation (2.20).

Accordingly, by inverting the Laplace transform (2.23) and using the convolution theorem (see Appendix C), the stress-relaxation equation corresponding to (2.22) can be written as

$$\sigma(t) = \int_0^t R(t - \tau) \frac{d}{d\tau} \varepsilon(\tau) \, d\tau. \tag{2.24}$$

Similarly, in the creep phase, one may solve equation (2.21) for the strain which would correspond to a constant stress applied to the Maxwell mode at time $t = 0^+$.

Then,

$$\bar{\varepsilon}(s) = d_t \bar{F}(s) \bar{\sigma}(s) \tag{2.25}$$

in which

$$\bar{F}(s) = E^{-1}\left(\frac{1}{d_t} + \frac{1}{\lambda \, d_t^2}\right) \tag{2.26}$$

is the Laplace transform of the creep function $F(t)$ expressed earlier by (2.16). Inverting (2.26) and using the convolution theorem, the creep response equation for the Maxwell model is expressed as

$$\varepsilon(t) = E^{-1}\sigma(t) + \eta^{-1}\int_0^t \sigma(\tau) \, d\tau. \tag{2.27}$$

(b) The Kelvin (Voigt) model

The Kelvin (Voigt) model is built up of a linear spring and a dashpot in parallel (Fig. 2.11(a).

Because of the parallel arrangement of the spring and the dashpot, this model will exhibit primary (decelerating) creep when first loaded. This is because the spring can only extend as rapidly as the dashpot. Thus, the model cannot by itself exhibit steady-state creep. For the same reason, it also cannot demonstrate steady-state stress relaxation. On the other hand, if, after a period of uniaxial tension, the stress is released, the spring would then attempt to return to its unstressed length, hence exerting compression on the dashpot during the process. The dashpot would, then, slowly retract, under the stress, to its original length permitting the spring to contract.

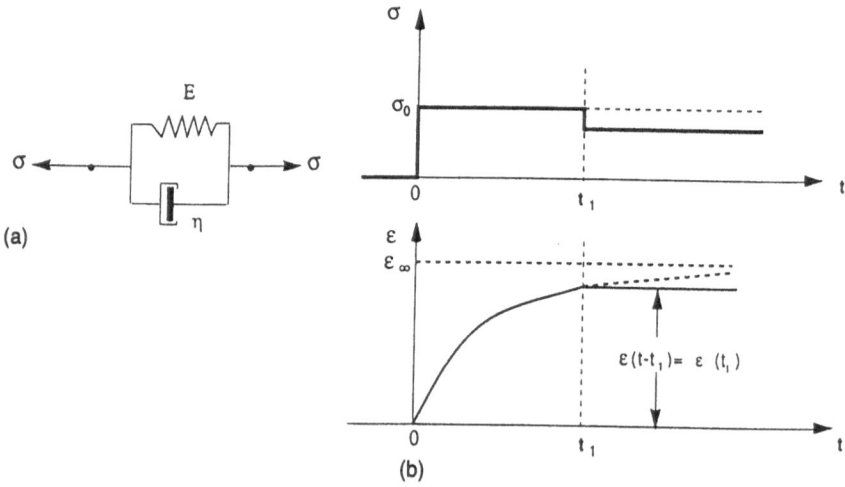

Fig. 2.11 The Kelvin (Voigt) model: (a) the model; (b) creep response (incomplete relaxation of the model is shown for $t > t_1$).

Thus, compressive creep under zero external stress will occur and will eventually, after infinite time, allow all of the prior tensile creep strain to be recovered. Such a property (viscoelastic contraction) can be significant in the creep behaviour of a large class of viscoelastic materials. As one may recall, viscoelastic contraction as described above does not occur in the Maxwell model.

For the spring 's' and dashpot 'd' in parallel arrangement, the Kelvin–Voigt model, the total stress is given by

$$\sigma = \sigma_s + \sigma_d$$
$$= E\varepsilon_s + \eta\dot{\varepsilon}_d. \tag{2.28}$$

This response relation is connected to the differential response equation (2.5) through

$$p = 0, \quad a_0 = 1, \quad q = 1, \quad b_0 = E, \quad b_1 = \eta.$$

In a creep experiment, we let $\sigma = \sigma_0$ for $t \geq 0^+$; thus, equation (2.28) has the solution

$$\varepsilon(t) = \frac{\sigma_0}{E} + C \exp(-t/\lambda), \quad \lambda = \eta/E, \tag{2.29}$$

whereby the constant of integration is given, subject to the initial conditions $\varepsilon_0 = \varepsilon(0) = 0$ at $t = 0$, by $C = -\sigma_0/E$. Hence, the creep constitutive equation (2.29) becomes

$$\varepsilon(t) = \frac{\sigma_0}{E} [1 - \exp(-t/\lambda)] = \sigma_0 F(t) \tag{2.30}$$

where

$$F(t) = E^{-1}[1 - \exp(-t/\lambda)] \tag{2.31}$$

is the creep function for the Kelvin (Voigt) model.

The creep response of the Kelvin model is illustrated in Fig. 2.11(b). As the time t tends to ∞, the strain approaches gradually a final limit. The latter is proportional to the stress, with an asymptotic modulus E_∞, whereby

$$\varepsilon_\infty = \varepsilon(\infty) = \frac{\sigma_0}{E_\infty}. \tag{2.32}$$

Such response behaviour is described as 'delayed elastic' and, hence, the Kelvin–Voigt model does represent the creep behaviour of real materials to a first approximation.

An alternative form of the creep response equation (2.30) for the Kelvin–Voigt model may be obtained by considering the Laplace transformation of (2.28). In this case, provided that, at $t < 0$, both $\sigma = 0$ and $\varepsilon = 0$, the Laplace transformation of (2.28) yields the following relation:

$$\bar{\sigma}(s) = E\bar{\varepsilon}(0) + \eta[d_t\bar{\varepsilon}(s) - \bar{\varepsilon}(0)] \tag{2.33}$$

where $d_t = d/dt$.

Thus, by inverting the Laplace transform (2.33) and using the convolution theorem, the creep response for the Kelvin–Voigt model is obtained as

$$\varepsilon(t) = \eta^{-1} \int_0^t \exp[-(t - \tau)]\sigma(\tau)\, d\tau, \quad t \geq 0. \tag{2.34}$$

On the other hand, during a relaxation experiment, the applied strain is constant, i.e. $\varepsilon(t) = \varepsilon_0 = \varepsilon(0^+)$ for $t \geq 0^+$. In view of the parallel arrangement of the elements of the Kelvin (Voigt) model, it is apparent that the model cannot portray stress relaxation when the external strain is applied at $t = 0^+$ and maintained constant afterwards. As shown in Fig. 2.11(b), when the strain is fixed at $t = t_1$ the stress is immediately relaxed by a certain amount and then remains constant at this value. In other words, the relaxation of the Kelvin model is incomplete.

(c) The three-element model

The three-element model consists of a linear spring in series with a Kelvin–Voigt element (Fig. 2.12(a)). It is sometimes referred to as the **standard linear model**.

With reference to Fig. 2.12(a), the responses of both parts of the model are expressed as

$$\sigma = E\varepsilon_s, \quad \sigma = E'\varepsilon_s' + \eta\dot{\varepsilon}_d \tag{2.35}$$

From equation (2.35), and using the Laplace transformation, the following equation is obtained:

$$(E + E')\sigma + \eta\dot{\sigma} = EE'\varepsilon + E\eta\dot{\varepsilon}. \tag{2.36}$$

In the creep phase, it can be shown that

$$\varepsilon(t) = \frac{\sigma_0}{\xi_1}\{\lambda[1 - \exp(-t/\lambda)] + \xi_2\exp(-t/\lambda)]\} \tag{2.37}$$

where

$$\xi_1 = \frac{E\eta}{E + E'}$$

$$\xi_2 = \frac{\eta}{E + E'}. \tag{2.38}$$

The creep response of the three-element model is demonstrated in Fig. 2.12(b). The model portrays an instant elasticity with

$$\varepsilon_0 = \varepsilon(0^+) = \frac{\sigma_0\xi_2}{\xi_1} = \frac{\sigma_0}{E} \tag{2.39}$$

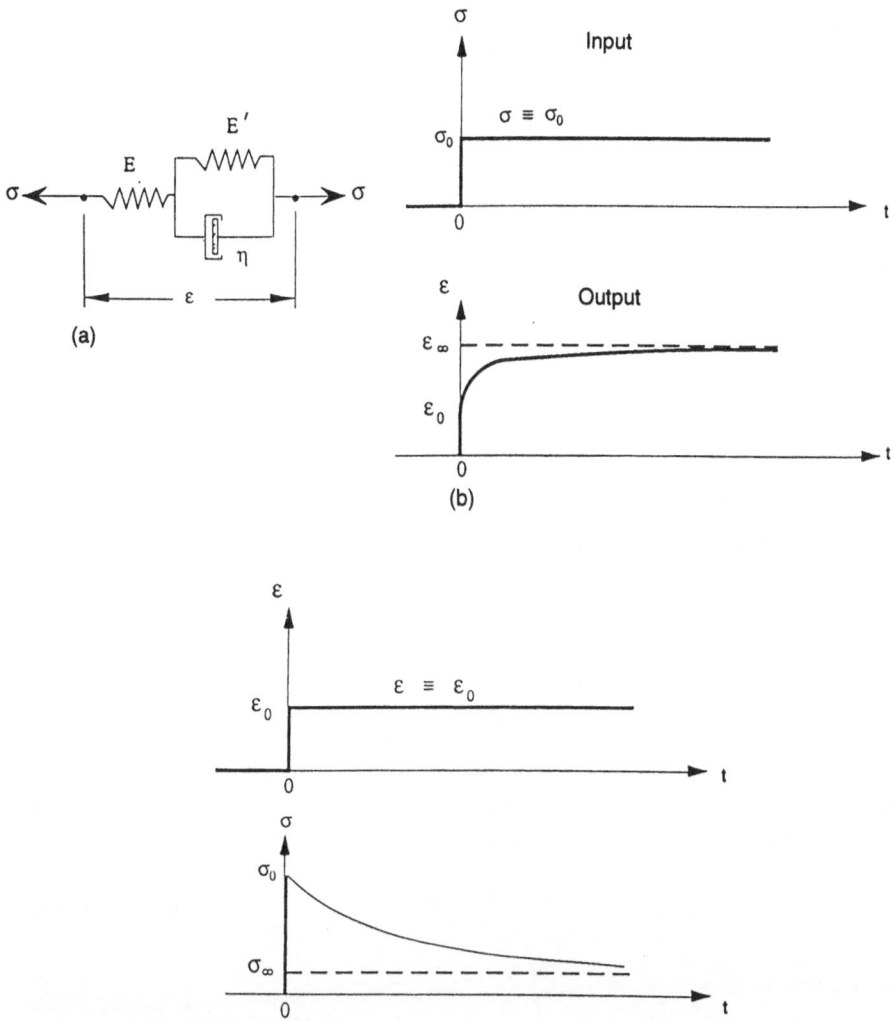

Fig. 2.12 The three-element model (standard linear solid): (a) the model; (b) creep response; (c) relaxation response.

in addition to an asymptotic elastic behaviour given by

$$\varepsilon_\infty = \varepsilon(\infty) = \frac{\sigma_0}{E_\infty} \tag{2.40}$$

where

$$E_\infty = \frac{EE'}{E + E'}. \tag{2.41}$$

For the relaxation phase, equation (2.35) leads, using the Laplace transform, to

$$\sigma(t) = E_x \varepsilon_0[1 - \exp(-t/\lambda')] + \sigma_0 \exp(-t/\lambda') \tag{2.42}$$

where

$$\lambda' = \frac{\eta}{E + E'}. \tag{2.43}$$

As demonstrated in Fig. 2.12(c), the model relaxes gradually to

$$\sigma_x = \sigma(\infty) = E_x \varepsilon \tag{2.44}$$

2.2.2 Generalized mechanical models

Two types of generalized mechanical models are of particular interest.

(a) The generalized Maxwell model

In this case, the basic Maxwell units are connected in parallel as shown in Fig. 2.13.

With reference to Fig. 2.13, consider a generalized Maxwell model of N different Maxwell elements arranged in parallel. Let i ($i = 1, 2, \ldots, N$) denote an individual Maxwell unit. Thus, the total strain in the generalized model is given by

$$\varepsilon = \varepsilon_i \tag{2.45}$$

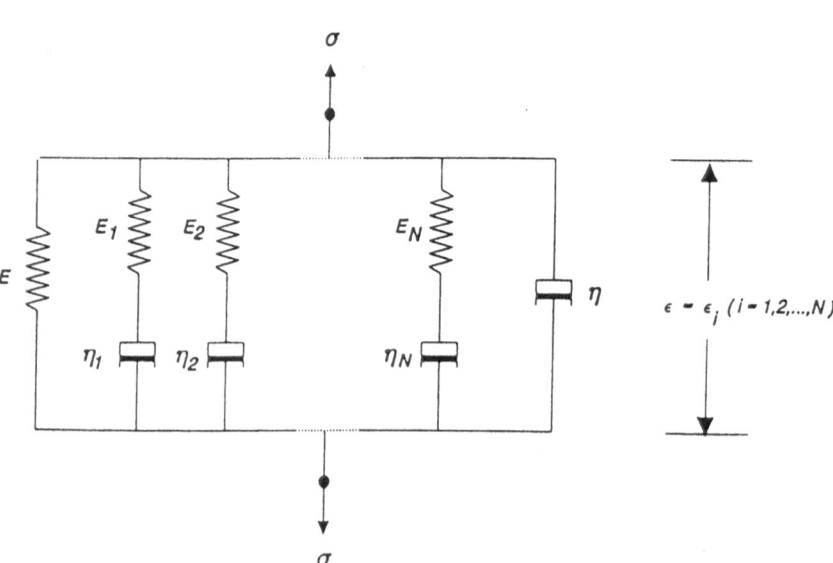

Fig. 2.13 Generalized Maxwell model.

and the total stress is

$$\sigma = \sum_{i=1}^{N} \sigma_i. \tag{2.46}$$

This is with the understanding that the stress is not shared equally between all the elements, that is $\sigma_1 \neq \sigma_2 \neq \cdots \neq \sigma_N$. In view of equations (2.10) and (2.45), it can be shown that

$$\dot{\varepsilon}_i(t) = E_i^{-1}(d_t + \lambda_i^{-1})\sigma_i = \dot{\varepsilon}(t) \tag{2.47}$$

and, thus,

$$\sigma(t) = \sum_{i=1}^{N} E_i(d_t + \lambda_i^{-1})^{-1}\dot{\varepsilon}(t) \tag{2.48}$$

where $\lambda_i = \eta_i/E_i$ and $d_t = d/dt$.

Further, with reference to (2.20), it can be shown that the relaxation function of the generalized Maxwell model is given by

$$R(t) = \sum_{i=1}^{N} E_i \exp(-t/\lambda_i). \tag{2.49}$$

(b) The generalized Kelvin (Voigt) model

In Fig. 2.14, a number $i = 1, 2, \ldots, N$ of Kelvin (Voigt) elements are attached in series to form the generalized Kelvin (Voigt) model. In this case, the stress in each

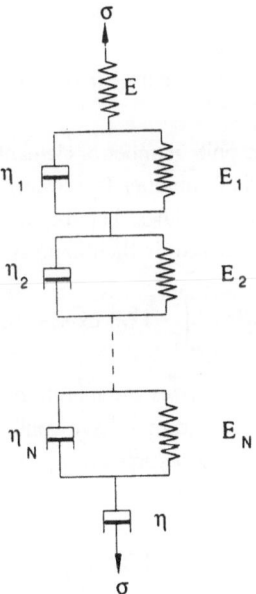

Fig. 2.14 Generalized Kelvin (Voigt) model.

element is the same, i.e. $\sigma_1 = \sigma_2 = \cdots = \sigma_N$. Thus, it can be shown, with reference to (2.28), that

$$\sigma(t) = E_i(1 + d_i\lambda_i)\varepsilon_i(t) \tag{2.50}$$

and

$$\varepsilon(t) = \sigma(t) \sum_{i=1}^{N} E_i^{-1}(1 + d_i\lambda_i)^{-1} \tag{2.51}$$

where, in the above two equations, $\lambda_i = \eta_i/E_i$.

Further, with reference to (2.31), the creep function of the generalized Kelvin (Voigt) model can be written as

$$F(t) = \sum_{i=1}^{N} E_i^{-1} [1 - \exp(-t/\lambda_i)], \quad \lambda_i = \eta_i/E_i. \tag{2.52}$$

2.2.3 Relaxation and retardation spectra

The stress relaxation response of a generalized Maxwell model is expressed, with reference to (2.24), as

$$\sigma(t) = \int_{-\infty}^{t} R(t - \tau)\dot{\varepsilon}(\tau) \, d\tau \tag{2.53}$$

where $R(\cdot)$ is the relaxation function of the model expressed previously by (2.49), i.e.

$$R(t - \tau) = \sum_{i=1}^{N} E_i \exp[-(t - \tau)/\lambda_i], \quad \lambda_i = \eta_i/E_i. \tag{2.54}$$

Now, we consider that the finite number of elements N in the generalized Maxwell model is replaced by an infinite number (spectrum) of elements characterized by a probability density of relaxation times. Let the latter be designated by $\Gamma(\lambda)$. The relaxation function for the spectrum is then expressed in view of (2.54) as

$$R(t) = \int_{0}^{\infty} \Gamma(\lambda) \exp(-t/\lambda) \, d\lambda. \tag{2.55}$$

The function $\Gamma(\lambda)$ which characterizes the infinite number of parallel Maxwell units is nonnegative for $0 \le \lambda < \infty$ and may have continuous and discontinuous parts.

The relaxation function is often expressed (e.g. Gross, 1947, 1953), as a generalization of (2.55) as follows:

$$R(t) = \int_{0}^{\infty} \beta\Gamma(\lambda) \exp(-t/\lambda) \, d\lambda \tag{2.56}$$

where $\Gamma(\lambda) \, d\lambda$ is the distribution of relaxation times and where β is a normalization factor such that

$$\int_0^\infty \Gamma(\lambda) \, d\lambda = 1, \tag{2.57a}$$

i.e.

$$\beta = R(0). \tag{2.57b}$$

Similarly, for a generalized Kelvin (Voigt) model, the creep function $F(\cdot)$ is identified, for a finite number of Kelvin–Voigt elements arranged in series, by equation (2.52), i.e.

$$F(t - \tau) = \sum_{i=1}^N E_i^{-1}\{1 - \exp[-(t - \tau)/\lambda_i]\}, \quad \lambda_i = \eta_i/E_i. \tag{2.58}$$

For an infinite number (spectrum) of Kelvin (Voigt) elements in the generalized model, the creep function may be expressed, in terms of the probability density of retardation times $f(\lambda)$, as

$$F(t) = \int_0^\infty f(\lambda)[1 - \exp(-t/\lambda)] \, d\lambda. \tag{2.59}$$

The creep function $F(t)$ may also be expressed, as a generalization of the expression (2.59), as

$$F(t) = \int_0^\infty \alpha f(\lambda)[1 - \exp(-t/\lambda)] \, d\lambda \tag{2.60}$$

where $f(\lambda) \, d\lambda$ is the distribution function of retardation times and α is a normalization factor. The latter is determined such that

$$\int_0^\infty f(\lambda) \, d\lambda = 1; \tag{2.61a}$$

hence,

$$\alpha = F(\infty). \tag{2.61b}$$

With reference to equations (2.60) and (2.61), the function $f(\lambda)$ is a nondecreasing function for $0 \le \lambda < \infty$ and may have continuous and discontinuous parts.

2.3 PHENOMENOLOGICAL FRAMEWORK OF THE LINEAR VISCOELASTIC THEORY

The assumption that a viscoelastic solid is linear is sufficient to establish explicit single integral expressions connecting responses to inputs without having to specify *a priori* its physical makeups and the physical significance of input and response quantities. If one designates an input stimulus on a viscoelastic specimen by I, then, in view of the hereditary effect, the response at time t, say $\Upsilon(t)$, would be in general a function of the time history of the input $I(t, \tau)$, including its current value $I(t)$. That is,

$$\Upsilon(t) = f[I(t, \tau), \quad I(t)], \quad \tau \le t. \tag{2.62}$$

Let σ represent the one-dimensional Cauchy stress and ε designate the corresponding one-dimensional infinitesimal strain defined at every material point x in a rectangular Cartesian reference frame within the time interval $-\infty < t < \infty$. One seeks the response (constitutive) relationship between σ and ε for the linear viscoelastic model of the material system. Such a linear constitutive relation would be generally consistent with the smallness assumptions of the infinitesimal deformation theory (Chapter 1). In general, the resulting viscoelastic constitutive equation would be a function of both position and time. However, on the assumption that the material specimen is homogeneous, the dependence of the constitutive formulism on position is withheld at present.

If one designates an input stress on the material specimen by $\sigma(t)$, then the resulting strain would be, in general, a function of the history $\sigma(t, \tau)$ and its current value $\sigma(t)$ in the manner expressed by (2.62), i.e.

$$\varepsilon(t) = f[\sigma(t, \tau), \quad \sigma(t)], \quad \tau \le t. \tag{2.63}$$

This equation is again a statement of the concept of the hereditary effect in the viscoelastic material specimen: that is the resulting strain is a function of the history of the stress input and not just its current value. The function $f(\cdot)$ appearing in (2.63) is defined to be linear if, and only if, it satisfies the following two criteria (Schapery, 1974):

1. the criterion of homogeneity (or proportionality), expressed by

$$f[c\sigma(t)] = cf[\sigma(t)] \tag{2.64}$$

where c is a constant;

2. the criterion of superposition of input histories, i.e.

$$f[\sigma_1(t_1), \sigma_2(t_2), \ldots, \sigma_n(t_n)] = f[\sigma_1(t_1)] + f[\sigma_2(t_2)] + \cdots + f[\sigma_n(t_n)]. \tag{2.65}$$

For a viscoelastic material to be identified as linear viscoelastic, equations (2.64) and (2.65) must be satisfied. In this regard, although the superposition criterion (2.65) is not implied by the homogeneity equation (2.64), it is possible to show that for all practical situations the opposite to this statement is true. Many viscoelastic materials exhibit, to a satisfactory degree of accuracy, the linearity requirements (1) and (2) above. This would be particularly true between certain limiting stress or strain values. For example, the creep curves of a composite solid propellant subjected to different levels of loading are shown in Fig. 2.5 (due to Blatz (1956)) to satisfy these requirements approximately. Linear viscoelastic behaviour is demonstrated in Fig. 2.5 by the fact that at any instant of time the strains are approximately proportional to the stress level (e.g. Lee, 1960). A similar example can be demonstrated for the case of relaxation of stress in stretched cellulose fibres (cotton and flax at 65% RH and 25°C) as shown in Fig. 2.7 (due to Meredith (1954)). Here, linearity is indicated by the fact that at any time t the stresses are approximately proportional to the strain input level. On the other hand, the stress-relaxation response of both viscous rayon (Fig. 2.8) and cellulose acetate (Fig. 2.9) is apparently nonlinear.

In a creep experiment, we consider a viscoelastic material specimen which is subjected to zero stress at time $t < 0$ and a constant stress σ_0 for $t \geq 0^+$. Following expression (2.63), the corresponding strain is given by

$$\varepsilon(t) = f(\sigma_0, t). \tag{2.66}$$

Further, in the case where linearity is assumed, equation (2.66) can be written, with reference to Fig. 2.15, as

$$\varepsilon(t) = F(t)\sigma_0 = F(0)\sigma_0 + [F(t) - F(0)]\sigma_0. \tag{2.67}$$

The first term on the right-hand side of equation (2.67) is the instantaneous strain which occurs on application of the load at time $t = 0$. The second term represents the delayed (retarded) part of strain which is a function of time. The function $F(t)$, introduced earlier, is the creep function. It is a function of time only for the homogeneous material specimen and characterizes the rheological properties of the linear viscoelastic material for the one-dimensional stress–strain situation. For $t < 0$, $F(t) = 0$ and, for $t \geq 0$, $F(t)$ is usually a monotonically increasing function of time. Figure 2.16 demonstrates the function $F(t)$ by comparison with corresponding response functions for elastic and viscous specimens.

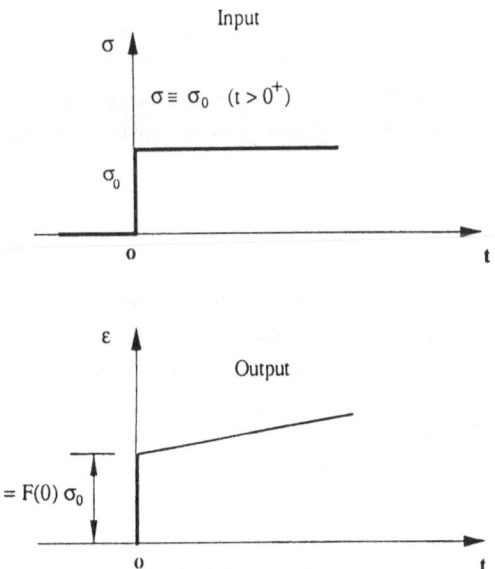

Fig. 2.15 Linear creep response.

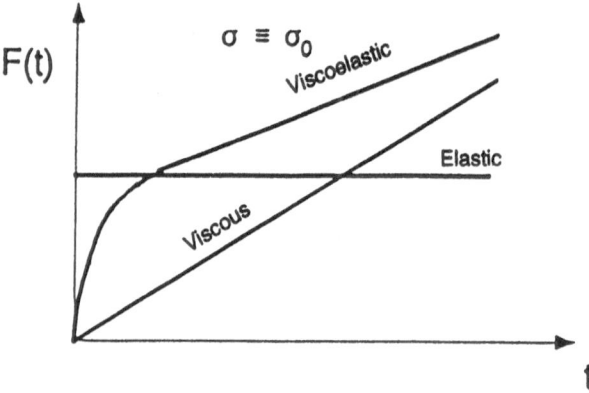

Fig. 2.16 Comparison of the strain response function $F(t)$ for elastic, viscous and viscoelastic material specimens when subjected to a constant unit stress.

Similarly, in a relaxation experiment, if one designates an input strain on the material specimen by $\varepsilon(t)$, then the resulting stress would be, in general, a function of the history of $\varepsilon(t)$, i.e.

$$\sigma(t) = r[\varepsilon(t, \tau), \quad \varepsilon(t)]. \tag{2.68}$$

Further, if a constant strain ε_0 is applied to the material specimen at time $t \geq 0$, the resulting stress can be expressed as

$$\sigma(t) = r(\varepsilon_0, t). \tag{2.69}$$

Such a relaxation experiment may be accomplished by applying at time $t = 0$ whatever stress is required to produce very rapidly the amount of strain required, ε_0, and then fixing the ends of the test specimen and recording the variation of stress with time, i.e. $\sigma(t)$. Equation (2.69) can be written, with reference to Fig. 2.17, where linear viscoelasticity is assumed, as

$$\sigma(t) = R(t)\varepsilon_0 = R(0)\varepsilon_0 + [R(t) - R(0)]\varepsilon_0 \tag{2.70}$$

in which the function $R(t)$ is the relaxation function. It is usually a positive decreasing function of time. Experimentally, the one-dimensional relaxation function of a linear viscoelastic material is found to be a monotonically decreasing function of time (Fig. 2.18). This property is not, however, a consequence of compatibility with thermo-dynamics or of the dissipative property of the viscoelastic material (Day, 1972). In this context, Gurtin and Herrera (1965) presented a dissipative one-dimensional relaxation function which is not monotonically decreasing.

In the study of the linear response of viscoelastic materials, we use the criterion of superposition of input histories, equation (2.65), to determine the output produced by common action of several inputs. This may translate, for a tensile test, for instance, into tensile stresses applied successively with different magnitudes.

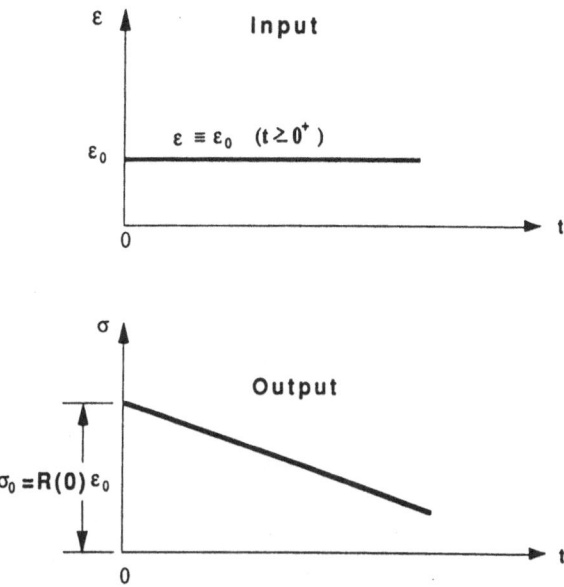

Fig. 2.17 Linear relaxation response.

In a creep experiment, if at $t = 0$ a stress σ_0 is applied suddenly, a creep strain $\varepsilon(t) = F(t)\sigma_0$ is produced whereby $F(t)$ is the creep function. Further, if the input stress σ_0 is maintained unchanged, then the response formula, identified previously by equation (2.67), may be used to describe the strain for the entire range of time considered. This is illustrated in Fig. 2.15. Consider now the case that, at time $t = \tau$, an additional amount of stress $\Delta\sigma_\tau$ is applied. Thus, for $t > \tau$, an additional amount of strain will be produced which will be proportional to $\Delta\sigma_\tau$ via the same creep relationship (2.67) with the time measured from $t = \tau$. Hence, the total strain for

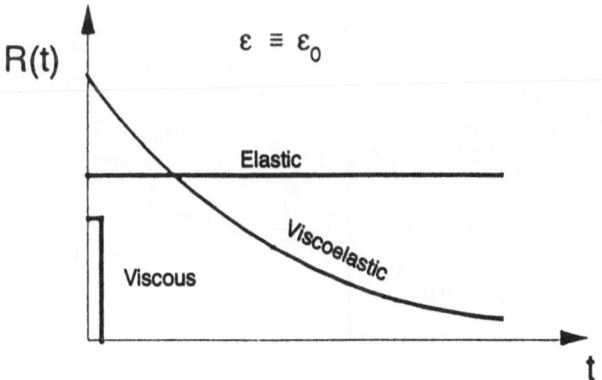

Fig. 2.18 Comparison of the stress response function $R(t)$ for elastic, viscous and viscoelastic material specimens when subjected to a constant unit strain.

$t > \tau$ is expressed by

$$\varepsilon(t) = F(t)\sigma_0 + F(t - \tau)\,\Delta\sigma_\tau. \tag{2.71}$$

As illustrated in Fig. 2.19, the first term on the right-hand side of (2.71) is the original amount of strain at any time t due to the initial value of stress σ_0 and the second term identifies the added amount of strain starting from $t = \tau$ produced by the additional increment of stress $\Delta\sigma_\tau$ applied at that time.

The procedure illustrated above may be generalized to treat cases of discontinuous input histories. This can be accomplished through the use of unit step functions (Appendix B). In this regard, it is always convenient to select the one input as a unit step function $H(t - \tau)$ defined by

$$H(t - \tau) = \begin{cases} 0, & t \leq \tau, \\ 1, & t \geq \tau. \end{cases} \tag{2.72}$$

Thus, in terms of the unit step function, the stimulus $\sigma(t) = 0$ for $t < 0$ and $\sigma(t) = \sigma_0$ for $t \geq 0$ can be written as

$$\sigma(t) = \sigma_0 H(t), \quad -\infty < t < \infty, \tag{2.73}$$

which has the response identified previously by the constitutive equation (2.67). Further, the application of a constant stress σ_τ for a duration of time $t \geq \tau$ can be

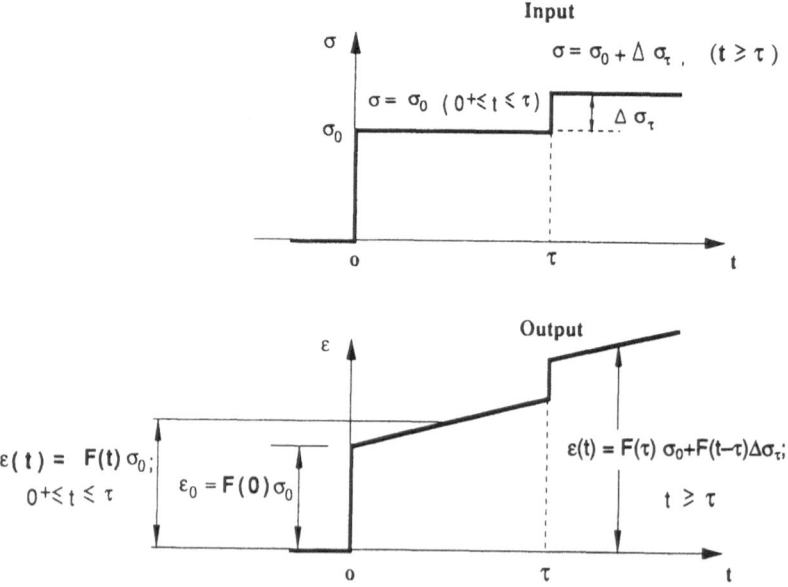

Fig. 2.19 Linear creep response with an additional increment of stress $\Delta\sigma_\tau$ applied at time $t = \tau$.

expressed by employing the unit step function as

$$\sigma(t) = H(t - \tau)\sigma_\tau. \tag{2.74}$$

Hence, the identity

$$\sigma(t) = \int_{-\infty}^{\infty} H(t - \tau) \, d\sigma(\tau) \tag{2.75}$$

can be thought of as a superposition of step functions with $d\sigma(\tau)$ being the height of the step at $t = \tau$ (e.g. Flügge, 1975). Since the linearity assumption guarantees the application of the superposition principle, then, with reference to (2.67), the strain function is a superposition of functions $F(t - \tau) \, d\sigma(\tau)$ with $-\infty < t < \infty$. Hence, the creep response is given by

$$\varepsilon(t) = \int_{-\infty}^{\infty} F(t - \tau) \, d\sigma(\tau) = \int_{-\infty}^{\infty} F(t - \tau)\dot{\sigma}(\tau) \, d\tau. \tag{2.76}$$

The constitutive relations in (2.76) are forms of the 'principle of superposition'. They were introduced as purely empirical laws in the theory of after-effect by Boltzmann (1874). The principle of superposition also appears in the laws of dielectrics (Hopkinson, 1876; Froehlich, 1949) and in the laws of linear electrical networks (Carson, 1926). The integrals in (2.76) were first introduced in pure mathematics where they are called Duhamel's integrals (Duhamel, 1833); Gross (1953) should also be referred to.

For the general case, when a stress σ_0 is applied at time $t = 0$, but that the stress then varies as an arbitrary function of time $\sigma(t)$, the response equation corresponding to (2.76) can be written as

$$\varepsilon(t) = F(t)\sigma_0 + \int_0^t F(t, \tau)\dot{\sigma}(\tau) \, d\tau. \tag{2.77}$$

The creep response equation (2.77) illustrates how the strain at any time t depends on all that has happened before and at present, i.e. on the entire input history. The latter, in the present case of creep, is the stress history $[\sigma(t, \tau), \sigma(t)]$. Hence, the integral in the response equation (2.77) is called the 'hereditary integral' following the introduction by Volterra (1913) of the term 'hereditary law' for functional relations similar to (2.77).

In many applications, the response of a linear viscoelastic material depends only on the time elasped since the application of the load. In this case, equation (2.77) can be written as

$$\varepsilon(t) = F(t)\sigma_0 + \int_0^t F(t - \tau)\dot{\sigma}(\tau) \, d\tau. \tag{2.78}$$

Viscoelastic materials which obey the time differential postulate indicated in (2.78) are referred to as '**non-aging**' or '**time translation invariant**' materials. Equation (2.78) can be further modified to read as

$$\varepsilon(t) = F(0)\sigma(t) + \int_0^t \sigma(\tau) \frac{dF(t-\tau)}{d(t-\tau)} d\tau. \tag{2.79}$$

While equation (2.78) separates the strains caused by the initial stress σ_0 and by the latter stress increase, equation (2.79) expresses the strain that would occur if the total stress σ were applied at the present time t and the additional strain resulting from the fact that much or all of the stress has been applied earlier and had time to produce creep. Equation (2.79), after some changes, can lead to the following form of the Stieltjes integral:

$$\varepsilon(t) = \int_{-\infty}^{\infty} F(t-\tau) \, d\sigma(\tau). \tag{2.80}$$

This brings us back to the form of equation (2.76).

In the relaxation case, a treatment similar to the above may be carried out based on the concept of the relaxation function $R(t)$ in equation (2.70). In this case, if an input strain $\varepsilon(t) = \varepsilon_0 H(t)$ is maintained in the material, the resulting response may be identified by the constitutive equation (2.70). Further, the application of a constant strain ε_τ for a duration of time $t \geq \tau$ can be expressed as

$$\varepsilon(t) = H(t-\tau)\varepsilon_\tau.$$

Hence, the identity

$$\varepsilon(t) = \int_{-\infty}^{\infty} H(t-\tau) \, d\varepsilon(\tau) \tag{2.81}$$

can be thought of as superposition of step functions with $d\varepsilon(\tau)$ being the height of the step at $t = \tau$. Accordingly, in this case, the stress-relaxation function would be thought of as a superposition of functions $R(t-\tau) \, d\varepsilon(\tau)$ with $-\infty < t < \infty$. Thus, a stress-relaxation response equation corresponding to the creep response equation (2.76) can be written as

$$\sigma(t) = \int_{-\infty}^{\infty} R(t-\tau)\dot{\varepsilon}(\tau) \, d\tau. \tag{2.82}$$

Hence, the stress-relaxation constitutive equation would assume, for a non-aging material, one of the following forms:

$$\sigma(t) = R(t)\varepsilon_0 + \int_0^t R(t-\tau)\dot{\varepsilon}(\tau) \, d\tau.$$

$$= R(0)\varepsilon(t) + \int_0^t \varepsilon(\tau) \frac{dR(t-\tau)}{d(t-\tau)} d\tau$$

$$= \int_{\tau=-\infty}^{\tau=+\infty} R(t-\tau) \, d\varepsilon(\tau). \tag{2.83}$$

The response equations (2.78) and (2.82) are Boltzmann's formulation (Boltzmann, 1874) of the constitutive equation for a linear viscoelastic material specimen under uniaxial loading. Hence, such material is often referred to as a 'Boltzmann's solid'. Alternatively, as mentioned earlier, it is called a 'linear hereditary material' as it was named about the same time by Volterra (1913).

The creep response relation (2.76) and the corresponding relaxation equation (2.82) are typical aftereffect relations. Also, the creep function $F(t)$ and the relaxation function $R(t)$ are aftereffect functions or memory functions. They are not quite arbitrary functions for thermodynamic reasons. They must be so that for all applied stress functions $\sigma(t)$ and all times t the following inequality holds (König and Meixner, 1958; Meixner, 1965; Axelrad, 1970):

$$\int_{-\infty}^{t} \sigma(\tau)\dot{\varepsilon}(\tau) \; d\tau \geq 0. \tag{2.84}$$

This inequality is referred to as the '**passivity property**'.

2.4 INTERCONVERSION OF CREEP AND RELAXATION DATA

In linear viscoelasticity, the creep function $F(t)$ and the relaxation function $R(t)$ are interrelated and each would permit the construction of the viscoelastic constitutive relation for an arbitrary given behaviour. This is under the condition that the principle of time invariance is applicable; that is the material is influenced only by $\sigma(t)$ and $\varepsilon(t)$ and no other stimuli at any time being present.

Recalling the creep constitutive equation (2.76), namely

$$\varepsilon(t) = \int_{-\infty}^{t} F(t - \tau) \frac{d\sigma(\tau)}{d\tau} \; d\tau,$$

one may consider (e.g. Ward, 1983) a loading programme starting at time $t = 0$ in which the stress would decay exactly as the relaxation function $R(\tau)$. Thus, the corresponding strain would remain constant as in a typical stress-relaxation experiment. That is, if

$$\frac{d\sigma(\tau)}{dt} = \frac{dR(\tau)}{d\tau}$$

then

$$\int_{0}^{t} \frac{dR(\tau)}{dt} F(t - \tau) \; d\tau = \text{constant}. \tag{2.85}$$

For simplicity, one may further normalize the definition of $R(\tau)$ and $F(\tau)$ so that the constant in the above equation becomes unity. Accordingly, one has

$$\int_{0}^{t} \frac{dR(\tau)}{d\tau} F(t - \tau) \; d\tau = 1 \tag{2.86}$$

which is sometimes interpreted to give

$$\int_0^t R(\tau)F(t - \tau)\, d\tau = t. \tag{2.87}$$

On the other hand, for the interconversion of the creep and relaxation time rates, the following relation is sometimes used:

$$m = \frac{d \ln F(t)}{d \ln t} \approx -\frac{d \ln R(t)}{d \ln t} = n \tag{2.88}$$

where m and n are the values of the corresponding shown derivatives.

The accuracy of formula (2.88) is, however, still unknown for a large class of viscoelastic materials (Struik, 1987). According to the latter reference, it is found that

$$0 \le 1 - R(t)F(t) \le \tfrac{1}{6}(\pi\mu)^2, \quad \mu \ll 1, \tag{2.89}$$

and

$$\left|\frac{m}{n} - 1\right| < Bm \quad \text{with} \quad B \approx 1. \tag{2.90}$$

The approximate equality of $F(t)$ and $[R(t)]^{-1}$, corresponding to (2.89), follows from the smallness of the slopes m and n for time $\tau < t$. It was concluded by Struik (1987) that (2.89) can be applied without restriction for the case of glassy polymers whereby m is generally less than 0.05.

Equations (2.88) and (2.90), however, are much less general. They are derived from the assumption that $dF(\tau)/d \ln \tau$ and $-dR(\tau)/dt$ are either increasing with τ for $\tau < t$ or only slowly decreasing. Although for polymers these conditions are always nearly fulfilled, equations (2.88) and (2.90) may fail in some applications even for small values of m and n. In this context, Struik (1987) demonstrated the limitations of the latter two equations for the case of a simple mechanical model consisting of a linear spring in parallel with a Maxwell unit. Struik (1987) concluded that these two equations, (2.88) and (2.90), may be valid only under the condition that the dispersion regions of response of viscoelastic materials are sufficiently broad; however, for dispersions with only a single relaxation time, the two equations (2.88) and (2.90) would fail.

The above condition that the dispersion regions of the material must be sufficiently broad is nearly always fulfilled by glassy polymers. This is illustrated in Struik's paper by two examples concerning PMMAL and CHMA (polycyclohexylmethacrylate).

It is seen from the previous discussion that the creep and relaxation functions of a linear viscoelastic system are mutually connected in a simple manner. This permits the calculation of the distribution function of relaxation times of stress and the distribution function of retardation times of strain, when the relaxation function or the creep function is given. A transformation can be also established for the conversion of one distribution function into another. The reader is referred, in this context, to Gross (1947). Some of the results concerning the above are dealt with in Chapter 3.

PROBLEMS

2.1 Determine the stress–strain relations for the mechanical models shown in the figure below.

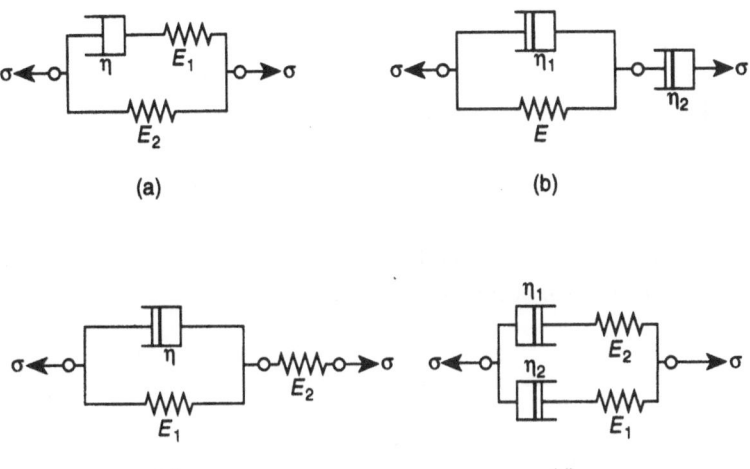

(a)

(b)

(c)

(d)

2.2 Comment on the use of the Maxwell model for the prediction of the response behaviour of real linear viscoelastic materials.

2.3 Verify the constitutive equations for the three-element model given by (2.37) and (2.42).

2.4 Determine the creep-recovery response of the following mechanical models, if a constant stress of magnitude σ_0 is applied suddenly at time $t = 0$ and maintained afterwards for a duration $t = 2\tau$, when it is completely removed:

1. the Maxwell model;
2. the Kelvin–Voigt model;
3. the standard linear solid.

2.5 Determine, by employing the superposition principle, the response of the Maxwell model and the Kelvin–Voigt model to the stress loadings shown in the figure below.

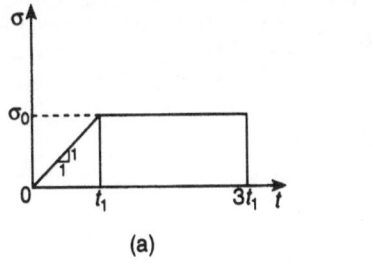

(a)

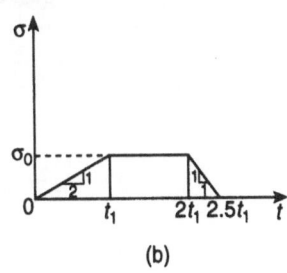

(b)

2.6 Comment on the use of both the Kelvin and three-element models for the prediction of the linear viscoelastic response illustrated in the figure below.

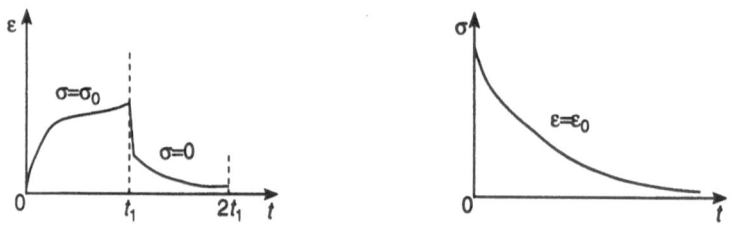

2.7 Determine the creep function $F(t)$ and the relaxation function $R(t)$ for the mechanical models shown in the figure below.

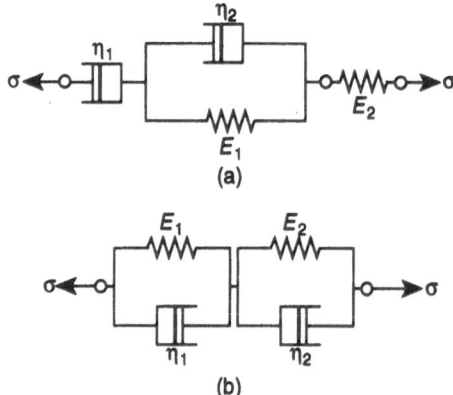

2.8 Use of the superposition principle to determine the creep recovery response of the models shown in the figure below.

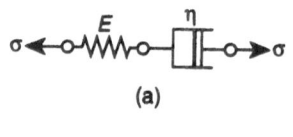

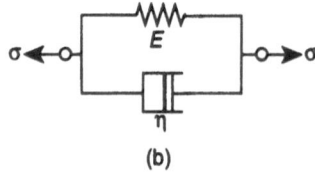

REFERENCES

Alfrey, T. (1948) *Mechanical Behaviour of High Polymers*, Interscience, New York.

Axelrad, D. R. (1970) Mechanical models of relaxation phenomena. *Adv. Mol. Relaxation Processes*, **2**, 41–68.

Bland, D. R. (1960) *The Theory of Linear Viscoelasticity*, Pergamon, New York.

Blatz, P. J. (1956) Rheology of composite solid propellants. *J. Ind. Eng. Chem.*, **48**(4), 727–9.

Boltzmann, L. (1874) Zür Theorie der elastichen Nachwirkung, Sitzungsber, Kaiserlich. *Akad. Wiss. Wien, Math. Naturwiss. Kl.*, **70**(2), 275–306.

Carson, J. R. (1926) *Electrical Circuit Theory and Operational Calculus*, McGraw-Hill, New York.

Christensen, R. M. (1971) *Theory of Viscoelasticity*, Academic Press, New York.

Coleman, B. D. and Noll, W. (1961) Foundations of linear viscoelasticity. *Rev. Mod. Phys.*, **33**, 239–49.

Day, W. A. (1972) *The Thermodynamics of Simple Materials with Fading Memory*, Springer, New York.

Duhamel, J. M. C. (1833) Sur la méthode générale relative au mouvement de la chaleur dans les corps solides plongés dans des milieux dont la température varie avec le temps. *J. Ec. Polytech., Paris*, **14**(22), 20–77.

Eirich, F. R. (1956, 1957) *Rheology Theory and Applications*, Vols 1 and 2, Academic Press, New York.

Ferry, J. D. (1961) *Viscoelasticity Properties of Polymers*, Wiley, New York.

Flügge, W. (1975) *Viscoelasticity*, Springer, New York.

Froehlich, H. (1949) *Theory of Dielectrics*, Clarendon, Oxford.

Gittus, J. (1975) *Creep, Viscoelasticity and Creep Fracture in Solids*, Wiley, New York.

Gross, B. (1947) On creep and relaxation. *J. Appl. Phys.*, **18** (February), 212–21.

Gross, B. (1953) *Mathematical Structure of the Theories of Viscoelasticity*, Hermann, Paris.

Gurtin, M. E. and Herrera, R. I. (1965) On dissipation inequalities and linear viscoelasticity. *Q. Appl. Math.*, **23**, 235–45.

Gurtin, M. E. and Sternberg, E. (1962) On the linear theory of viscoelasticity. *Arch. Ration. Mech. Anal.*, **11**, 291–356.

Halpin, J. C. (1968) Introduction to viscoelasticity, in *Composite Materials Workshop* (eds S. W. Tsai, J. C. Halpin and N. J. Pagano), Technomic, Stamford, CT, pp. 87–152.

Hopkinson, J. (1876) The presidual charge of the Leyden jar. *Philos. Trans. R. Soc. London*, **166**, 489–94.

König, H. and Meixner, J. (1958) Linear Systeme und Lineare Transformationen. *Math. Nachr.*, **19**, 256–322.

Leaderman, H. (1943) *Elastic and Creep Properties of Filamentous and Other High Polymers*, Textile Foundation, Washington, DC.

Leaderman, H. (1958) In *Viscoelasticity Phenomena in Amorphous High Polymeric Systems*, Vol. II (ed. F. Eirich), Academic Press, New York, pp. 1–6.

Lee, E. H. (1960) *Viscoelastic Stress Analysis*, First Symposium on Naval Structural Mechanics, Pergamon, New York, pp. 456–82.

Leitman, M. J. and Fisher, G. M. C. (1973) The linear theory of viscoelasticity, in *Handbuch der Physik*, Vol. VI a/3 (ed. S. Flügge), Springer, Berlin, pp. 1–123.

Lethersich, W. (1950) The rheological properties of dielectric polymers. *Br. J. Appl. Phys.*, **1** (November), 294–301.

Lockett, F. J. (1972) *Nonlinear Viscoelastic Solids*, Academic Press, New York.

Markovitz, H. (1977) Boltzmann and the beginnings of linear viscoelasticity. *Trans. Soc. Rheol.*, **21**(3), 381–98.

Meixner, J. (1965) Linear passive systems, in Proc. Int. Symp. on Statistical Mechanics and Thermodynamics, 1964, North-Holland, Amsterdam, pp. 52–68.

Schapery, R. A. (1974) Viscoelastic behaviour and analysis of composite materials, in *Mechanics of Composite Materials*, Vol. 2 (ed. G. Sendeckj), Academic Press, New York, pp. 86–168.

Scott-Blair, G. W. (1949) *Survey of General and Applied Rheology*, Pitman, London.

Struik, L. C. E. (1987) The accuracy of some formulae for the interconversion of creep and relaxation data. *Rheol. Acta*, **26**, 7–13.

Stuart, H. A. (1956) *Die Physik der Hochpolymeren*, Vol. 4, Springer, Berlin.

Tschoegl, N. W. (1989) *The Phenomenological Theory of Linear Viscoelastic Behaviour, An Introduction*, Springer, New York.

Volterra, V. (1913) *Fonctions de Lignes*, Gauthier-Villard, Paris.

Volterra, V. and Peres, J. (1936) *Théorie Générale des Fonctionelles*, Gauthier-Villard, Paris.

Ward, I. M. (1983) *Mechanical Properties of Solid Polymers*, 2nd edn, Wiley, New York.

FURTHER READING

Aklonis, J. J. (1972) *Introduction to Polymer Viscoelasticity*, Wiley, New York.

Alexander, R. L. (1964) *Limits of Linear Viscoelastic Behaviour of an Asphalt Concrete in Tension and Compression*, Institute of Transportation and Traffic Engineering, University of California, Berkeley, CA.

Andrade, E. N. (1910) The viscous flow in metals and allied phenomena. *Proc. R. Soc. London, Ser. A*, **84**, 1–12.

Alfrey, T. and Doty, P. M. (1945) Methods of specifying the properties of viscoelastic materials. *J. Appl. Phys.*, **16**, 700–13.

Bazant, Z. P. (1975) Theory of creep and shrinkage in concrete structures: a précis of recent developments. *Mech. Today*, **2**, 1–93.

Bernal, J. D. (1958) Structure arrangements of macromolecules. *Discuss. Faraday Soc.*, **25**, 7–18.

Bland, D. R. and Lee, E. H. (1956) On the determination of a viscoelastic model for stress analysis of plastics. *J. Appl. Mech.*, **23**, 416–20.

Berry, D. S. and Hunter, S. C. (1956) The propagation of dynamic stresses in viscoelastic rods. *J. Mech. Phys. Solids*, **4**, 72–95.

Breuer, S. (1969) Lower bounds on work in linear viscoelasticity. *Q. Appl. Math.*, **27**(2), 139–46.

Creus, G. J. (1986) *Viscoelasticity: Basic Theory and Applications to Concrete Structures*, Springer, Berlin.

Findley, W. N. (1944) Creep characteristics of plastics, in Symposium on Plastics, ASTM, pp. 118–34.

Findley, W. N. and Khosla, G. (1955) Application of the superposition principle and theories of mechanical equation of state, strain and time hardening to creep of plastics under changing loads. *J. Appl. Phys.*, **26**(7), 821–32.

Finnie, I. and Heller, W. R. (1959) *Creep of Engineering Materials*, McGraw-Hill, New York.

Glauz, R. D. and Lee, E. H. (1954) Transient wave analysis in linear time-dependent material. *J. Appl. Phys.*, **25**, 947–53.

Haddad, Y. M. (1988) On the theory of the viscoelastic solid. *Res Mech.*, **25**, 225–59.

Hill, R. L. (1967) The creep behaviour of individual pulp fibers under tensile stress. *Tappi*, **50**(8), 432–40.

Hilton, H. H. (1964) Viscoelastic analysis, in *Engineering Design for Plastics* (ed. E. Baer), Reinhold, New York.

Hopkins, I. L. and Hamming, R. W. (1957) On creep and relaxation. *J. Appl. Phys.*, **28**(8), 906–9.

Hunter, S. C. (1960) Viscoelastic waves, in *Progress in Solid Mechanics*, Vol. I (eds I. N. Sneddon and R. Hill), North-Holland, Amsterdam, pp. 1–57.

Kauman, W. G. (1966) On the deformation and setting of the wood cell wall. *Holz Roh Werkst.*, **24**(11), 551–6.

Kolsky, H. (1965) Experimental studies of the mechanical behaviour of linear viscoelastic solids, in Proc. 4th Symposium on Naval Structural Mechanics, Pergamon, London, pp. 381–442.

Lee, E. H. (1956) Special issues on rheology of polymers. *J. Appl. Phys.*, **27**, 665–72.

Lee, E. H. (1960) In *Viscoelasticity: Phenomenological Aspects* (ed. J. T. Bergen), Academic Press, New York, pp. 1–150.

Lee, E. H. (1962) Viscoelasticity, in *Handbook of Engineering Mechanics* (ed. W. Flügge), McGraw-Hill, New York, pp. 53/1–53/22.

Lubliner, J. and Salkman, J. L. (1967) On uniqueness in general linear viscoelasticity. *Q. Appl. Math.*, **25**, 129–38.

Mark, H. and Tobolosky, A. V. (1950) *Physical Chemistry of High Polymeric Materials*, Interscience, New York.

Mazilu, P. (1973) On the constitutive law of Boltzmann–Volterra. *Rev. Roum. Math. Pures Appl.*, **18**, 1067–9.

Meredith, R. (1954) Relaxation of stress in stretched cellulose fibers. *J. Text. Inst.*, **45**, T438–T460.

McHenry, D. (1943) A new aspect of creep in concrete and its application to design. *Proc. ASTM*, **43**, 1064–86.

Odeh, F. and Tadjbakhsh, I. (1965) Uniqueness in the linear theory of viscoelasticity. *Arch. Ration. Mech. Anal.*, **18**, 244–50.

Pipkin, A. C. (1972) *Lectures on Viscoelasticity Theory*, Springer, New York.

Roesler, F. C. and Twyman, W. A. (1955) An iteration method for the determination of relaxation spectra. *Proc. Phys. Soc. B*, **68**(2), 97–105.

Roscoe, R. (1950) Mechanical models for the representation of viscoelastic properties. *Br. J. Appl. Phys.*, **1**, 171–3.

Roy, M. (1966). *Milieux Continus*, Dunod, Paris.

Swindeman, R. W. and Bolling, E. (1989) Relaxation response of A533B Steel from 25 to 600°C, in Proceedings, the 1989 ASME Pressure Vessels and Piping Conference–JSME Cosponsorship, PVP, Vol. 172, Collection and Uses of Relaxation Data in Design, Honolulu, Hawaii, July 23–27, 1989, pp. 21–8.

Tapsell, H. J. and Johnson, A. E. (1940) Creep under combined tension and torsion. *Engineering*, **150**, 24–8.

Tobolosky, A. and Eyring, H. J. (1943) Mechanical properties of polymeric materials. *J. Chem. Phys.*, **11**, 125–34.

Volterra, V. (1909) Sulle equazioni integro diffenziali della teoria dell'elasticita. *Att. Reale Accad. Lincei*, **18**, 295–301.

3

Transition to dynamic viscoelasticity

3.1 INTRODUCTION

In Chapter 2, we introduced the idealized theory of isothermal linear viscoelasticity under the conditions of quasi-static stress or strain. Within the scope of the presented formulations, the linear viscoelastic behaviour of a material may be described by one of the following functions

1. the creep function $F(t)$;
2. the relaxation function $R(t)$;
3. the retardation spectrum $f(\lambda)$;
4. the relaxation spectrum $\Gamma(\lambda)$.

Of the four functions above, the first two can be determined by experiment. That is, the creep function $F(t)$ is determined by a creep experiment and the relaxation function $R(t)$ is determined from a relaxation experiment. The other two functions $f(\lambda)$ and $\Gamma(\lambda)$, however, may be determined by integral transformation. That is, the retardation spectrum $f(\lambda)$ is determined by integral transformation, via equation (2.60), from the creep function $F(t)$ in (1) above. Also, the relaxation spectrum $\Gamma(\lambda)$ is determined by integral transformation, via equation (2.56), from the relaxation function $R(t)$ in (2) above. In view of the interrelation between the functions $F(t)$ and $R(t)$, as discussed in section 2.4, any one of the functions (1)–(4) above may be used to characterize the linear viscoelastic behaviour of the material completely with no other restrictions imposed other than linearity of the response behaviour (Boltzmann's superposition principle) and the constancy of temperature (isothermal) condition.

An alternative approach to the viscoelastic response of the material is to subject the specimen to an alternating stimulus and simultaneously measuring the output. Hence, in the present chapter, we extend the formulation of Chapter 2 to include the possibility of characterizing the linear viscoelastic behaviour of materials under dynamic loading or deformations. This is particularly suitable for gaining information about processes which take place rapidly, e.g. wave propagation. Here, we follow the sequence of development of Chapter 2. First, we present the material functions that characterize the linear viscoelastic material under dynamic conditions. Then, we introduce the interrelations between these functions and their connection to the material functions characterizing the linear viscoelastic response under static input conditions.

3.2 CHARACTERIZATION OF THE LINEAR VISCOELASTIC RESPONSE UNDER DYNAMIC LOADING CONDITIONS

For stresses and strains which are not too large, the linear viscoelastic properties exhibited under dynamic loading are described by a complex modulus $E^*(i\omega)$ which is a function of the frequency of the loading ω. As derived below, a linear relation exists between the stresses and strains in question in the form

$$\sigma(\omega, t) = E^*(i\omega)\varepsilon(\omega, t). \tag{3.1}$$

In the case of periodic loading, both the stress and strain appearing in the above equation are harmonic functions of time and the modulus $E^*(i\omega)$ is given by real and imaginary parts as

$$E^*(i\omega) = E_1(\omega) + iE_2(\omega). \tag{3.2}$$

Because of the effect of delayed elasticity and viscous flow in the viscoelastic material, the stress and strain will be generally out of phase. Thus, $\sigma(t)$ and/or $\varepsilon(t)$ will be complex numbers.

3.2.1 Complex modulus and complex compliance

In the case of linear viscoelastic material subjected to a sinusoidal stress, when equilibrium is reached, both the stress and the strain will vary sinusoidally, but the strain will lag behind the stress, i.e.

$$\varepsilon = \varepsilon_0 \sin(\omega t) \tag{3.3}$$

and

$$\sigma = \sigma_0 \sin(\omega t + \delta) \tag{3.4}$$

where ω is the angular frequency and δ is the phase lag. Expanding equation (3.4),

$$\sigma = \sigma_0 \sin(\omega t) \cos \delta + \sigma_0 \cos(\omega t) \sin \delta. \tag{3.5}$$

We introduce the moduli E_1 and E_2 where

$$E_1 = \frac{\sigma_0}{\varepsilon_0} \cos \delta, \quad E_2 = \frac{\sigma_0}{\varepsilon_0} \sin \delta \tag{3.6a}$$

and

$$\tan \delta = E_2/E_1. \tag{3.6b}$$

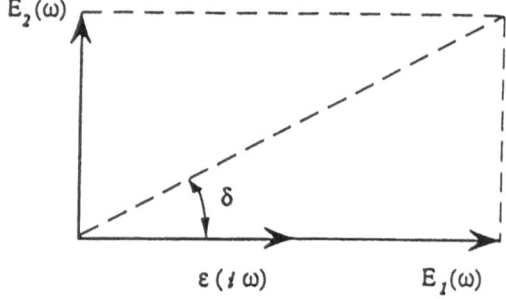

Fig. 3.1 Storage modulus $E_1(\omega)$, loss modulus $E_2(\omega)$ and phase lag δ for a linear viscoelastic material (Gittus, 1975).

Thus, in terms of E_1 and E_2 as expressed in (3.6a), equation (3.5) can be written in the following form:

$$\sigma = \varepsilon_0 E_1 \sin(\omega t) + \varepsilon_0 E_2 \cos(\omega t). \tag{3.7}$$

In other words, the stress–strain relationship (3.7) is defined, in view of (3.6), by the modulus E_1 in phase with the strain and by the modulus E_2 which is $90°$ out of phase with the strain (Fig. 3.1).

Meantime, if we use a complex representation of the input and output, i.e.

$$\varepsilon = \varepsilon_0 \exp(i\omega t) \tag{3.8}$$

and

$$\sigma = \sigma_0 \exp[i(\omega t + \delta)], \tag{3.9}$$

then

$$\frac{\sigma}{\varepsilon} = E^* = \frac{\sigma_0}{\varepsilon_0} \exp(i\delta)$$

$$= \frac{\sigma_0}{\varepsilon_0} (\cos \delta + i \sin \delta)$$

$$= E_1 + iE_2 \tag{3.10}$$

where equation (3.6a) has been used and E^* is referred to as the 'complex modulus'. In the case of a linear viscoelastic response, however, the amplitude ratio, σ_0/ε_0, and the phase difference remain constant for fixed frequency and are generally independent of the amplitude of excitation (e.g. Lee, 1960). In (3.10), the real part E_1 of the complex modulus, which is in phase with the strain, is often referred to as the 'storage modulus' or, simply, the 'dynamic modulus'. It is associated with the energy stored in the specimen due to the applied strain. The imaginary part E_2 of the complex modulus, which is out of phase with the strain, defines the dissipation of energy. It is often called the 'loss modulus' or 'internal friction modulus'. It forms part of the energy dissipated per cycle. The latter is denoted below by ΔW and is expressed as

$$\Delta W = \int \sigma \, d\varepsilon = \int_0^{2\pi/\omega} \sigma \frac{d\varepsilon}{dt} \, dt$$

$$= \omega \varepsilon_0^2 \int_0^{2\pi/\omega} [E_1 \sin(\omega t) \cos(\omega t)$$

$$+ E_2 \cos^2(\omega t)] \, dt$$

$$= \pi E_2 \varepsilon_0^2 \tag{3.11}$$

where equations (3.3) and (3.7) have been used. Typical values of E_1, E_2 and $\tan \delta$ for a polymer would be around 10^{10} dyn cm^{-2}, 10^8 dyn cm^{-2} and 0.01 respectively (Ward, 1983). The angle δ, expressed by (3.6b), is of particular interest. It represents, as mentioned earlier, the phase angle by which the strain lags behind the stress. The tangent of δ is conventionally employed as a measure of 'internal friction' of a linear viscoelastic material. Comparing (3.6b) and (3.10), it follows that

$$\tan \delta = \frac{\text{imaginary part of } E^*}{\text{real part of } E^*} = \frac{E_2}{E_1}.$$

A similar treatment can also be followed to define the 'complex compliance'. The latter is denoted by $J^*(i\omega)$ and it is related to the 'complex modulus' by the relation

$$J^*(i\omega) = \frac{1}{E^*(i\omega)} \quad \text{and} \quad E^*(i\omega) = \frac{1}{J^*(i\omega)}. \tag{3.12}$$

That is, by definition, the complex compliance is the reciprocal of the complex modulus and conversely. As discussed later in this chapter, equation (3.12) allows one to establish correlations between the creep phase and the relaxation phase of viscoelastic behaviour. Thus, the variation of either $E^*(i\omega)$ or $J^*(i\omega)$ as a function of frequency is an alternative form for defining the viscoelastic response of the material.

With reference to (3.12), the complex compliance is decomposed into its real and imaginary parts as

$$J^*(i\omega) = J_1(\omega) + iJ_2(\omega). \tag{3.13}$$

Combining equations (3.10), (3.12) and (3.13), it follows that

$$(E_1 + iE_2)(J_1 + iJ_2) = 1. \tag{3.14}$$

For a meaningful description of the dynamic viscoelastic response of the material, it is necessary to determine the moduli E_1 and E_2 or, alternatively, the compliances J_1 and J_2 in terms of frequency. In this case, we refer back to the basic mechanical models introduced earlier in Chapter 2 (section 2.2).

3.3 MECHANICAL MODELS OF VISCOELASTIC RESPONSE UNDER DYNAMIC LOADING

3.3.1 Maxwell model

We recall equation (2.10), characterizing the response behaviour of the Maxwell model,

$$\sigma + \lambda \frac{d\sigma}{dt} = E\lambda \frac{d\varepsilon}{dt}, \quad \lambda = \eta/E, \tag{3.15}$$

where η and E are the characteristics of the elements of the Maxwell model.

Now, we let

$$\sigma = \sigma_0 \exp(i\omega t) \tag{3.16}$$

which, from equation (3.10), can be also expressed as

$$\sigma = \sigma_0 \exp(i\omega t) = (E_1 + iE_2)\varepsilon. \tag{3.17}$$

Thus, by combining equations (3.15) and (3.17), it follows that

$$E_1 + iE_2 = E \frac{i\omega\lambda}{1 + i\omega\lambda}. \tag{3.18}$$

Equating real and imaginary parts in both sides of (3.18),

$$E_1 = E \frac{\omega^2\lambda^2}{1 + \omega^2\lambda^2}, \quad E_2 = E \frac{\omega\lambda}{1 + \omega^2\lambda^2}, \tag{3.19}$$

and, with reference to (3.6b), the phase lag is expressed, for the Maxwell model, as

$$\tan \delta = \frac{1}{\omega\lambda}. \tag{3.20}$$

Figure 3.2 illustrates schematically the variation with frequency of the moduli E_1 and E_2 and the phase lag for the Maxwell model.

With reference to Fig. 3.2, the following points (Gittus, 1975) may be made.

1. At very small frequency of vibrations (i.e. when ω tends to zero), the energy loss in the Newtonian fluid contained in the dashpot would be extremely small. Hence E_2, which is proportional to the energy loss (equation (3.11)), would also be very small. At low frequencies, only very small displacements will be produced in the Maxwell element. Consequently, the modulus E_1, which is in phase with the displacement (equation (3.7)), would also be very small at low frequencies.
2. At moderate to high frequency (i.e. when ω approaches the reciprocal of the relaxation time λ of the system; ω tends to $1/\lambda$ and $\ln(\lambda\omega)$ tends to zero), the fluid in the dashpot flows relatively rapidly. Thus the viscous force becomes larger and, accordingly, both the energy loss in the system and the modulus E_2 will increase. At the same time, the spring must respond accordingly through large extensions, and hence the modulus E_1 will rise as ω approaches $1/\lambda$.

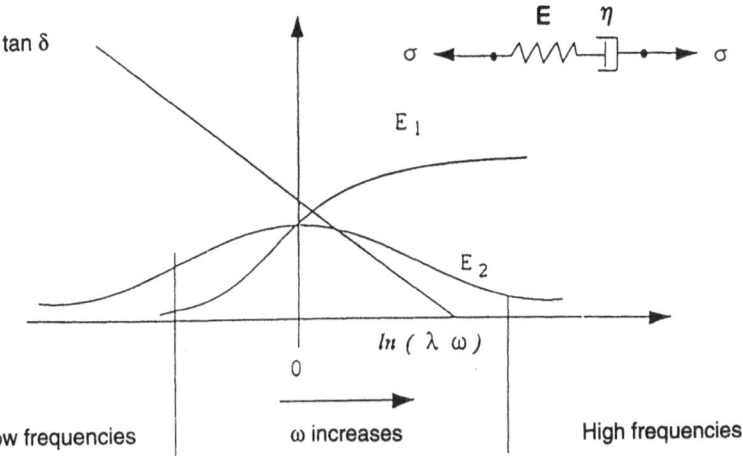

Fig. 3.2 Variation with frequency of moduli E_1, E_2 and phase lag δ for the Maxwell model (Gittus, 1975).

3. At very high frequencies, the spring would support most of the elongation while the dashpot will hardly respond. Accordingly, in view of (3.19), the modulus E_1 will approach its maximum value while the value of the modulus E_2 would approach zero value. Thus, at high frequencies, E_2 is negligible and, in view of (3.10), the value of the complex modulus E^* will be approximately equal to E_1.

4. Following point (3), since $\tan \delta = E_2/E_1$ (equation (3.6b)), $\tan \delta$ decreases monotonically as frequency increases.

3.3.2 Kelvin (Voigt) model

An analytical treatment similar to that carried out above for the Maxwell model leads to the following expressions for the real (storage) compliance J_1 and the imaginary (loss) compliance of a Kelvin model subjected to sinusoidal input. That is,

$$J_1 = \frac{1}{E(1 + \omega^2 \lambda^2)}, \quad J_2 = \frac{\omega\lambda}{E(1 + \omega^2 \lambda^2)} \tag{3.21}$$

and

$$\tan \delta = \omega\lambda. \tag{3.22}$$

The frequency dependence of the compliances J_1 and J_2 and the phase lag for the Kelvin model is illustrated in Fig. 3.3.

1. At very small frequencies of vibration, the viscous drag of the dashpot is small. Accordingly, both the energy loss and loss compliance J_2 are expected to be small.

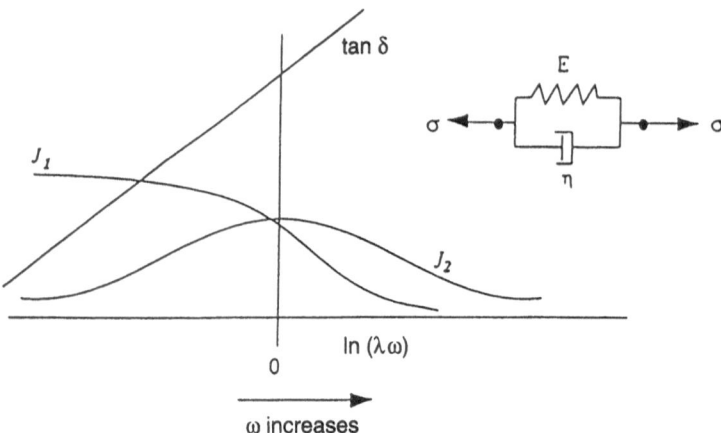

Fig. 3.3 Variation with frequency of compliances J_1, J_2 and phase lag δ for the Kelvin model (Gittus, 1975).

2. At moderate to high frequency (i.e. when ω approaches $1/\lambda$), the dashpot moves more rapidly. Hence the viscous force exerted by the dashpot increases. Thus the applied external force is unable to deform the spring as it counterbalances, at the same time, a large viscous force from the dashpot. Accordingly, the spring elongation and the storage compliance J_1 decrease, while J_2 increases.
3. At very high frequencies, the viscous resistance of the fluid in the dashpot becomes very high, preventing the spring from elongating, and hence J_1 will approach zero. At the same time, the displacement of the dashpot becomes so small that the energy loss per cycle and hence J_2 approach zero.

3.3.3 Three-element model (Standard Linear Solid)

The response of the standard linear solid under static loading is governed by equation (2.36). Solution of this equation for the case of a sinusoidal stimulus follows the same treatment presented earlier for the Maxwell model. This yields, for the moduli E_1 and E_2 and the phase lag δ, the following expressions:

$$E_1 = \frac{E(1 + \omega^2\lambda\lambda')}{1 + \omega^2\lambda^2} = \frac{E + \omega^2\lambda^2E'}{1 + \omega^2\lambda^2}; \qquad (3.23)$$

$$E_2 = \frac{E(\lambda' - \lambda)\omega}{1 + \omega^2\lambda^2} = \frac{(E' - E)\omega\lambda}{1 + \omega^2\lambda^2}; \qquad (3.24)$$

$$\tan \delta = \frac{(\lambda' - \lambda)\omega}{1 + \omega^2\lambda\lambda'} = \frac{(E' - E)\omega\lambda}{1 + \omega^2\lambda^2E'}; \qquad (3.25)$$

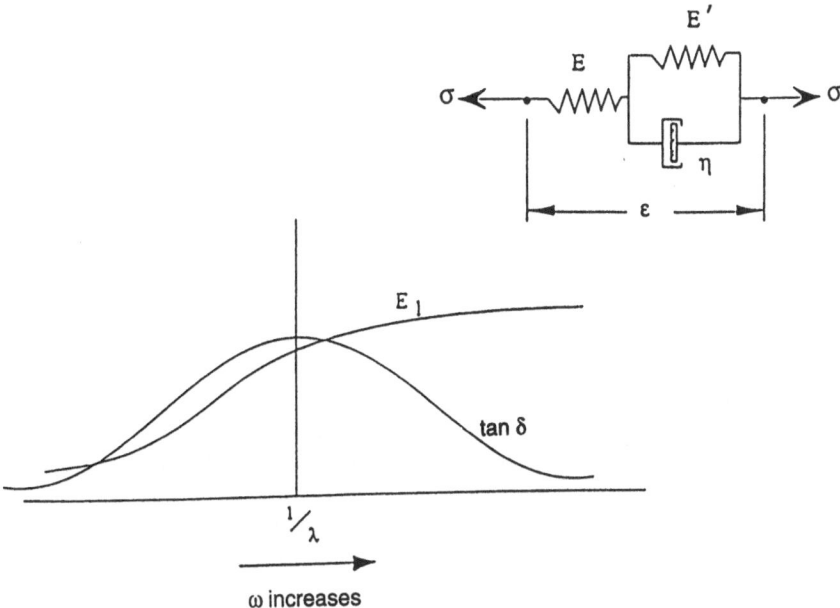

Fig. 3.4 Variation with frequency of the storage modulus E_1 and phase lag δ for the three-element model (Gittus, 1975).

The identification of the parameters appearing in equations (3.23)–(3.25) are presented in conjunction with this model in section 2.2 and shown in Fig. 3.4. In this figure, the variation of moduli and phase lag with frequency is also illustrated. As governed by (3.25), it can be seen in Fig. 3.4 that the phase lag δ goes through a maximum when $\omega^2 = 1/\lambda\lambda'$, instead of varying in a straight line fashion with the frequency, as it does in the case of the Maxwell and Kelvin models. This is, of course, provided that $E = E'$.

As an illustration of the dynamic response of a real viscoelastic material, Fig. 3.5 shows double-logarithmic plots of storage and loss moduli (G' and G'') in shear as functions of circular frequency ω for melts of a standard polystyrene and a binary blend with a low molar mass function (Schausberger, Knoglinger and Janeschitz-Kriegl, 1987).

Figures 3.6 and 3.7, both due to Kolsky (1960), give a comparison between the observed behaviour of four viscoelastic materials, namely polyethylene, polymethyl methacrylate, ebonite and polystyrene, and the predicted behaviour of the two- and three-element mechanical models. The experimental values in the two figures are due to Lethersich (1950) and give the response behaviour of the mentioned materials to time-dependent sinusoidal shear deformation. In Fig. 3.6, it is seen that, whereas the Maxwell and Voigt models are completely inadequate in predicting the variation of tan δ with frequency for real materials, the standard linear solid seems to give a reasonable approximation of the observed behaviour of such materials over at least

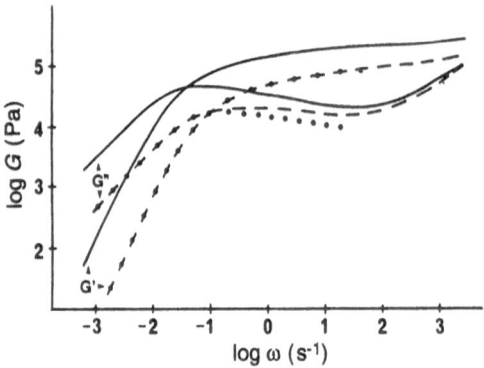

Fig. 3.5 Double-logarithmic plots of storage and loss moduli G' and G'' as functions of circular frequency ω (reference temperature $T_r = 180°C$) for the melts of a standard polystyrene and a binary blend with a low molar mass fraction: measured values for a standard of $M_W = 670$ kg mol^{-1}; –––, values for a binary mixture of this standard with 35 wt% of a standard with $M_W = 8$ kg mol^{-1}; ●, positions of the full lines after appropriate shifting. (Source: Schausberger, A., Knoglinger, H. and Janeschitz-Kriegl, H. (1987) The role of short chain molecules for the rheology of polystyrene melts. *Rheol. Acta*, **26**, 468–473, Steinkopff Verlag Darmstadt Reprinted by permission of Steinkopff Verlag Darmstadt.)

one decade of frequency. In Fig. 3.7, the plot of the complex modulus against logarithm of frequency again shows that the predicted behaviour of the two-element model is very inadequate for describing the response of real materials; the standard linear solid seems to be much more suitable, but only over a limited frequency range.

3.4 DYNAMIC VISCOELASTIC BEHAVIOUR

3.4.1 Creep response under alternating stress

Consider a linear viscoelastic solid subjected in the steady state to a sinusoidal stress. This will result, as mentioned earlier, in a sinusoidal strain partly in phase and partly in quadrature with the stress. Thus, using a complex representation of the stress input, the creep response equation can be written as

$$\varepsilon(t) = J^*(i\omega)\sigma(t), \quad \sigma(t) = \sigma_0 \exp(i\omega t). \tag{3.26}$$

In the above equation, $J^*(i\omega)$ is the complex viscoelastic compliance. The latter may be interpreted as the complex strain due to a sinusoidal stress input of unit magnitude. Decomposing the complex compliance into real and imaginary parts, then

$$J^*(i\omega) = J_1(\omega) + iJ_2(\omega), \tag{3.27}$$

where $J_1(\omega)$ is the dynamic storage compliance and $J_2(\omega)$ is the dynamic friction compliance previously introduced in section 3.2.1.

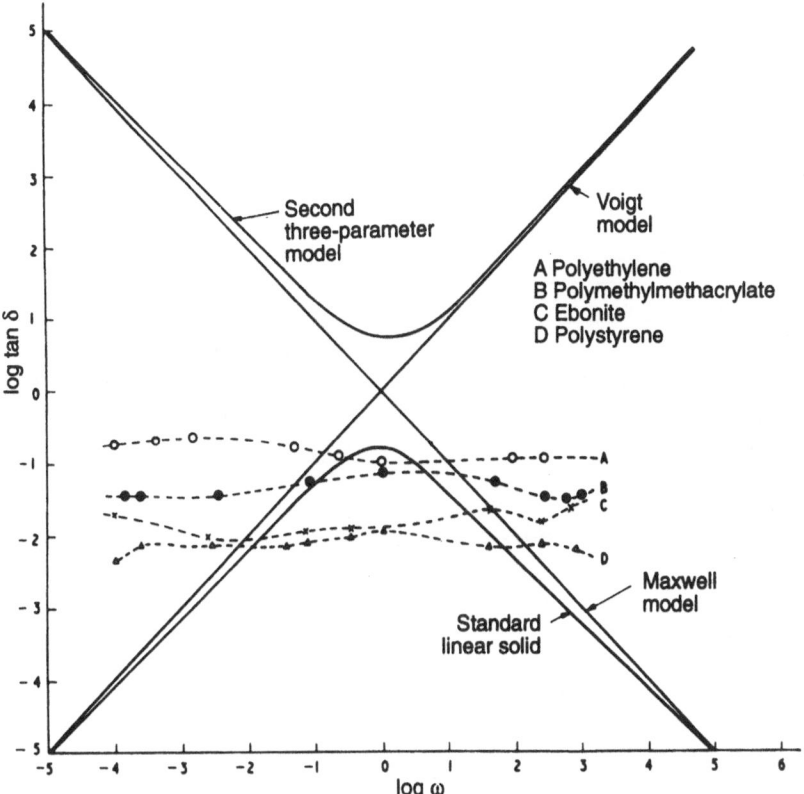

Fig. 3.6 Comparison between response of model solids and measured values of tan δ in shear (Lethersich, 1950). The second three-parameter model is constructed with a dashpot in parallel with a Maxwell model. (Source: Kolsky, H. (1960) Viscoelastic waves, in Int. Symp. on Stress Wave Propagation in Materials (Ed. N. Davids), Interscience Publishers, New York, pp. 59–90. Reprinted with permission.)

Recalling the creep constitutive equation (2.76) in the form of the hereditary integral, one can write

$$\varepsilon(t) = \int_0^t F(t - \tau) \frac{d\sigma(\tau)}{d\tau} \, d\tau. \tag{3.28}$$

We replace, for convenience, in the above equation the variable $t - \tau$ by ξ; then

$$\varepsilon(t) = \int_0^t F(\xi) \frac{d\sigma(t - \xi)}{d\xi} \, d\xi. \tag{3.29}$$

Assuming periodic stress input and substituting in the above equation for $\sigma(t) = \sigma_0 \exp(i\omega t)$, then

$$\varepsilon(t) = \int_0^t F(\xi) i\omega \sigma_0 \exp[i\omega(t - \xi)] \, d\xi \tag{3.30}$$

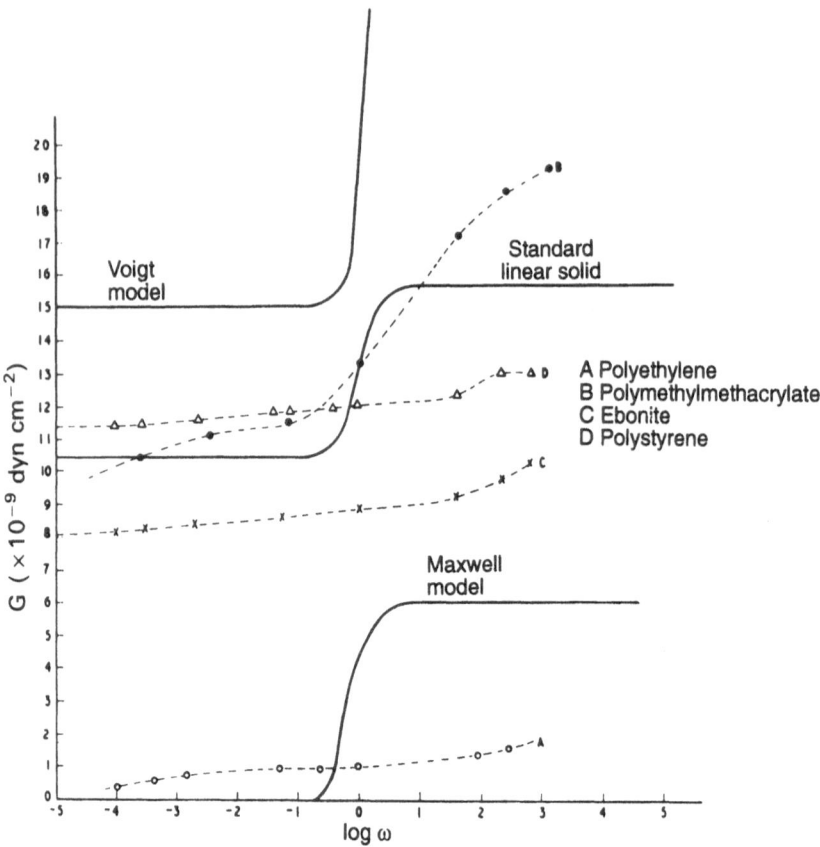

Fig. 3.7 Comparison between response of model solids and measured values of complex modulus in shear (Lethersich, 1950). (Source: Kolsky, H. (1960) Viscoelastic waves, in Int. Symp. on Stress Wave Propagation in Materials (Ed. N. Davids), Interscience Publishers, New York, pp. 59–90. Reprinted with permission.)

which can be written as

$$\varepsilon(t) = i\omega\sigma_0 \exp(i\omega t) \int_0^\infty F(\xi) \exp(-i\omega\xi) \, d\xi. \tag{3.31}$$

On the assumption that the creep function $F(t) = 0$ for $t < 0$, one can replace the lower limit of the integral above by $-\infty$ and write this integral in the following form of Fourier transformation (Appendix D)

$$\bar{F}(\omega) = \int_{-\infty}^\infty F(\tau) \exp(-i\omega\tau) \, d\tau \tag{3.32}$$

where $\bar{F}(\omega)$ is the Fourier transform of the creep function under static stress.

Combining equations (3.31) and (3.32), it follows that

$$\varepsilon(t) = i\omega\sigma_0\bar{F}(\omega) \exp(i\omega t). \tag{3.33}$$

Equation (3.33) illustrates a basic characteristic concerning the linearity in response; that is, when the stress is periodic, the resulting strain will also be periodic. Hence, if

$$\varepsilon(t) = \varepsilon_0 \exp(i\omega t),$$

then, by comparison with (3.33), it follows that

$$\varepsilon_0 = i\omega\sigma_0 \bar{F}(\omega) \tag{3.34}$$

or

$$\frac{\varepsilon_0}{\sigma_0} = i\omega\bar{F}(\omega). \tag{3.35}$$

In the above equation, the ratio ε_0/σ_0 is a complex number which is expressed, in view of (3.10), as

$$\frac{\varepsilon_0}{\sigma_0} = i\omega\bar{F}(\omega) = \frac{\exp(i\delta)}{E^*}. \tag{3.36}$$

As mentioned in section 3.2, the tangent of δ is often used as a measure of 'internal friction' of a linear viscoelastic material; hence, in view of (3.36), the internal friction can be determined when the Fourier transform of the creep function is known.

3.4.2 Relaxation response under alternating strain

Consider the complex presentation of the alternating strain input. Then the relaxation response equation is expressed in an analogous manner to (3.26) by

$$\sigma(t) = E^*(i\omega)\varepsilon(t), \quad \varepsilon(t) = \varepsilon_0 \exp(i\omega t), \tag{3.37}$$

where $E^*(i\omega)$ is the complex viscoelastic modulus; this is interpreted as the complex stress resulting from the application of a sinusoidal strain of unit magnitude.

Recalling the stress-relaxation constitutive equation (2.82) in the form of the hereditary integral, one writes

$$\sigma(t) = \int_0^\infty R(t - \tau) \frac{d\varepsilon(\tau)}{d\tau} \, d\tau. \tag{3.38}$$

Equation (3.38), with a similar procedure to that carried out on the creep constitutive equation (3.28), leads to

$$\frac{\sigma_0}{\varepsilon_0} = i\omega\bar{R}(\omega) = E^* \exp(-i\delta) \tag{3.39}$$

where equation (3.10) has been used and $\bar{R}(\omega)$ is the Fourier transform of the relaxation function under static strain, i.e.

$$\bar{R}(\omega) = \int_{-\infty}^\infty R(\tau) \exp(-i\omega\tau) \, d\tau. \tag{3.40}$$

Further, by combining (3.35) and (3.39), a relationship between the Fourier transform of the creep function and that of the relaxation function is expressed as

$$-\omega^2 \bar{F}(\omega)\bar{R}(\omega) = 1, \tag{3.41}$$

which again is a correlation between the creep and relaxation phases of the viscoelastic response in the linear case.

3.5 RELATIONS BETWEEN MATERIAL FUNCTIONS ASSOCIATED WITH CREEP AND RELAXATION

The material functions characterizing the response behaviour of a linear viscoelastic material are interrelated. That is, one material function is analytically deduced from another. We present below the different relations between the different material functions associated with both creep and stress relaxation of a linear viscoelastic material under static and dynamic loading conditions. For more information, the reader is referred, in this context, to Gross (1947), Alfrey (1948), Leaderman (1954a, b), Schwarzl and Staverman (1952), Fung (1965) and Ferry (1970), among others.

3.5.1 Material functions associated with creep

(a) Complex compliance function and creep function

In order to establish a relationship between the dynamic complex compliance $J^*(i\omega)$ and the creep function $F(\tau)$, pertaining to the static response, we employ the creep constitutive equation (3.28). Substituting $\sigma(t) = \sigma_0 \exp(i\omega t)$ in (3.28), it can be shown that

$$J^*(i\omega) = \int_0^\infty \exp(-i\omega\tau) \frac{dF(\tau)}{d\tau} \, d\tau. \tag{3.42}$$

Decomposing this equation into real and imaginary components gives

$$J_1(\omega) = \int_0^\infty \frac{dF(\tau)}{d\tau} \cos(\omega\tau) \, d\tau \tag{3.43a}$$

and

$$J_2(\omega) = -\int_0^\infty \frac{dF(\tau)}{d\tau} \sin(\omega\tau) \, d\tau \tag{3.43b}$$

where $J^*(i\omega) = J_1(\omega) + iJ_2(\omega)$.

It is apparent from equations (3.43) that $J_1(\omega)$ and $J_2(\omega)$ are Fourier transforms of the creep function $F(\tau)$. Thus, by inversion of (3.43a) or (3.43b), the creep function can be determined in terms of the dynamic compliance components. The simultaneous existence of the two relations (3.43a) and (3.43b) implies that a relationship between

$J_1(\omega)$ and $J_2(\omega)$ exists (Gross, 1953). Benbow (1956), for instance, considered the determination of dynamic moduli and internal friction of high polymers from static creep measurements with reference to some measurements made of polythene. This was carried out using an integral transformation as based on a procedure put forward by Roesler and Pearson (1954) and Roesler and Twyman (1955).

(b) Complex compliance function and retardation spectrum

Consider, in view of (2.60), the rate of creep function as

$$\frac{dF(t)}{dt} = \int_0^\infty \frac{\alpha f(\lambda)}{\lambda} \exp(-t/\lambda)\, d\lambda. \tag{3.44}$$

Following Gross (1953), one introduces a creep frequency $s = 1/\lambda$ and a frequency distribution

$$N(s) = \alpha f(1/s)/s^2. \tag{3.45}$$

Hence, $N(s)$ represents also a retardation spectrum. Combining equations (3.44) and (3.45), the former becomes

$$\frac{dF(t)}{dt} = \int_0^\infty sN(s)\, \exp(-ts)\, ds \tag{3.46}$$

which is a form of a Laplace integral.

Substituting expression (3.46) into (3.42), it can be shown that the complex compliance function can be expressed in terms of the retardation spectrum (the distribution function of retardation times) as

$$J^*(i\omega) = \int_0^\infty \frac{sN(s)}{s + i\omega}\, ds. \tag{3.47}$$

Decomposing the above equation into real and imaginary components, gives

$$J_1(\omega) = \int_0^\infty N(s) \frac{s^2}{s^2 + \omega^2}\, ds \tag{3.48a}$$

and

$$J_2(\omega) = -\int_0^\infty N(s) \frac{\omega s}{s^2 + \omega^2}\, ds. \tag{3.48b}$$

In order to determine the retardation spectrum explicitly from the creep compliance, one substitutes in equation (3.47) $\omega \exp(\pm i\pi)$ for $i\omega$; then, this equation is transformed to (Gross, 1953)

$$J^*[\omega \exp(\pm i\pi)] = \int_0^\infty \frac{sN(s)}{s - \omega}\, ds \mp i\pi\omega N(\omega) \tag{3.49}$$

which results in the determination of the retardation spectrum

$$N(\omega) = \pm \frac{1}{\pi\omega} \text{ Im } J^* \, \omega[\exp(\mp i\pi)] \tag{3.50}$$

or in terms of J_1 and J_2 as

$$N(\omega) = \pm \frac{2}{\pi\omega} \text{ Im } J_1[\exp(\mp i\pi/2)] \tag{3.51a}$$

and

$$N(\omega) = \frac{2}{\pi\omega} \text{ Re } J_2[\exp(\mp i\pi/2)] \tag{3.51b}$$

where Im(·) and Re(·) refer, respectively, to imaginary and real parts of the complex variable or function (·). Thus, in terms of the formulations above, one can determine the retardation frequency spectrum from the complex compliance function or from its real and imaginary parts. Approximation formulae for the determination of the retardation spectrum were given by Alfrey and Doty (1945), Schwarzl (1951), Andrews (1952) and Marvin (1952). For a review and evaluation of such approximation formulae, the reader is referred to Leaderman (1954a, b).

3.5.2 Material functions associated with relaxation

(a) Complex modulus function and relaxation function

Substituting the alternating stress input $\sigma = \sigma_0 \exp(i\omega t)$ into the relaxation constitutive equation (3.38), it can be concluded that

$$E^*(i\omega) = i\omega \int_0^\infty \exp(-i\omega\tau)R(\tau) \, d\tau. \tag{3.52}$$

Further, the complex modulus function, as represented by the above equation, can be decomposed into real and imaginary parts, i.e.

$$E^*(i\omega) = E_1(\omega) + iE_2(\omega)$$

where

$$E_1(\omega) = \omega \int_0^\infty R(\tau) \, \sin(\omega\tau) \, d\tau \tag{3.53a}$$

and

$$E_2(\omega) = \omega \int_0^\infty R(\tau) \, \cos(\omega\tau) \, d\tau \tag{3.53b}$$

Equations (3.52) and (3.53) are forms of Fourier integrals. Both equations establish a connection between the dynamic response and the static response in linear viscoelasticity. As shown, these equations permit one to determine the dynamic complex modulus function or its components in terms of the relaxation function in the static case. Conversely, these equations may be inverted to determine the relaxation function explicitly in terms of the dynamic moduli. Considering, for instance, the inversion of (3.53), the relaxation function is expressed as

$$R(t) = \frac{2}{\pi} \int_0^\infty \frac{E_1(\omega)}{\omega} \sin(\omega t) \, d\omega \tag{3.54a}$$

or, alternatively,

$$R(t) = \frac{2}{\pi} \int_0^\infty \frac{E_2(\omega)}{\omega} \cos(\omega t) \, d\omega. \tag{3.54b}$$

That is, the relaxation function pertaining to the viscoelastic response under static strain can be determined from dynamic data.

(b) Complex modulus function and relaxation spectrum

Consider the expression (2.56) for the relaxation function, i.e.

$$R(t) = \int_0^\infty \beta \Gamma(\lambda) \exp(-t/\lambda) \, d\lambda. \tag{3.55}$$

We introduce now a relaxation frequency $s = 1/\lambda$ and a corresponding relaxation frequency spectrum $N'(s) \, ds$ where

$$N'(s) = \beta \Gamma(1/s)/s^2, \quad s = 1/\lambda. \tag{3.56}$$

Combining (3.55) and (3.56), the relaxation function is expressed in terms of the frequency spectrum as

$$R(t) = \int_0^\infty N'(s) \exp(-ts) \, ds. \tag{3.57}$$

The integral in this equation is a Laplace transform. Thus, when the relaxation function $R(t)$ is given, the inversion of (3.57) results in the determination of the relaxation frequency spectrum $N'(s)$ (e.g. Gross, 1947, 1953; Macey, 1948; Pol and Bremmer, 1951).

Combining (3.52) and (3.57), it can be shown through an iteration procedure that the complex modulus is

$$E^*(i\omega) = i\omega \int_0^\infty \frac{N'(s)}{s + i\omega} \, ds. \tag{3.58}$$

Meantime, separating $E^*(i\omega)$ into its real and imaginary parts, the dynamic storage modulus and the dynamic friction modulus are expressed, respectively by

$$E_1(\omega) = \int_0^\infty N'(s) \frac{\omega^2}{\omega^2 + s^2} \, ds \qquad (3.59a)$$

and

$$E_2(\omega) = \int_0^\infty N'(s) \frac{\omega s}{\omega^2 + s^2} \, ds. \qquad (3.59b)$$

Further, by substituting $\omega \exp(\pm i\pi)$ for $i\omega$ in equation (3.58), it can be shown that the latter equation will transform into

$$E^*[\omega \exp(\pm i\pi)] = -\omega \int_0^\infty \frac{N'(s)}{s - \omega} \, ds \pm i\pi\omega N'(\omega). \qquad (3.60)$$

Inversion of equation (3.60) results in the determination of the relaxation spectrum. That is,

$$N'(\omega) = \pm \frac{1}{\pi\omega} \, \text{Im} \, E^*[\omega \exp(\pm i\pi)] \qquad (3.61)$$

or, alternatively, in terms of the dynamic moduli E_1 and E_2,

$$N'(\omega) = \pm \frac{2}{\pi\omega} \, \text{Im} \, E_1[\exp(\pm i\pi/2)] \qquad (3.62a)$$

and

$$N'(\omega) = \frac{2}{\pi\omega} \, \text{Re} \, E_2[\exp(\pm i\pi/2)]. \qquad (3.62b)$$

Thus, one is able to determine the relaxation frequency spectrum from the complex modulus as well as from its real and imaginary components (Roesler and Pearson, 1954; Roesler, 1955; Roesler and Twyman, 1955).

3.5.3 Interrelation between the retardation spectrum and the relaxation spectrum

The retardation and relaxation spectra are time distribution functions of the compliance and the elastic moduli, respectively. As we have discussed in sections 3.5.1(b) and 3.5.2(b), both spectra are determinable from the creep and relaxation data respectively, but with the involvement of the numerical solution of the corresponding integral equations.

As shown earlier, equations (3.12) allow us to establish a correlation between the material functions associated with the creep phase and those pertaining to the relaxation phase of viscoelastic behaviour. In this section, we seek along the same lines a relationship between the retardation spectrum and the relaxation spectrum.

Following Gross (1953), one substitutes $\omega \exp(\pm i\pi)$ for $i\omega$ in (3.12); then

$$E^*[\omega \exp(\pm i\pi)] = 1/J^*[\omega \exp(\pm i\pi)] \tag{3.63a}$$

and

$$J^*[\omega \exp(\pm i\pi)] = 1/E^*[\omega \exp(\pm i\pi)]. \tag{3.63b}$$

Separating into real and imaginary parts and recalling the creep distribution function $f(\lambda)$ and the relaxation distribution function $\Gamma(\lambda)$, it can be shown that (Gross, 1953)

$$\alpha f(\lambda) = \frac{1}{\pi \lambda^2} \frac{\pi \beta \Gamma(\lambda)}{[k(\lambda)]^2 + [\pi \beta \Gamma(\lambda)]^2} \tag{3.64a}$$

and

$$\beta \Gamma(\lambda) = \frac{1}{\pi \lambda^2} \frac{\pi \alpha f(\lambda)}{[K(\lambda)]^2 + [\pi \alpha f(\lambda)]^2} \tag{3.64b}$$

where

$$k(\lambda) = \int_0^\infty \beta \Gamma(u) \frac{u}{\lambda(\lambda - u)} \, du \tag{3.65a}$$

and

$$K(\lambda) = \int_0^\infty \alpha f(u) \frac{du}{\lambda - u}. \tag{3.65b}$$

Relations (3.64) express the interrelations between the distribution functions in creep and relaxation. These relations are of particular interest for the discussion of the variations in the spectra when one shifts from creep to relaxation and vice versa. Figure 3.8 gives schematic illustration of relaxation and retardation spectra (e.g. Gross and Pelzer, 1951). It should be emphasized that equations (3.64a) and (3.64b) are based on the assumption of continuous spectra which might include, as special cases, both line and mixed type spectra (Gross, 1953).

In the theory of viscoelasticity, both the retardation and the relaxation spectra are of interest. This may be due to the fact that either of the two spectra provides an overall reflection of the mechanical properties of the viscoelastic material over the extent of time or frequency considered. Further, there is a general expectation that these spectra could be correlated, in some manner, with the molecular structure of the viscoelastic material.

The interrelations between material functions associated with creep and relaxation, as dealt with in this section, are demonstrated in Fig. 3.9.

As can be recognized from Fig. 3.9, the linear viscoelastic behaviour of material can be determined by one of the functions in the following two groups.

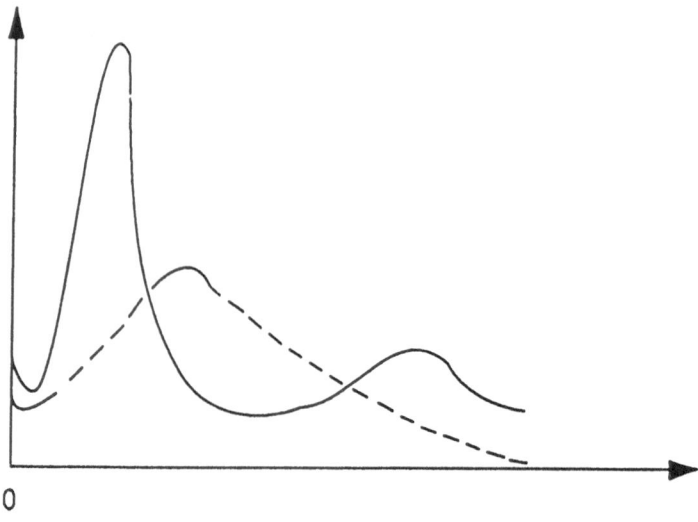

Fig. 3.8 Schematic representation of relaxation (——) and retardation (– – –) spectra (Gross, 1953).

1. Material functions determined by experiment:

   (a) the retardation function $F(t)$;
   (b) the relaxation function $R(t)$;
   (c) the dynamic complex compliance $J^*(i\omega)$;
   (d) the dynamic complex modulus $E^*(i\omega)$.
2. Material functions determined by integral transformation
   of experimental functions (1):

   (a) the retardation spectrum under static loading, $f(\lambda)$, or the retardation spectrum under dynamic loading, $N(s)$;
   (b) the relaxation spectrum under static loading, $\Gamma(\lambda)$, or the relaxation spectrum under dynamic loading, $N'(s)$.

Both the retardation function $F(t)$ and the relaxation function $R(t)$ in (1) above can be determined directly from an experiment under static loading. That is, $F(t)$ is determined from a creep experiment and $R(t)$ is determined from a relaxation experiment. The complex compliance $J^*(i\omega)$ and the complex modulus $E^*(i\omega)$ may be, however, determined from dynamic oscillation measurements with prescribed stress or strain respectively. On the other hand, in group (2) above, the retardation spectrum $f(\lambda)$ is obtained by integral transformation from the retardation function $F(t)$ and the retardation spectrum $N(s)$ is obtained by integral transformation from the complex compliance $J^*(i\omega)$. Similarly, the relaxation spectrum under static loading $\Gamma(\lambda)$ or that under dynamic loading $N'(s)$ is obtained by integral transformation from the relaxation function $R(t)$ or the complex modulus $E^*(i\omega)$. Thus, any one of the

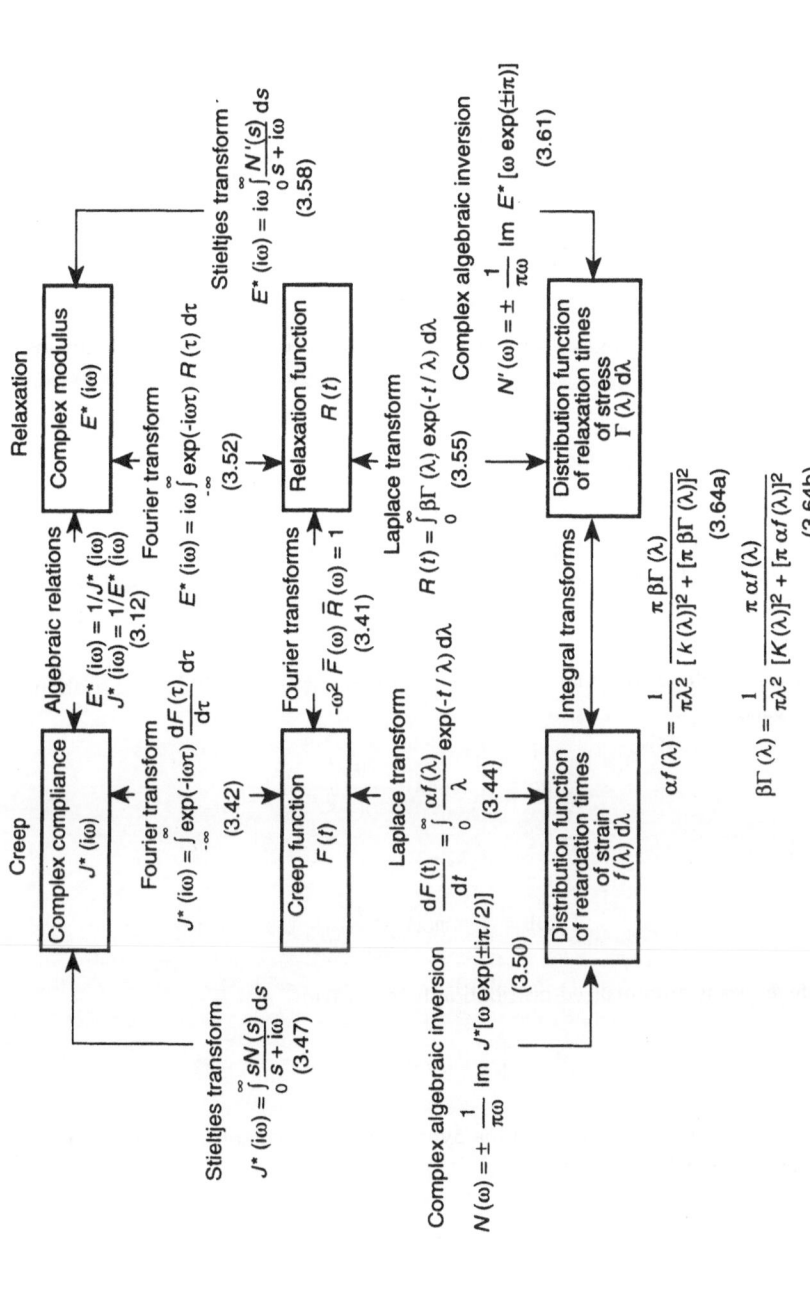

Fig. 3.9 Interrelations between material functions associated with creep and relaxation in the two cases of quasi-static and dynamic loading.

material functions stated above would suffice for the characterization of the linear viscoelastic response behaviour under isothermal conditions. In practice, however, a difficulty arises from the fact that creep and relaxation experiments under static testing conditions cannot be performed in very short times (i.e. instantaneously) and dynamic measurements cannot be easily extended to very long times (Schwarzl and Staverman, 1952). Accordingly, the mathematical transformation formulism mentioned above may not be directly applicable. Thus, approximate methods of transforming operator representation could be required (Lee, 1960). Alternatively, one would be inclined to use the appropriate type of measurement and determine the viscoelastic response of the material by the particular material function pertaining directly to such measurement.

PROBLEMS

3.1 Determine the storage and loss moduli as functions of $\ln(\omega\tau)$ for the following models:

1. the Maxwell model;
2. the Kelvin–Voigt model.

Illustrate, by a sketch, the shape of these functions for each model.

3.2 Determine the energy dissipated per cycle in terms of the loss compliance $J_2(\omega)$.

3.3 For a material with a single relaxation time λ show that E_1 is proportional to $\omega^2\lambda^2/(1 + \omega^2\lambda^2)$ and E_2 is proportional to $\omega\lambda/(1 + \omega^2\lambda^2)$ where ω is the frequency.

3.4 If $\phi(\lambda)$ denotes a spectrum of relaxation times, show that the following relations are valid:

$$E_1(\omega) = \int_0^\infty \phi(\lambda)\, \frac{\omega^2\lambda^2}{1 + \omega^2\lambda^2}\, d\lambda$$

and

$$E_2(\omega) = \int_0^\infty \phi(\lambda)\, \frac{\omega\lambda}{1 + \omega^2\lambda^2}\, d\lambda$$

where $\phi(\cdot)$ is a normalized distribution function with

$$\int_0^\infty \phi(\lambda)\, d\lambda = 1.$$

3.5 Prove that the relation between the dynamic complex compliance $J^*(i\omega)$ and the creep function $F(t)$ is given by

$$J^*(i\omega) = \int_0^\infty \exp(-i\omega\tau)\, \frac{dF(\tau)}{d\tau}\, d\tau.$$

3.6 Derive the expression (3.58) for the complex modulus from expression (3.52).

3.7 Following problem 3.6, derive the expressions for the components of the complex modulus in terms of the relaxation spectrum.

3.8 For a creep frequency $s = 1/\lambda$ and a frequency distribution $N(s) = \alpha f(1/s)/s^2$ where α is a constant, prove that the time derivative of the creep function is given by

$$\frac{dF(t)}{dt} = \int_0^\infty sN(s) \exp(-ts)\, ds.$$

3.9 Following problem 3.8, show that the complex compliance function can be expressed in terms of the distribution function of retardation times $N(s)$ as

$$J^*(i\omega) = \int_0^\infty \frac{sN(s)}{s + i\omega}\, ds.$$

Express the components of the complex compliance function $J_1(\omega)$ and $J_2(\omega)$ for this case.

REFERENCES

Alfrey, T. (1948) *Mechanical Behaviour of High Polymers*, Interscience, New York.

Alfrey, T. and Doty, P. (1945) The methods of specifying the properties of viscoelastic materials. *J. Appl. Phys.*, **16**, 700–13.

Andrews, R. D. (1952) Correlation of dynamic and static measurements on rubber-like materials. *Ind. Eng. Chem.*, **44**, 707–15.

Benbow, J. J. (1956) The determination of dynamic moduli and internal friction of high polymers from creep measurements. *Proc. Phys. Soc. B*, **69**(9), 885–92.

Ferry, J. D. (1970) *Viscoelastic Properties of Polymers*, 2nd edn, Wiley, New York.

Fung, Y. C. (1965) *Foundations of Solid Mechanics*, Prentice-Hall, New York.

Gittus, J. (1975) *Creep, Viscoelasticity and Creep Fracture in Solids*, Wiley, New York.

Gross, B. (1947) On creep and relaxation. *J. Appl. Phys.*, **18**, 212–21.

Gross, B. (1953) *Mathematical Structure of the Theories of Viscoelasticity*, Hermann, Paris.

Gross, B. and Pelzer, H. (1951) On creep and relaxation III. *J. Appl. Phys.*, **22**, 1035–9.

Kolsky, H. (1960) Viscoelastic waves, in Int. Symp. on Stress Wave Propagation in Materials (ed. N. Davids), Interscience, New York, pp. 59–90.

Leaderman, H. (1954a) Approximations in linear viscoelasticity theory: delta function approximation. *J. Appl. Phys.*, **25**, 294–6.

Leaderman, H. (1954b) Rheology of polyisobutylene, IV. Calculation of the retardation time function and dynamic response from creep data, in Proc. 2nd Int. Congr. on Rheology, Butterworth, London, pp. 203–13.

Lee, E. H. (1960) Viscoelastic stress analysis, in 1st Symp. on Naval Structural Mechanics, Pergamon, New York, pp. 456–82.

Lethersich, W. (1950) The rheological properties of dielectric polymers. *Br. J. Appl. Phys.*, **1** (November), 294–301.

Macey, B. (1948) On the application of Laplace pairs to the analysis of relaxation curves. *J. Sci. Instrum.*, **25**, 251–3.

Marvin, R. S. (1952) A new approximate conversion method for relating stress relaxation and dynamic modulus. *Phys. Rev.*, **86**, 644–5.

Pol, B. Van Der and Bremmer, H. (1951) *Operational Calculus*, Cambridge University Press, Cambridge.

Roesler, F. C. (1955) Some applications of Fourier series in the numerical treatment of linear behaviour. *Proc. Phys. Soc. B*, **68**, 89–96.

Roesler, F. C. and Pearson, J. R. A. (1954) Determination of relaxation spectra from damping measurements. *Proc. Phys. Soc. B*, **67**, 338–47.

Roesler, F. C. and Twyman, W. A. (1955) An iteration method for the determination of relaxation spectra. *Proc. Phys. Soc. B*, **68**(2), 97–105.

Schausberger, A., Knoglinger, H. and Janeschitz-Kriegl, H. (1987) The role of short chain molecules for the rheology of polystyrene melts, II. Linear viscoelastic properties. *Rheol. Acta*, **26**, 468–73.

Schwarzl, F. (1951) Näherungsmethoden in der Theorie des viscoelastischen Verhaltens I. *Physica*, **17**, 830–40.

Schwarzl, F. and Staverman, A. J. (1952) Time–temperature dependence of linear viscoelastic behaviour. *J. Appl. Phys.*, **23**(8), 838–43.

Ward, I. M. (1983) *Mechanical Properties of Solid Polymers*, 2nd edn, Wiley, New York.

FURTHER READING

Barker, L. M. and Hollenbach, R. E. (1964) System for measuring the dynamic properties of materials. *Rev. Sci. Instrum.*, **35**(6), 742–6.

Christensen, R. M. (1972) Restrictions upon viscoelastic relaxation functions and complex moduli. *Trans. Soc. Rheol.*, **16**, 603–14.

Davis, J. L. (1987) *Introduction to Dynamics of Continuous Media*, Macmillan, New York.

Day, M. A. (1970) Some results on the least work needed to produce a given strain in a given time in a viscoelastic material and a uniqueness theorem for dynamic viscoelasticity. *Q.J. Mech. Appl. Math.*, **23**, 469–79.

Doi, M. and Edwards, S. F. (1986) *The Theory of Polymer Dynamics*, Clarendon, Oxford.

Edelstein, W. S. and Gurtin, M. E. (1964) Uniqueness theorems in the linear dynamic theory of anisotropic viscoelastic solids. *Arch. Ration. Mech. Anal.*, **17**, 47–60.

Gibson, R. F., Hwang, S. J. and Sheppard, C. H. (1990) Characterization of creep in polymer composites by the use of frequency–time transformations. *J. Compos. Mater.*, **24** (April), 441–53.

Hunter, S. C. (1960) Viscoelastic waves, in *Progress in Solid Mechanics* (eds I. N. Sneddon and R. Hill), North-Holland, Amsterdam, pp. 1–57.

Kolsky, H. (1949) An investigation of the mechanical properties of materials at very high rates of loading. *Proc. Phys. Soc. B*, **62**, 676–700.

Lethersich, W. (1950) The measurement of the coefficient of internal friction of solid rods by a resonance method. *Br. J. Appl. Phys.*, **1**, 18–22.

Magrab, E. B. and Blomquist, D. S. (1971) *The Measurement of Time-varying Phenomena*, Wiley, New York.

Nashif, A. D., Jones, D. I. G. and Henderson, J. P. (1965) *Vibration Damping*, Wiley, New York.

Orbey, N. and Ded, J. M. (1991) Determination of the relaxation system from oscillatory-shear data. *J. Rheol.*, **35**, 1035–49.

Pandit, S. M. (1991) *Model and Spectrum Analysis: Data Dependent Systems in State Space*, Wiley, New York.

Pindera, J. T. and Straka, P. (1974) On physical measures of rheological response of some materials in wide ranges of temperature and spectral frequency. *Rheol. Acta*, **13**, 338–51.

Ricketts, T. E. and Goldsmith, W. (1970) Dynamic properties of rocks and composite structural materials. *Int. J. Rock Mech. Min. Sci.*, **7**, 313–35.

Yu, P. and Haddad, Y. M. (1993) A dynamic system method for the characterization of the rheological response of a class of viscoelastic materials. *Int. J. Pressure Vessels Piping*, in press.

4

Three-dimensional linear viscoelasticity

4.1 INTRODUCTION

In Chapter 2 we introduced the basic formulations concerning the one-dimensional response behaviour of a linear viscoelastic material under static loading. In the present chapter, we extend the presentation to include the generalization of the formulations to the three-dimensional case. The most direct approach in this regard is to replace the one-dimensional constitutive relationships by their corresponding tensorial equivalents.

4.2 THREE-DIMENSIONAL CONSTITUTIVE EQUATIONS

Let $\varepsilon_{ij}(x_i, t)$ denote the strain field defined at every point x_i of the material at time t $(-\infty < t < \infty)$. The latter is derived within the infinitesimal deformation theory from the deformation field $u_i(x_i, t)$ by equation (1.73), i.e.

$$\varepsilon_{ij}(x_i, t) = \tfrac{1}{2}[u_{i,j}(x_i, t) + u_{j,i}(x_i, t)].\tag{4.1}$$

Consider now the Cauchy stress tensor field $\sigma_{ij}(x_i, t)$ corresponding to $\varepsilon_{ij}(x_i, t)$. For a non-aging viscoelastic material with an arbitrary degree of anisotropy, the hypothesis that the current value of the creep strain depends on the complete stress history is expressed by

$$\varepsilon_{kl}(t) = f_{kl}[\sigma_{ij}(t - \tau), \quad \sigma_{ij}(t)], \quad 0 \le (t, \tau) < \infty,\tag{4.2}$$

where, for notational simplicity, explicit dependence of the stresses, strains and material properties on coordinates x_i is omitted. In equation (4.2), $f_{kl}[\cdot]$ is a linear tensor-valued function which transforms each stress history $\sigma_{ij}(t, \tau)$, $0 \le (t, \tau) < \infty$, into a corresponding strain history $\varepsilon_{kl}(t)$. The functional $f_{kl}[\cdot]$, as seen from equation (4.2), depends also on the current state of the stress $\sigma_{ij}(t)$. In an analogous form to (2.76) the response relation (4.2) above can be written (e.g. Axelrad, 1970) in the following form of the Stieltjes integral:

$$\varepsilon_{kl}(t) = \int_{-\infty}^{x} F_{klij}(t - \tau) \frac{\partial \sigma_{ij}(\tau)}{\partial \tau}\, d\tau\tag{4.3}$$

.In this equation, $F_{klij}(\cdot)$ is a fourth-order tensorial function referred to as the 'tensorial creep function'. $F_{klij}(t) = 0$ for $-\infty < t < 0$ and each of its components is of bounded variation in every closed subinterval $-\infty < t < \infty$. The fourth-order tensor components $F_{klij}(t)$ are known as the creep compliances. The creep function $F_{klij}(\cdot)$ is interpreted as the strain response at $t \geq 0$ to a constant unit stress applied at $t = 0$, so that $F_{klij}(t) = 0$ for $t < 0$ and $F_{klij}(0)$ is the instantaneous response at time $t = 0$. It is assumed that $F_{klij}(t)$ and at least its first derivative are continuous for $t \geq 0$, under the understanding that stress and strain histories with discontinuities at $t = 0$ are mathematically permissible (Gurtin and Sternberg, 1962). In the constitutive equation (4.3), it is noticed that the lower limit of the hereditary integral is taken as $-\infty$ which allows for the possibility of input histories prior to the time origin $t = 0$. This could be of benefit in particular engineering applications such as, for example, the study of steady-state vibrational response (Flügge, 1975).

With reference to the constitutive creep equation (4.3), one may also consider the case of discontinuous stress histories. This can be accomplished in equation (4.3), as shown previously in the one-dimensional case (Chapter 2, section 2.3), through the use of unit step functions. Thus, by resolving in equation (4.3) the stress histories $\sigma_{ij}(t)$, which have a step discontinuity at $t = 0$, into a uniformly convergent sequence of continuous functions, the constitutive relation corresponding to (4.3) can be written as

$$\varepsilon_{kl}(t) = F_{klij}(t)\sigma_{ij}(0) + \int_0^t F_{klij}(t - \tau)\,\frac{\partial\sigma_{ij}(\tau)}{\partial\tau}\,d\tau \tag{4.4}$$

or, alternatively,

$$\varepsilon_{kl}(t) = F_{klij}(0)\sigma_{ij}(t) + \int_0^t \dot{F}_{klij}(\tau)\sigma_{ij}(t - \tau)\,d\tau \tag{4.5}$$

where the fourth order tensor $F_{klij}(0)$ is usually referred to as the 'instantaneous elastic compliance' of the material. Equations (4.4) and (4.5) can be seen to follow the superposition of input histories principle (Boltzmann's) that the current output is determined by the superposition of the response to the complete spectrum of increments of inputs.

Any discontinuity in the form of a jump in the stress input $\sigma_{ij}(t)$ will contribute a corresponding instantaneous response similar to the first term in equation (4.4). For instance, if $\sigma_{ij}(t)$ experiences another jump, say $\Delta\sigma_{ij}(t_1)$ at time t_1, $0 < t_1 < t$, while $F_{klij}(t)$ and $\partial\sigma_{ij}(t)/\partial t$ are continuous elsewhere, equation (4.4) would take the following form:

$$\varepsilon_{kl}(t) = F_{klij}(t)\sigma_{ij}(0) + F_{klij}(t - t_1)\,\Delta\sigma_{ij}(t_1)H(t - t_1)$$

$$+ \int_0^t F_{klij}(t - \tau)\,\frac{\partial\sigma_{ij}(\tau)}{\partial\tau}\,d\tau \tag{4.6}$$

where $H(t - t_1)$ is the Heaviside unit step function (Appendix B).

Similarly, in analogous manner to equation (4.2), the relaxation stress can be expressed as a function of the entire strain history in the viscoelastic, non-aging material as

$$\sigma_{ij}(t) = r_{ij}[\varepsilon_{ij}(t - \tau), \quad \varepsilon_{ij}(t)], \quad 0 \le (t, \tau) < \infty. \tag{4.7}$$

Consequently, the stress-relaxation equation can be written, in a corresponding form to equation (4.3), as

$$\sigma_{ij}(t) = \int_{-\infty}^{t} R_{ijkl}(t - \tau) \frac{\partial \varepsilon_{kl}(t - \tau)}{\partial \tau} \, d\tau \tag{4.8}$$

in which $R_{ijkl}(t - \tau)$ is the fourth-order tensorial relaxation function. The components of $R_{ijkl}(t)$ are known as the relaxation moduli.

Further, if the strain is applied to the material specimen at $t = 0$, i.e. if $\varepsilon_{kl}(t) = 0$ for $t < 0$, relation (4.8) can be written, in a similar fashion to equation (4.4), as

$$\sigma_{ij}(t) = R_{ijkl}(t)\varepsilon_{kl}(0) + \int_{0}^{t} R_{ijkl}(t - \tau) \frac{\partial \varepsilon_{kl}(\tau)}{\partial \tau} \, d\tau. \tag{4.9}$$

In the stress-relaxation constitutive relation (4.9), $R_{ijkl}(t, \tau) = 0$ for $-\infty < t < 0$. Both $R_{ijkl}(t, \tau)$ and its first derivative are assumed to be continuous functions of time during $0 \le t < \infty$. An alternative form to (4.9) is the constitutive relaxation relation

$$\sigma_{ij}(t) = R_{ijkl}(0)\varepsilon_{kl}(t) + \int_{0}^{t} \dot{R}_{ijkl}(\tau)\varepsilon_{kl}(t - \tau) \, d\tau \tag{4.10}$$

where the fourth-order tensor $R_{ijkl}(0)$ is the 'instantaneous elastic modulus' of the material, i.e. at time $t = 0$.

Further, it is often assumed (Day, 1972) that

$$\int_{0}^{+\infty} \| \dot{R}_{ijkl}(t) \| \, dt < +\infty, \quad \int_{0}^{+\infty} t \, \| \dot{R}_{ijkl}(t) \| \, dt < +\infty, \tag{4.11}$$

where $\| \cdot \|$ the norm. These assumptions ensure that the 'equilibrium elastic modulus'

$$R_{ijkl}(+\infty) = \lim_{t \to +\infty} R_{ijkl}(t)$$

exists.

Any discontinuity in the form of a jump of the strain history $\varepsilon_{ij}(t)$ will again contribute an immediate discontinuity in the response and, hence, an additional term similar to the first term in equation (4.9). Consider, for instance, that $\varepsilon_{ij}(t)$ has a jump $\Delta\varepsilon_{ij}(t_1)$ at time t_1, $0 < t_1 < t$, while $R_{klij}(t)$ and $\partial \varepsilon_{ij}(t)/\partial t$ are continuous elsewhere; then, equation (4.9) becomes

$$\sigma_{ij}(t) = R_{ijkl}(t)\varepsilon_{kl}(0)$$
$$+ R_{ijkl}(t - t_1) \, \Delta\varepsilon_{kl}(t_1)H(t - t_1)$$
$$+ \int_{0}^{t} R_{ijkl}(t - \tau) \frac{\partial \varepsilon_{kl}(\tau)}{\partial \tau} \, d\tau \tag{4.12}$$

where $H(t - t_1)$ is the Heaviside unit step function.

The constitutive equations (4.3) and (4.8) may also be expressed in the form of a convolution integral which is frequently used in the mathematical branch of abstract algebra. In the creep phase, for instance, consider the two tensorial functions $F_{klij}(t)$ and $\sigma_{kl}(t)$ to be defined in the interval $-\infty < t < \infty$. The convolution of the two functions is established by the convolution product defined in the time range $0 \le t < \infty$ as follows:

$$F_{ijkl}(t) * d\sigma_{kl}(t) = \int_0^t F_{ijkl}(t - \tau) \frac{d\sigma_{kl}(t)}{d\tau} d\tau$$

$$+ F_{ijkl}(t)\sigma_{kl}(0) \tag{4.13}$$

which is identical to equation (4.4). The integral in (4.13) is often referred to as a Riemannian or Stietjes' integral and has been used by Gurtin and Sternberg (1962) in their development of the theory of viscoelasticity. An important characteristic of the convolution of functions is the commutativity property. Hence, in view of the latter property, the creep constitutive equation (4.3) can be expressed, with reference to (4.13), as

$$\varepsilon_{ij}(t) = F_{ijkl}(t) * d\sigma_{kl}(t) = \sigma_{kl}(t) * dF_{ijkl}(t) \tag{4.14}$$

and, in analogous manner, the relaxation constitutive equation (4.8) may be written as

$$\sigma_{ij}(t) = R_{ijkl}(t) * d\varepsilon_{kl}(t) = \varepsilon_{kl}(t) * dR_{ijkl}(t). \tag{4.15}$$

For other properties concerning the convolution of functions, Appendix C should be referred to.

4.3 CREEP AND RELAXATION FUNCTIONS

With reference to the constitutive equations (4.3) and (4.8), both the creep and relaxation functions, $F_{ijkl}(t)$ and $R_{ijkl}(t)$ respectively, are components of completely symmetric fourth-order tensors. In this context, symmetry of the stress and strain tensors implies that

$$F_{ijkl}(t) = F_{jikl}(t) = F_{ijlk}(t)$$

and

$$R_{ijkl}(t) = R_{jikl}(t) = R_{ijlk}(t) \tag{4.16}$$

so that at most 36 of the 81 components of $F_{ijkl}(t)$ (or $R_{ijkl}(t)$) are independent. This is in addition to symmetry with respect to an interchange of the first two and last two indices, i.e.

$$F_{ijkl}(t) = F_{klij}(t) \quad \text{and} \quad R_{ijkl}(t) = R_{klij}(t) \tag{4.17}$$

as considered by Biot (1954) in his study of the thermodynamics of stable, irreversible systems based on the first and second laws of thermodynamics and on Onsager's principle (Fung, 1965; Day, 1971a, 1972). In view of the above, one may consider the creep and relaxation tensors to be completely symmetric (Schapery, 1974). Rogers and Pipkin (1963), however, argued that the symmetry conditions (4.16) cannot be stated in a clear-cut fashion and discussed some additional restrictions imposed by equations (4.16) (Halpin and Pagano, 1968).

A series of tests has been proposed by Hayes and Morland (1969) to determine the possible 36 independent relaxation tensor components of a homogeneous linear anisotropic viscoelastic body. The set of tests comprises 3 simple shear tests and 21 tests of pure compression. Both types of tests employ, essentially, the same testing apparatus.

For each component of the viscoelastic stress–strain relations (4.3) and (4.8) to satisfy the fading memory type of behaviour, it is required that the magnitude of the slope of each component of both the creep and the relaxation function tensors is a continuously decreasing function of time. Thus,

$$\left|\frac{dF_{ijkl}(t)}{dt}\right|_{t=t_2} \leq \left|\frac{dF_{ijkl}(t)}{dt}\right|_{t=t_1}, \quad t_2 \geq t_1 \geq 0, \tag{4.18a}$$

and

$$\left|\frac{dR_{ijkl}(t)}{dt}\right|_{t=t_2} \leq \left|\frac{dR_{ijkl}(t)}{dt}\right|_{t=t_1}, \quad t_2 \geq t_1 \geq 0. \tag{4.18b}$$

This is with the understanding that

$$F_{ijkl}(t) = 0 \quad \text{and} \quad R_{ijkl}(t) = 0 \quad \text{for} \quad -\infty < t < 0. \tag{4.19}$$

The requirement implied by equations (4.18) is often referred to as the 'axiom of nonretroactivity'.

Considering that $F_{ijkl}(0) \neq 0$, $R_{ijkl}(0) \neq 0$ and both are finite, the integrals (4.3) and (4.8) may be inverted to give relations between the two sets of creep and stress-relaxation response functions. One relationship can be written in the form of the following Volterra integral equation:

$$R_{ijkl}(0)F_{klmn}(t) + \int_0^t \frac{\partial R_{ijkl}(t-\tau)}{\partial \tau} F_{ijkl}(\tau)\, d\tau = H(t)\delta_{im}\delta_{jn} \tag{4.20}$$

where $H(t)$ is the Heaviside unit step function and δ_{ij} is the Kronecker delta. On the other hand, applying Laplace transformation to the creep constitutive equation (4.3) and to the stress-relaxation constitutive equation (4.8) with the lower limit of the integral in both equations is taken as zero, it can be shown that

$$[\bar{F}_{ijkl}(s)] = [\bar{R}_{ijkl}(s)]^{-1} \tag{4.21}$$

where s is the Laplace transformation variable.

This equation provides the means for relating the creep tensorial function F_{ijkl} and the relaxation function R_{ijkl} in both the s and the t planes irrespective of the degree of anisotropy. Existence of the inverse tensors of (4.21) is guaranteed by thermodynamics because of the positive–definite and semidefinite characters of these tensors (Schapery, 1974). A similar relation to (4.21) can be shown to be valid for the complex compliances and moduli. This is established by substituting $i\omega$ for s in (4.21), that is

$$[\bar{F}_{ijkl}(i\omega)] = [\bar{R}_{ijkl}(i\omega)]^{-1} \tag{4.22}$$

The corresponding quasi-static approximation to equations (4.21) and (4.22) is

$$F_{ijkl}(t) = [R_{ijkl}(t)]^{-1} \tag{4.23}$$

Equations (4.21)–(4.23) are identical to those for linear elastic media at a particular instant of time or if the time effect is completely eliminated. This is due to the complete symmetry of the tensorial material functions appearing in these equations.

4.4 REDUCTION TO THE ISOTROPIC CASE

In the case of an isotropic medium, the invariance of properties under rotation of the coordinate axes would reduce the constitutive equation of such a medium to two pairs of response operators: one associated with the stress and strain deviators which pertain to pure shear and the other for average hydrostatic compression (dilatation).

For an isotropic viscoelastic material, both the material functions $F_{ijkl}(t)$ and $R_{ijkl}(t)$ are isotropic, i.e. invariant with respect to any rotation of the Cartesian reference coordinates. Consider the creep tensorial function $F_{ijkl}(t)$. The most general isotropic representation of a fourth-order tensor is given by

$$F_{ijkl}(t) = \tfrac{1}{3}[F_2(t) - F_1(t)]\delta_{ij}\delta_{kl}$$
$$+ \tfrac{1}{2}[F_1(t)](\delta_{ik}\delta_{jl} + \delta_{il}\delta_{jk}) \tag{4.24}$$

where F_1, F_2 are scalar functions such that $F_1 = 0$ and $F_2 = 0$ for $-\infty < t < 0$ and δ_{ij} is the Kronecker delta. We recall the definition of the deviatoric stress σ'_{ij} (equation 1.42)), i.e.

$$\sigma'_{ij} = \sigma_{ij} - \tfrac{1}{3}\delta_{ij}\sigma_{kk} \qquad \text{with} \qquad \sigma'_{ii} = 0. \tag{4.25}$$

Meantime, the deviatoric strain ε'_{ij} is expressed as

$$\varepsilon'_{ij} = \varepsilon_{ij} - \tfrac{1}{3}\delta_{ij}\varepsilon_{kk} \qquad \text{with} \qquad \varepsilon'_{ii} = 0. \tag{4.26}$$

Accordingly, the constitutive creep equation (4.3) subject to (4.24)–(4.26) reduces to

$$\varepsilon'_{ij}(t) = \int_{-\infty}^{t} F_1(t - \tau) \frac{d\sigma'_{ij}(\tau)}{d\tau} \tag{4.27a}$$

and

$$\varepsilon_{kk}(t) = \int_{-\infty}^{t} F_2(t - \tau) \frac{d\sigma_{kk}(\tau)}{d\tau} \tag{4.27b}$$

where $F_1(t)$ and $F_2(t)$ are the independent creep functions appearing in (4.24). $F_1(t)$ is the creep function in shear and $F_2(t)$ is referred to as the creep function in isotropic compression (dilatation). On the other hand, if (4.24) is substituted in the convolution constitutive relation (4.14), one obtains the following creep response equations for an isotropic material:

$$\varepsilon'_{ij}(t) = \sigma'_{ij}(t) * dF_1(t) = F_1(t) * d\sigma'_{ij}(t); \tag{4.28a}$$

$$\varepsilon_{kk}(t) = \sigma_{kk}(t) * dF_2(t) = F_2(t) * d\sigma_{kk}(t). \tag{4.28b}$$

In a similar manner, the relaxation stress components can be expressed, in the isotropic case, as

$$\sigma'_{ij}(t) = \int_{-\infty}^{t} R_1(t - \tau) \frac{d\varepsilon'_{ij}(\tau)}{d\tau} d\tau \tag{4.29a}$$

and

$$\sigma_{kk}(t) = \int_{-\infty}^{t} R_2(t - \tau) \frac{d\varepsilon_{kk}(\tau)}{d\tau} d\tau. \tag{4.29b}$$

In the constitutive relations above, $R_1(t)$ and $R_2(t)$ are the independent relaxation functions. $R_1(t)$ is referred to as the relaxation function in shear and $R_2(t)$ is the relaxation function in isotropic compression (dilatation).

Equations (4.29) can be further expressed in a convolution form respectively as

$$\sigma'_{ij}(t) = \varepsilon'_{ij}(t) * dR_1(t) = R_1(t) * d\varepsilon'_{ij}(t) \tag{4.30a}$$

and

$$\sigma_{kk}(t) = \varepsilon_{kk}(t) * dR_2(t) = R_2(t) * d\varepsilon_{kk}(t). \tag{4.30b}$$

The relations between the functions $F_\alpha(t)$ and $R_\alpha(t)$ ($\alpha = 1, 2$) can be obtained (Christensen, 1971), using the Laplace transformation, in the form

$$\lim_{t \to \infty} F_\alpha(t) = \lim_{t \to \infty} [R_\alpha(t)]^{-\infty}, \quad \alpha = 1, 2, \tag{4.31a}$$

and

$$\lim_{t \to \infty} R_\alpha(t) = \lim_{t \to \infty} [F_\alpha(t)]^{-1}, \quad \alpha = 1, 2. \tag{4.31b}$$

For a particular viscoelastic material, the above properties are determined from creep or stress-relaxation experiments (Ferry, 1950; Kolsky, 1967).

4.5 REMARKS ON THE THERMODYNAMIC RESTRICTIONS ON ISOTHERMAL LINEAR VISCOELASTICITY

We present below a few remarks concerning the restrictions imposed by thermo-dynamics on the formulation of idealized linear viscoelasticity. The reader is referred to the cited references and Chapter 5 of this text for more detailed treatment of the subject matter.

1. Coleman (1964) discussed the thermodynamic restrictions imposed on the relaxation functions in linear viscoelasticity. To this effect, Coleman (1964) advanced, based on the minimal property of free energy, that the difference between the instantaneous and equilibrium elastic moduli, $\mathbf{R}(0) - \mathbf{R}(+\infty)$, must be positive semidefinite and symmetric. However, it can be argued that, based on the symmetry of the stress and strain tensors, that $\mathbf{R}(\infty)$ is symmetric (Day, 1972); thus, Coleman's work (1964) implies that the two moduli $\mathbf{R}(0)$ and $\mathbf{R}(+\infty)$ are both symmetric and that

$$\mathbf{R}(0) \geq \mathbf{R}(+\infty) \tag{4.32}$$

2. The relaxation function $\mathbf{R}(t)$ is compatible with thermodynamics if the following relation is satisfied (Day, 1972);

$$\int_{-\infty}^{+\infty} \boldsymbol{\sigma}[t, \boldsymbol{\varepsilon}(t)]\dot{\boldsymbol{\varepsilon}}(t) \, \mathrm{d}t \geq \mathbf{0} \tag{4.33}$$

whenever $\boldsymbol{\varepsilon}(t)$ is a closed strain path in the sense that there is a symmetric tensor $\mathbf{A}$ and times t_0, t_1 with $\boldsymbol{\varepsilon}(t) = \mathbf{A}$ for every $t \leq t_0$ and for every $t \geq t_1$. Relation (4.33) implies that the work done around any closed isothermal strain path starting from equilibrium cannot be negative.

A different restriction, which also has a thermodynamic character (Day, 1972), was proposed by König and Meixner (1958). This restriction implies the requirement

$$\int_{-\infty}^{t_1} \boldsymbol{\sigma}[t, \boldsymbol{\varepsilon}(t)]\dot{\boldsymbol{\varepsilon}}(t) \, \mathrm{d}t \geq \mathbf{0} \tag{4.34}$$

for every time t_1 and for every path starting from the state of zero strain in the sense that $\boldsymbol{\varepsilon}(-\infty) = \mathbf{0}$, i.e. there is a time t_0 such that $\boldsymbol{\varepsilon}(t) = \mathbf{0}$ for every $t \leq t_0$. Relation (4.34) asserts that the relaxation function $\mathbf{R}(t)$ is **dissipative**.

3. The following implications of dissipativity for linear viscoelastic materials should be noted (Day, 1972).

   (a) Shu and Onat (1965) showed that, if $\mathbf{R}(t)$ is dissipative, then the instantaneous elastic modulus $\mathbf{R}(0)$ is symmetric. This is in accordance with the implications of Coleman's work (1964) discussed under (1) above.

   (b) Gurtin and Herrera (1965) concluded that, if $\mathbf{R}(t)$ is dissipative, then, the moduli $\mathbf{R}(0)$ and $\mathbf{R}(+\infty)$ are both symmetric and positive semidefinite and

$$\mathbf{R}(0) \geq \pm \mathbf{R}(t) \tag{4.35}$$

   for every $t \geq 0$. In addition, relation (4.32) can be rewritten as

$$\mathbf{R}(0) \geq \mathbf{R}(+\infty) \geq \mathbf{0} \tag{4.36}$$

   and

$$\dot{\mathbf{R}}(0) \geq \mathbf{0}. \tag{4.37}$$

(c) Day (1972) established, in addition, the following connections concerning the compatability between dissipative relaxation functions and thermodynamics.

   (i) The relaxation function $\mathbf{R}(t)$ is dissipative if and only if it is compatible with thermodynamics and the equilibrium elastic modulus $\mathbf{R}(\infty)$ is positive semidefinite.

   (ii) The relaxation function $\mathbf{R}(t)$ is compatible with thermodynamics if and only if $\mathbf{R}(t) - \mathbf{R}(+\infty)$ is dissipative and the equilibrium elastic modulus $\mathbf{R}(+\infty)$ is symmetric.

(d) It follows, based on (b) and (c) above, that, if $\mathbf{R}(t)$ is compatible with thermodynamics and in particular if $\mathbf{R}(t)$ is dissipative, then the elastic moduli $\mathbf{R}(0)$ and $\mathbf{R}(+\infty)$ are both symmetric and

$$\mathbf{R}(0) - \mathbf{R}(+\infty) \geq +[\mathbf{R}(t) - \mathbf{R}(+\infty)] \tag{4.38}$$

for every $t \geq 0$. Further, relations (4.32) and (4.34) will hold.

(e) Relation (4.38) reflects, in essence, the connection between the elastic and viscoelastic responses of the material (Day, 1971a, b, 1972):

   (i) A linear elastic material is a linear viscoelastic material whose relaxation function is identically time independent, that is $\mathbf{R}(t) = \mathbf{R}(0)$ for every $t \geq 0$. Thus, the elastic moduli $\mathbf{R}(0)$ and $\mathbf{R}(+\infty)$ coincide. Further, in view of (4.38), if $\mathbf{R}(0) = \mathbf{R}(+\infty)$, then $\mathbf{R}(t) = \mathbf{R}(+\infty) = \mathbf{R}(0)$ for $t \geq 0$. Accordingly, a linear viscoelastic material which is compatible with thermodynamics is elastic if, and only if, its equilibrium and instantaneous elastic moduli coincide.

   (ii) For a linear viscoelastic material which is compatible with thermodynamics and which is not elastic, $\mathbf{R}(0) \neq \mathbf{R}(+\infty)$. The response of such a linear viscoelastic material to rapid changes in strain will be always different from its response at equilibrium.

(f) The positive definiteness of $\mathbf{R}(0)$, relation (4.36), and the negative definiteness of $\dot{\mathbf{R}}(0)$, relation (4.37), together imply the decay of shock and acceleration waves in viscoelastic materials (Coleman, Gurtin and Herrera, 1965; Day, 1972). The subject of wave propagation in viscoelastic materials is dealt with in Chapter 8.

PROBLEMS

4.1 Comment on the symmetry of both the creep function $F_{ijkl}(t)$ and the relaxation function $R_{ijkl}(t)$. What are the implications of such symmetry?

4.2 Comment on the validity of the following two assumptions within the context of linear viscoelasticity (equation (4.11)):

$$\int_0^\infty \|\dot{R}_{ijkl}(t)\| \, dt < +\infty, \quad \int_0^\infty t \, \|\dot{R}_{ijkl}(t)\| \, dt < +\infty,$$

where $\|\cdot\|$ indicates the norm and $R_{ijkl}(t)$ is the relaxation tensor.

4.3 Show that, if the assumptions mentioned in problem 4.2 are valid, then the equilibrium elastic modulus

$$R_{ijkl}(+\infty) = \lim_{t \to \infty} R_{ijkl}(t)$$

exists.

4.4 Considering that, if the creep function $F_{ijkl} \neq 0$, the relaxation function $R_{ijkl} \neq 0$ and both are finite, show that the integrals (4.3) and (4.8) may be inverted to give the following relation between the creep and relaxation functions:

$$R_{ijkl}(0)F_{klmn}(t) + \int_0^t \frac{\partial R_{ijkl}(t - \tau)}{\partial \tau} F_{ijkl}(\tau) \, d\tau = H(t)\delta_{im}\delta_{in}$$

where $H(t)$ is the Heaviside function and δ_{ij} is the Kronecker delta.

4.5 Prove the validity of expressions (4.28) and (4.30).

4.6 Comment on the remarks presented in this chapter concerning the restrictions imposed by thermodynamics on the stress-relaxation function $\mathbf{R}(t)$ in idealized linear viscoelasticity.

REFERENCES

Axelrad, D. R. (1970) Mechanical models of relaxation phenomena. *Adv. Mol. Relaxation Processes*, **2**, 41–68.

Biot, M. A. (1954) Theory of stress–strain relations in anisotropic viscoelasticity and relaxation phenomena. *J. Appl. Phys.*, **25**(11), 1385–91.

Christensen, R. M. (1971) *Theory of Viscoelasticity*, Academic Press, New York.

Coleman, B. D. (1964) On thermodynamics, strain impulses and viscoelasticity. *Arch. Ration. Mech. Anal.*, **17**, 230–54.

Coleman, B. D., Gurtin, M. E. and Herrera, R. I. (1965) *Wave Propagation in Dissipative Materials*, Springer, Berlin.

Day, W. A. (1971a) Time-reversal and the symmetry of the relaxation function of a linear viscoelastic material. *Arch. Ration. Mech. Anal.*, **40**, 155–9.

Day, W. A. (1971b) When is a linear viscoelastic material elastic? *Mathematica*, **18**, 134–37.

Day, W. A. (1972) *Thermodynamics of Simple Materials with Fading Memory*, Springer, New York.

Ferry, J. D. (1950) Mechanical properties of substances of high molecular weight, VI. Dispersion in concentrated polymer solutions and its dependence on temperature and concentration. *J. Am. Chem. Soc.*, **72**, 3746–52.

Flügge, W. (1975) *Viscoelasticity*, Springer, New York.

Fung, Y. C. (1965) *Foundations of Solid Mechanics*, Prentice-Hall, Englewood Cliffs, NJ.

Gurtin, M. E. and Herrera, I. (1965) On dissipation inequalities and linear viscoelasticity. *Q. Appl. Math.*, **23**, 235–45.

Gurtin, M. E. and Sternberg, E. (1962) On the linear theory of viscoelasticity. *Ration. Mech. Anal.*, **11**, 291–356.

Halpin, J. C. and Pagano, N. J. (1968) Observations on linear anisotropic viscoelasticity. *J. Compos. Mater.*, **2**(1), 68–80.

Hayes, M. A. and Morland, L. W. (1969) The response function of an anisotropic linear viscoelastic material. *Trans. Soc. Rheol.*, **13**(2), 231–40.

Kolsky, H. (1967) Experimental studies of the mechanical behaviour of linear viscoelastic solids, in Proc. 4th Symp. on Naval Structural Mechanics, April 1965, Pergamon, Oxford, pp. 357–79.

König, H. and Meixner, J. (1958) Linear Systeme und Linear Transformationen. *Math. Nachr.*, **19**, 256–322.

Rogers, T. G. and Pipkin, A. C. (1963) Asymmetric relaxation and compliance matrices in linear viscoelasticity. *J. Appl. Math. Phys.*, **14**, 334–43.

Schapery, R. A. (1974) Viscoelastic behaviour and analysis of composite materials, in *Mechanics of Composite Materials*, Vol. 2 (ed. G. Sendeckj), Academic Press, New York, pp. 86–168.

Shu, L. S. and Onat, E. T. (1965) On anisotropic linear viscoelastic solids, in Proc. 4th Symp. on Naval Structural Mechanics, April 1965, Pergamon, Oxford, pp. 203–215.

FURTHER READING

Bland, D. R. (1960) *The Theory of Linear Viscoelasticity*, Pergamon, Oxford.

Breuer, S. (1969) Lower bounds on work in linear viscoelasticity. *Q. Appl. Math.*, **27**(2), 139–46.

Brown, R. L. and Sidebottom, A. M. (1971) A comparison of creep theories for multiaxial loading of polyethylene. *Trans. Soc. Rheol.*, **15**, 3–23.

Coleman, B. D. and Noll, W. (1960) An approximation theorem for functionals with applications in continuum mechanics. *Arch. Ration. Mech. Anal.*, **5**, 355–81.

Ewing, P. D., Turner, S. and Williams, J. G. (1973) Combined tension–torsion creep of polyethylene with abrupt changes of stress. *J. Strain Anal.*, **8**, 83–9.

Findley, W. N., Reed, R. M. and Stern, P. (1967) Hydrostatic creep of solid plastics. *J. Appl. Mech.*, **34**, 895–904.

Gross, B. (1947) On creep and relaxation. *J. Appl. Phys.*, **18** (February), 213–21.

Gurtin, M. E. and Sternberg, (1963) A reciprocal theorem in the linear theory of viscoelastic solids. *J. Soc. Ind. Appl. Math.*, **11**, 607–13.

Hopkins, I. L. and Hamming, R. W. (1958) On creep and relaxation. *J. Appl. Phys.*, **28**(8), 906–9.

Landel, R. F. and Peng, S. T. J. (1986) Equations of state and constitutive equations. *J. Rheol.*, **30**(4), 741–65.

Leaderman, H. (1954) Approximations in linear viscoelasticity theory: delta function approximations. *J. Appl. Phys.*, **25**(3), 294–6.

Misoulis, E. (1988) A Heuristic Approach to Modeling Viscoelasticity in Polymer Processing, Society of Plastics Engineers, Technical papers, Vol. XXXIV, ANTEC '88, Atlanta, GA, pp. 140–44.

Nolte, K. G. and Findley, W. N. (1974) Approximation of irregular loading by intervals of constant stress rate to predict creep and relaxation of polyurethane by three integral representations. *Trans. Soc. Rheol.*, **18**(1), 123–43.

Nunziato, J. W., Schuler, K. W. and Walsh, E. K. (1972) The bulk response of viscoelastic solids. *Trans. Soc. Rheol.*, **16**(1), 15–32.

Oldroyd, J. G. (1950) On the formulation of rheological equations of state. *Proc. R. Soc. London, Ser. A*, **200**, 523–41.

Onaran, K. and Findley, W. N. (1963) Combined stress creep experiments on viscoelastic material with abrupt changes in state of stress, in Proc. Joint Int. Conf. on Creep, Institution of Mechanical Engineers, London, pp. 285–97.

Tapseh, H. J. and Johnson, A. E. (1963) Creep under combined tension and torsion. *Engineering*, **150**, 24–5.

5

Thermoviscoelasticity

5.1 INTRODUCTION

The viscoelastic response behaviour appropriate to the formulation of the constitutive equations as dealt with in Chapters 2–4 generally exhibits a very strong dependence on the service temperature. The simplest and most direct situation of such dependence occurs when the constitutive equation is to be related or connected with different base temperatures within the context of the isothermal theory. In this, the material functions characterizing the viscoelastic behaviour of the material would be identified in an ambient temperature environment equal to the base temperature at which the constitutive equation is to be applied. On the other hand, when the constitutive equation is to be used in a nonisothermal situation, the identification problem is more involved and must be dealt within the realm of an advanced topic of thermoviscoelasticity. Thus, one generally deals with the following two classifications of the theory of thermoviscoelasticity.

- *Classification I* refers to a thermoviscoelastic treatment which examines the performance of the material as related to a fixed reference temperature T_R. In this treatment, the effects due to infinitesimal temperature variation from T_R would be neglected.
- *Classification II* refers to an advanced thermoviscoelastic theory within which the dependence of the material performance on a transient temperature field is dealt with.

One is generally concerned with the thermomechanical process and its effect on the response behaviour of the viscoelastic medium. For this purpose, the increase of the total energy of the medium is seen to be due to the work done by the external forces as well as the supply of energy (heat) from other sources. In such a study, however, one must differentiate between reversible and irreversible effects. Reversible effects would include cases in which temporary changes are produced in the material. These may include geometrical changes such as expansion and contraction as well as changes in the values of the material parameters or functions characterizing the constitutive equations of the material, but the format of the constitutive equations would essentially remain unchanged. On the other hand, irreversible effects include situations in which permanent changes are occurring in the material. In the case of a viscoelastic medium, such changes may include, for instance, primary bond rupture and weight loss. Polymeric composite materials and their constituents, for instance, are often subject to irreversible changes under a variety of environmental and chemical aging influences. In the latter context, temperature, moisture and water vapour are considered as important factors during the manufacturing process and in service (e.g. Steel, 1965; Fried, 1970; Tsai, 1970; Schapery, 1974). For the purpose of viscoelastic analysis, one simple case is where the response is considered over times that are short compared with the time scale over which changes due to environmental and/or aging effects would occur (Schapery, 1974); otherwise, one must allow for the dependence of the material parameters and functions characterizing the constitutive equations on such changes.

5.2 THERMODYNAMICS OF THE DEFORMATION PROCESS

5.2.1 A thermodynamic process

We consider a solid body B occupying a regular domain of volume V with surface boundary S. A material particle of B is defined in the reference configuration by the position vector $\mathbf{X}$. A thermodynamic process in B is described by the following eight functions of $\mathbf{X}$ and the time parameter t (Coleman, 1963, 1964a, b):

1. the spatial particle position $\mathbf{x} = \mathbf{x}(\mathbf{X}, t)$, which describes the evolution of the deformation process in the medium;
2. the symmetric stress tensor, Cauchy's $\boldsymbol{\sigma} = \boldsymbol{\sigma}(\mathbf{X}, t)$ or Piola–Kirchhoff's $\boldsymbol{\Sigma} = \boldsymbol{\Sigma}(\mathbf{X}, t)$;
3. the body force (per unit mass) $\boldsymbol{\chi} = \boldsymbol{\chi}(\mathbf{X}, t)$, exerted on the body at $\mathbf{X}$ by outside bodies not intersecting B;
4. the specific internal energy (per unit mass) $e = e(\mathbf{X}, t)$;
5. the specific entropy (per unit mass) $S = S(\mathbf{X}, t)$;
6. the local absolute temperature $T = T(\mathbf{X}, t)$, assumed to be positive;
7. the heat flux vector $\mathbf{q} = \mathbf{q}(\mathbf{X}, t)$;
8. the heat supply $r = r(\mathbf{X}, t)$, which is the radiation energy (per unit mass and per unit time) that is absorbed by the body B at $\mathbf{X}$ and supplied by the environment or any other bodies not intersecting with B.

Couple stresses, body couples and other mechanical interactions not included in $\boldsymbol{\sigma}$ or $\boldsymbol{\chi}$ are assumed to be absent or at least negligible. In order for the above set of eight functions to be called a 'thermodynamic process', it must be compatible with both the law of balance of linear momentum and the law of balance of energy. This is with the understanding that the balance of moment of momentum is satisfied by the assumed symmetry of the stress tensor $\boldsymbol{\sigma}$ (Chapter 1, section 1.3.5). In a differential form, the laws of balance of linear momentum and balance of energy can be expressed, respectively, as

$$\text{div } \boldsymbol{\sigma} + \rho\boldsymbol{\chi} = \rho\ddot{\mathbf{x}} \tag{5.1}$$

and

$$\text{tr}(\boldsymbol{\sigma}\mathbf{L}) - \text{div}\,\mathbf{q} - \rho\dot{e} = -\rho r. \tag{5.2}$$

In the above two equations, ρ is the mass density, $\mathbf{L}$ is the velocity gradient ($\mathbf{L} = \text{grad}\,\dot{\mathbf{x}}$), 'tr' is the trace operator, a superimposed dot designates the material derivative, i.e. the derivative with respect to the time parameter t keeping the position vector $\mathbf{X}$ fixed, and the operators 'grad' and 'div' refer to spatial derivatives, i.e. the gradient and divergence with respect to the current position vector $\mathbf{x}$ with the time parameter t kept fixed. Thus, to specify a thermodynamic process, it would suffice to prescribe only the six functions $\mathbf{x}$, $\boldsymbol{\sigma}$, e, $\mathbf{q}$, S and T. The remaining functions χ and r are then determined by the two laws (5.1) and (5.2).

The law of balance of energy (5.2) is a form of the first law of thermodynamics. The latter states that the rate at which the total energy of the body increases is balanced by the power of the external forces and the rate at which heat is supplied to the body. In other words, the quantity of heat supplied to the body is measured as the difference between change of the total energy and work done by the external forces (Rivlin, 1975).

Consider that heat enters the body throughout its volume at a rate H per unit mass and through its surface at a rate $\dot{h}$ per unit area. Both $\dot{H}$ and $\dot{h}$ are measured in the reference configuration, whereby the superimposed dot indicates differentiation with respect to time. Thus, according to the first law of thermodynamics, the rate of change of the total energy is given by

$$\Xi = \int_V \rho\chi\cdot\dot{\mathbf{x}}\,dV + \int_S \mathbf{F}\cdot\dot{\mathbf{x}}\,dS + \int_V \rho\dot{H}\,dV + \int_S \dot{h}\,dS \tag{5.3}$$

where $\mathbf{F}$ is the force per unit area acting on the body and measured in the reference configuration. Let ψ denote the total energy of the body per unit mass; then, the specific internal energy e is defined by

$$e = \psi - \tfrac{1}{2}\dot{x}\dot{x}, \tag{5.4}$$

i.e. e, defined by the equation above, is the total energy per unit mass less the kinetic energy per unit mass.

5.2.2 Restrictions imposed by the Second Law of Thermodynamics

Regarding $\mathbf{q}/T$ to be a vectorial flux of entropy due to heat flow and r/T to be a scalar supply of entropy from radiation, Coleman (1964a, b) defined the time rate of entropy production in a part P of the body B to be

$$\Gamma = \frac{d}{dt}\int_P S\,dm - \int_P \frac{r}{T}\,dm + \int_{\partial P} \frac{1}{T}\,\mathbf{q}\cdot\mathbf{n}\,ds. \tag{5.5}$$

In (5.5), dm is an element of mass of the body B, $\mathbf{n}$ is the exterior unit normal to the surface ∂P of P and ds is the element of surface area in the current configuration of the body (i.e. at time t). Under an appropriate smoothness (Coleman, 1964a, b), equation (5.5) may be expressed as

$$\Gamma = \int_P \gamma \, dm \tag{5.6}$$

in which

$$\gamma = \dot{S} - \frac{r}{T} + \frac{1}{\rho} \, \mathrm{div}\!\left(\frac{\mathbf{q}}{T}\right)$$

$$= \dot{S} - \frac{r}{T} + \frac{1}{\rho T} \, \mathrm{div} \, \mathbf{q} - \frac{1}{\rho T^2} \, \mathbf{q} \cdot \mathbf{g}(T) \tag{5.7}$$

is the specific rate of entropy production and where $\mathbf{g}(T)$ is the gradient of the current temperature. In this context, Coleman and Noll (1963) and Coleman (1964a, b) gave the following postulate as a mathematical expression of the second law of thermodynamics.

Postulate 5.1

For every admissible thermodynamic process in a body B, the following 'Clausius–Duhem inequality' must hold for all t and all parts of P of B:

$$\Gamma \geq 0 \tag{5.8}$$

where Γ is the rate of production of entropy in a part P of the body B, equation (5.5).

It is evident that the postulate above places restrictions on the formats of the constitutive equations of materials.

Thermomechanical systems must satisfy the same general conservation laws, concerning mass and momentum, that were introduced earlier in Chapter 1. The law of conservation of energy, however, contains both mechanical and thermal energies. Since the change of thermal energy is related to the change of entropy, a description of the evolution of a thermomechanical system requires a knowledge of the entropy production.

Recalling the conservation of mass principle as expressed by the equation of continuity (1.6), that is

$$\frac{\partial \rho}{\partial t} + \frac{\partial(\rho v_i)}{\partial x_i} = 0, \tag{5.9}$$

the conservation of momentum is expressed by the equation of motion (1.22):

$$\rho \dot{v}_i = \sigma_{ij,i} + \rho \chi_i \quad \text{within} \quad V. \tag{5.10}$$

Meantime, Cauchy's formula (1.18) and (1.28) are respectively

$$T_i = \sigma_{ij} n_j$$

and $$\tag{5.11}$$

$$\sigma_{ij} = \sigma_{ji}.$$

The conservation of energy is given by

$$\rho \dot{e} = \sigma_{ij} v_{i,j} - q_{i,i} \tag{5.12}$$

where the superimposed dot indicates the material derivative defined earlier by (1.4), i.e.

$$(\cdot) = \frac{\partial}{\partial t} + v_j \frac{\partial}{\partial x_j} \tag{5.13}$$

and v_j are the velocity components.

In order to establish the entropy balance equation, one may assume (Fung, 1965) that the specific entropy S is a function of both the internal energy per unit mass e and the strain ε_{ij} irrespective of the equilibrium of the system. That is,

$$S = S(e, \varepsilon_{ij}). \tag{5.14}$$

This is in agreement with the expression for the total differential of the specific entropy S as given by Gibb's relation

$$\rho T \, dS = \rho \, de - \sigma_{ij} \, d\varepsilon_{ij}. \tag{5.15}$$

Thus, along the path of the motion, one may write that

$$\rho T \dot{S} = \rho \dot{e} - \sigma_{ij} v_{ij} \tag{5.16}$$

where the superimposed dot indicates the material derivative and v_{ij} is the rate of deformation tensor, i.e.

$$v_{ij} = \tfrac{1}{2}(v_{i,j} + v_{j,i}).$$

Combining (5.12) and (5.16), it follows that

$$\rho T \dot{S} = -q_{i,i} \tag{5.17}$$

which can be equivalently written as

$$\rho \dot{S} = -\frac{q_{i,i}}{T} = -\left(\frac{q_i}{T}\right)_{,i} - q_i \frac{T_{,i}}{T^2}. \tag{5.18}$$

The first term on the right-hand side of (5.18) is the divergence of the entropy flow and the second term is the entropy production which must be positive as previously discussed (equation (5.8)).

A constitutive law, defining the relationship between the stress tensor σ_{ij} and the strain tensor ε_{ij}, must be further included so that the strain field is uniquely defined. Thus, a sufficient number of differential equations are obtained for which a boundary value problem may be formulated.

5.3 RHEOLOGICAL EQUATIONS OF STATE

Significant research efforts have been undertaken in the last four decades towards the development of a rigorous thermomechanical theory concerning the response behaviour of materials with memory as based on phenomenological considerations. In this context, theories have been presented from different points of view by Biot

(1958, 1973), Coleman and Noll (1963), Coleman (1964a, b), Schapery (1964), Christensen and Naghdi (1967), Crochet and Naghdi (1974), Crochet (1975) and Rivlin (1975), among others. Other work of interest includes that of Coleman and Mizel (1963, 1964), Breuer and Onat (1964), Breuer (1969) and Day (1970) concerning the free energy concept, recoverable work and related work bounds. The reader is also referred to Müller (1967), Meixner (1969) and others for developments in the subject of continuum thermodynamics.

Coleman and Noll (1963) adopted the Clausius–Duhem inequality, as an expression for the second law of thermodynamics, to determine the validity of the constitutive equations of a body of material. In this, their paper demonstrates that the second law of thermodynamics requires the Clausius–Duhem inequality to be satisfied in a process that is compatible with the balance laws of mass, momentum, moment of momentum and energy. This translates into the requirement that the constitutive equations must be compatible with the Clausius–Duhem inequality in order for such constitutive relations to be able to describe the response behaviour of a material under the restrictions imposed by thermodynamics. Following the above approach, Coleman and Mizel (1963) studied heat conduction in rigid bodies and then (Coleman and Mizel, 1964) established the existence of caloric equations of state for materials of the rate type. For applications of the Coleman and Noll (1963) approach to different classes of materials, the reader is referred to the research works by Green and Naghdi (1965), Gurtin (1965), Gurtin and Williams (1966), Wang and Bowen (1966), Coleman and Gurtin (1967a, b), Green and Laws (1967), Laws (1967), Coleman and Mizel (1967, 1968), Owen (1968, 1970) and Coleman and Owen (1970), among others.

As a continuation of the work of Coleman and Noll (1963), Coleman (1964a, b) dealt with the foundations of a thermodynamic theory of materials with memory, from a macroscopic point of view and based on the principles of continuum physics. Again, Coleman takes the Clausius–Duhem inequality to be the expression of the second law of thermodynamics and establishes the restrictions for reducing the constitutive equations to forms compatible with thermodynamics. As the statement of the Clausius–Duhem inequality involves the entropy of the body, one must acquire, in Coleman's approach, the entropy from the beginning to deal with (Day, 1970, 1972).

5.3.1 Simple materials

Neglecting any thermodynamic effect, a substance for which the stress $\sigma(t)$ is determined by the history of a measure of strain is referred to, from a continuum mechanics point of view, as a 'simple material'. The response equation of such material may be written (Coleman, 1964a, b) as

$$\sigma(t) = \underset{\tau=0}{\overset{\infty}{\boldsymbol{\pi}}} \; [\mathbf{F}(t-\tau)]. \tag{5.19}$$

In this equation, $\mathbf{F}(t-\tau)$ is the deformation gradient at time $t-\tau$, $0 \le (t, \tau) < \infty$, and $\boldsymbol{\pi}$ is a functional mapping the function $\mathbf{F}(t - \tau)$ into tensor $\boldsymbol{\sigma}(t)$. $\boldsymbol{\pi}$ may be considered as a general functional subject to the requirements of material symmetry (Noll, 1958; Coleman and Noll, 1964; Coleman 1964a, b), the principle of material objectivity (Noll, 1958; Coleman, 1964a, b) and the principle of fading memory (Coleman and Noll, 1960, 1961; Coleman, 1964a, b).

In the more general case, i.e. when one includes thermodynamic effects, the stress $\boldsymbol{\sigma}(t)$ would depend on both the deformation gradient history $\mathbf{F}(t - \tau)$ as well as the temperature history $T(t - \tau)$. Thus a more generalized form of (5.19) is

$$\boldsymbol{\sigma}(t) = \overset{\scriptscriptstyle\infty}{\underset{\tau=0}{\boldsymbol{\pi}}} \; [\mathbf{F}(t - \tau), \; T(t - \tau)]. \tag{5.20}$$

One may also assume (Coleman, 1964a, b) that the specific internal energy per unit mass e is determined, similarly to $\boldsymbol{\sigma}(t)$, as illustrated above, by the histories $\mathbf{F}(t - \tau)$ and $T(t - \tau)$, i.e.

$$e(t) = \overset{\scriptscriptstyle\infty}{\underset{\tau=0}{e}} \; [\mathbf{F}(t - \tau), \; T(t - \tau)]. \tag{5.21}$$

On the other hand, the heat flux vector $\mathbf{q}$ is dependent on the temperature gradient $\mathbf{g}(T)$ during the thermodynamic process. Since, according to (5.20), the stress is assumed to depend on $\mathbf{F}(t - \tau)$ and $T(t - \tau)$, it is likely that these histories would influence the dependence of the heat flux $\mathbf{q}$ on the temperature gradient $\mathbf{g}(T)$. Accordingly, the constitutive equation for the heat flux may be expressed (Coleman, 1964a, b) as

$$\mathbf{q}(t) = \overset{\scriptscriptstyle\infty}{\underset{\tau=0}{\mathbf{q}}} \; [\mathbf{F}(t - \tau), \; T(t - \tau), \; \mathbf{g}(T)] \tag{5.22}$$

where $\mathbf{q}$ is a functional whose arguments are the histories $\mathbf{F}(t - \tau)$, $T(t - \tau)$ and the gradient of the current temperature, i.e. $\mathbf{g}(T)$.

We recall at this point the 'principle of equipresence' (Truesdell, 1951; Truesdell and Toupin, 1960; Coleman and Mizel, 1964; Coleman and Gurtin, 1967a, b) which, in its present form, reads (Coleman, 1964a, b) as follows.

An independent variable present in one constitutive equation of a material should be assumed to be so present in all, until its presence is shown to be in direct contradiction to the assumed symmetry of the material, the principle of material objectivity, or the laws of thermodynamics.

Thus, the constitutive equations (5.20) and (5.21) should also include in their argument the dependence on the temperature gradient $\mathbf{g}(T)$ present in (5.22). Accordingly, one replaces (5.20) and (5.21), respectively, by

$$\boldsymbol{\sigma}(t) = \overset{\scriptscriptstyle\infty}{\underset{\tau=0}{\boldsymbol{\pi}}} \; [\mathbf{F}(t - \tau), \; T(t - \tau), \; \mathbf{g}(T)] \tag{5.23}$$

and

$$e(t) = \overset{\propto}{\underset{\tau=0}{e}} \ [\mathbf{F}(t - \tau), \ T(t - \tau), \ \mathbf{g}(T)]. \tag{5.24}$$

In order to include the restrictions imposed by the Clausius–Duhem inequality on the above-mentioned constitutive relations (5.22)–(5.24), an expression for the specific entropy (per unit mass) is introduced in the form

$$S(t) = \overset{\propto}{\underset{\tau=0}{S}} \ [\mathbf{F}(t - \tau), \ T(t - \tau), \ \mathbf{g}(T)] \tag{5.25}$$

which also satisfies the principle of equipresence.

In Coleman's (1964a, b) theory, it is assumed that the four functionals $\mathbf{q}$, $\boldsymbol{\pi}$, e and S corresponding respectively to equations (5.22)–(5.25) are given at each point $\mathbf{X}$ of the material. These functions, in general, depend on the choice of the reference configuration. However, if there exists a reference configuration that would render these functionals independent of $\mathbf{X}$ for all material points in the body B, then one may consider B to be materially homogeneous.

(a) Admissibility

A thermodynamic process is said to be admissible in B if it is compatible with the constitutive relations (5.22)–(5.25) at each material point $\mathbf{X}$ of B and all times t. In this context, Coleman (1964a, b) showed the following remark to be valid.

Remark To every choice of the deformation function $\mathbf{x}(\mathbf{X}, t)$ and the temperature distribution $T(\mathbf{x}, t)$, $(\mathbf{x}, \mathbf{X}$ in B; $-\infty < t < \infty)$, there corresponds a unique admissible thermodynamic process in B.

Coleman (1964) also showed that the Clausius–Duhem inequality requires that the temperature gradient $\mathbf{g}(T)$ drops out from relations (5.23)–(5.25). Accordingly, the new set of constitutive equations are expressed as

$$\boldsymbol{\sigma}(t) = \overset{\propto}{\underset{\tau=0}{\boldsymbol{\pi}}} \ [\mathbf{F}(t - \tau), \ T(t - \tau)], \tag{5.26a}$$

$$e(t) = \overset{\propto}{\underset{\tau=0}{e}} \ [\mathbf{F}(t - \tau), \ T(t - \tau)], \tag{5.26b}$$

$$\mathbf{q}(t) = \overset{\propto}{\underset{\tau=0}{\mathbf{q}}} \ [\mathbf{F}(t - \tau), \ T(t - \tau), \ \mathbf{g}(T)] \tag{5.26c}$$

and

$$S(t) = \overset{\propto}{\underset{\tau=0}{S}} \ [\mathbf{F}(t - \tau), \ T(t - \tau)], \tag{5.26d}$$

Equations (5.26b) and (5.26d) may also be used to express a constitutive equation based on the specific Helmholtz free energy. Denoting the latter by A, it is defined by

$$A = e - TS.$$

Since both e and S are given in (5.26b) and (5.26d), respectively, by functionals of $\mathbf{F}(t - \tau)$ and $T(t - \tau)$, it follows that

$$A(t) = \underset{\tau=0}{\overset{\infty}{A}} \; [\mathbf{F}(t - \tau), \, T(t - \tau)], \tag{5.26e}$$

The reader is referred to Coleman (1964a, b) for theorems and remarks concerning the set of constitutive relations (5.26).

(b) Entropy as an independent variable

Recall the entropy constitutive equation (5.26d). That is,

$$S(t) = \underset{\tau=0}{\overset{\infty}{S}} \; [\mathbf{F}(t - \tau), \, T(t - \tau)].$$

Assume that the above functional transformation is invertible in the sense that there exists a functional $T(t)$ such that

$$T(t) = \underset{\tau=0}{\overset{\infty}{T}} \; [\mathbf{F}(t - \tau), \, S(t - \tau)]. \tag{5.27}$$

Accordingly, one may rewrite the other constitutive equations (5.26a)–(5.26c) and (5.26e), respectively, as

$$\boldsymbol{\sigma}(t) = \underset{\tau=0}{\overset{\infty}{\hat{\boldsymbol{\pi}}}} \; [\mathbf{F}(t - \tau), \, S(t - \tau)], \tag{5.28a}$$

$$e(t) = \underset{\tau=0}{\overset{\infty}{\hat{e}}} \; [\mathbf{F}(t - \tau), \, S(t - \tau)], \tag{5.28b}$$

$$\mathbf{q}(t) = \underset{\tau=0}{\overset{\infty}{\hat{\mathbf{q}}}} \; [\mathbf{F}(t - \tau), \, S(t - \tau), \, \mathbf{g}(T)] \tag{5.28c}$$

and

$$A(t) = \underset{\tau=0}{\overset{\infty}{\hat{A}}} \; [\mathbf{F}(t - \tau), \, S(t - \tau)]. \tag{5.28d}$$

(c) Internal energy as an independent variable

Consider the constitutive equation for internal energy (5.26b), i.e.

$$e(t) = \underset{\tau=0}{\overset{\infty}{e}} \; [\mathbf{F}(t - \tau), \, T(t - \tau)].$$

Assume now that the above functional transformation is invertible, i.e. there exists a functional $\check{T}(t)$ such that

$$T(t) = \overset{\infty}{\underset{\tau=0}{\check{T}}} \ [\mathbf{F}(t - \tau), \ e(t - \tau)]. \tag{5.29}$$

Accordingly, one may rewrite the rest of the constitutive equations (5.26), in sequence, as

$$\boldsymbol{\sigma}(t) = \overset{\infty}{\underset{\tau=0}{\check{\pi}}} \ [\mathbf{F}(t - \tau), \ e(t - \tau)], \tag{5.30a}$$

$$\mathbf{q}(t) = \overset{\infty}{\underset{\tau=0}{\check{q}}} \ [\mathbf{F}(t - \tau), \ e(t - \tau), \ \mathbf{g}(T)] \tag{5.30b}$$

$$S(t) = \overset{\infty}{\underset{\tau=0}{\check{S}}} \ [\mathbf{F}(t - \tau), \ e(t - \tau)] \tag{5.30c}$$

and

$$A(t) = \overset{\infty}{\underset{\tau=0}{\check{A}}} \ [\mathbf{F}(t - \tau), \ e(t - \tau)]. \tag{5.30d}$$

Coleman (1964) showed the following remarks to be valid.

Remark   In every admissible process

$$\dot{S} \geq \frac{1}{T}\left(\dot{e} - \frac{1}{\rho} \operatorname{tr} \boldsymbol{\sigma}\mathbf{L}\right) \tag{5.31a}$$

and

$$\dot{S} \geq \frac{1}{T}\left(r - \frac{1}{\rho} \operatorname{div} \mathbf{q}\right) \tag{5.31b}$$

where $\mathbf{L}$ is the velocity gradient, i.e. $\mathbf{L} = \operatorname{grad} \dot{x}$.

The inequality in (5.31) above is referred to as '**the principle of positive internal production of entropy**'.

Remark   Whenever the strain and internal energy are held constant, the entropy cannot decrease, regardless of the past history.

Remark   Whenever the strain and entropy are held constant, the internal energy cannot increase regardless of the past history.

Remark   In an admissible thermodynamic process, the material time derivative of the free energy obeys the inequality

$$\dot{A} \leq \frac{1}{\rho} \operatorname{tr} \boldsymbol{\sigma}\mathbf{L} - S\dot{T}. \tag{5.32}$$

Thus,

$$\text{if } \mathbf{L} = \mathbf{0}, \quad \dot{T} = 0, \quad \text{then} \quad \dot{A} \leq 0, \tag{5.33}$$

i.e. if, at a given instant of time, material point $\mathbf{X}$ is held at constant strain and temperature (e.g. isothermal stress relaxation), the free energy at $\mathbf{X}$ at that instant cannot increase regardless of the past history.

The significance of Coleman's (1964a, b) work in establishing the restrictions imposed by thermodynamics on the constitutive equations for materials with memory is apparent. However, as pointed out by Rivlin (1975), no prescription was given in Coleman's work for determining the actual form of the constitutive functionals either analytically or by deduction from experiment. Rivlin (1975) criticized Coleman's approach in that entropy cannot be regarded as a 'primitive quantity' since it is not in the same category as the primitive mass, length and time. Rivlin (1975), on the other hand, defined materials with memory as materials for which the Piola–Kirchhoff stress $\mathbf{\Sigma}$ and empirical temperature θ are functions of the histories of the specific internal energy $e(\tau)$ and the deformation gradient tensor $\mathbf{F}(\tau)$, with support $(-\infty, t^+)$, i.e.

$$\mathbf{\Sigma}(t) = \mathbf{\Sigma}[\mathbf{F}(\tau), \quad e(t)], \quad \theta(t) = \theta[\mathbf{F}(\tau), \quad e(\tau)]. \tag{5.34}$$

Alternatively, one may consider $\mathbf{\Sigma}(t)$ and $e(t)$ as functionals of the histories $\mathbf{F}(\tau)$ and $\theta(\tau)$ with support $(-\infty, t^+)$. That is,

$$\mathbf{\Sigma}(t) = \mathbf{\Sigma}[\mathbf{F}(\tau), \quad \theta(\tau)], \quad e(t) = e[\mathbf{F}(\tau), \quad \theta(\tau)]. \tag{5.35}$$

The support in (5.34) and (5.35) is taken by Rivlin (1975) to be $(-\infty, t^+)$, rather than $(-\infty, t)$, in order to include the possibility that $\mathbf{\Sigma}$ and θ may depend on the instantaneous values of the time derivatives of $\mathbf{F}(\tau)$ and $e(\tau)$ at the instant t, even though these may change discontinuously at time t.

Rivlin (1975) considered the material to have fading memory if the functionals in (5.34) and (5.35) are such that, for two histories which differ only up to time $t - \tau$, the differences in the functionals decrease to zero as τ increases to infinity. Coleman (1964a, b) considered also a similar assumption for the definition of the 'fading memory' of simple materials, that is the memory of such materials fades in time. Coleman's assumption implies the assertions that deformations and temperatures experienced in the distant past should have less effect on the present values of the entropy, energy, stress and heat flux than deformations and temperatures which occurred in the recent past. In this context, Coleman (1964a, b) introduced an 'influence function' $C(\tau)$, $0 \leq \tau < \infty$, which would characterize the rate at which the memory fades. The influence function $C(\tau)$ is assumed to be positive monotonic decreasing and continuous function for the time parameter τ (Coleman and Noll, 1960, 1961, 1964).

With reference to (5.34), a material is said (Rivlin, 1975) to be perfectly elastic if $\mathbf{\Sigma}$ and θ depend only on the instantaneous values of $\mathbf{F}$ and e. In this case, $\mathbf{\Sigma}$ and θ are ordinary functions of $\mathbf{F}$ and e, i.e.

$$\mathbf{\Sigma} = \mathbf{\Sigma}(\mathbf{F}, e), \quad \theta = \theta(\mathbf{F}, e). \tag{5.36}$$

Alternatively, Σ and e will be ordinary functions of $\mathbf{F}$ and θ and (5.35) will be replaced by

$$\Sigma = \Sigma(\mathbf{F}, \theta), \quad e = e(\mathbf{F}, \theta). \tag{5.37}$$

Accordingly, in the case of materials with fading memory, if we restrict ourselves to processes carried out quasi-statistically, the constitutive equations (5.34) and (5.35) will ensure the forms of the constitutive equations (5.36) and (5.37), respectively. In other words, materials with fading memory behave as perfectly elastic materials with respect to quasi-static processes.

Rivlin (1975) adopted Carathéodory's principle as a form of the second law of thermodynamics: there are states of a system, differing infinitesimally from a given state, which are unattainable from that state by any adiabatic process whatever. Here, 'state' is used in the sense of 'equilibrium state' and it is postulated that the materials considered can always be taken from any such state to another state by a quasi-static process. As a consequence of the above, Rivlin (1975) asserted the existence of the 'specific entropy' which is a function of the variables used to describe the state and of the absolute temperature which is a function of the empirical temperature. The function through which the specific entropy relates to the state variables depends on the material considered, while the function through which the absolute temperature is associated with the empirical temperature is independent of this material.

As an illustration of the above arguments, Rivlin (1975) considered a body of material with fading memory to be in equilibrium with uniform empirical temperature θ, specific internal energy e and deformation gradient $\mathbf{F}$. The constitutive equations describing the response of such material is assumed to be given by (5.34). Rivlin, then, assumed that the body is taken from a homogeneous equilibrium state by a homothermal quasi-static process to a neighbouring equilibrium state, in which θ, e and $\mathbf{F}$ are changed, respectively, to $\theta + d\theta$, $e + de$ and $\mathbf{F} + d\mathbf{F}$. Thus, letting dH be the amount of heat (per unit mass) which is absorbed by the body in this process, it can be shown, following the first law of thermodynamics, that

$$\rho \, dH = \rho \, de - \text{tr}(\Sigma \cdot d\mathbf{F}). \tag{5.38}$$

In (5.38), ρ is the material density in the fixed reference state with respect to which $\mathbf{F}$ is measured, and de is the increase in specific internal energy in the process.

With (5.34)–(5.37), equation (5.38) yields

$$\rho \, dH = \rho \left(\frac{\partial e}{\partial \theta} \right)_{\mathbf{F}} d\theta + \text{tr} \left[\rho \left(\frac{\partial e}{\partial \mathbf{F}} \right)_{\theta}^{+} - \Sigma \right] \mathbf{F}. \tag{5.39}$$

Based on the Carathéodory principle and on the assumption that the process to be quasi-static, then there must exist values of $d\theta$ and $d\mathbf{F}$ for which $dH \neq 0$ (Rivlin, 1975). In other words, the process must not be adiabatic for the transition between the two neighbouring states to take place. From this fact, Rivlin (1975) asserts, with the support of the work of Wilson (1957) and Kestin (1966), that there exists an integrating factor $1/T(\mathbf{F}, \theta)$ such that $\rho \, dH/T$, from (5.39), is a perfect differential.

Following Rivlin (1975), one may consider unit mass of a material with fading memory to be taken by a quasi-static homothermal process from an equilibrium state A to an equilibrium state B. Let the states A and B be identified, respectively, by the two sets of values $(\theta_A, \mathbf{F}_A, e_A, S_A)$ and $(\theta_B, \mathbf{F}_B, e_B, S_B)$. Let, also, dH denote the heat fed into the body in an infinitesimal step of the process. Since

$$dS = dH/T(\theta) \qquad (5.40)$$

and S is a function of θ and $\mathbf{F}$ only, then

$$S_B - S_A = \int_A^B dH/T(\theta) \qquad (5.41)$$

where the integration is carried out along the path in the ten-dimensional space $(\theta, \mathbf{F})$ followed by the process.

Consider, now, a body of a material with fading memory to be taken from an equilibrium state A to an equilibrium state B by a process which is not necessarily quasi-static. It can be shown (Rivlin, 1975; Kestin, 1966), by application of Carathéodory's principal, that if dH is the amount of heat (per unit mass) entering the system in an infinitesimal step of the process, at an instant at which the empirical temperature of the system is θ, then

$$\int_A^B \frac{dH}{T(\theta)} \leq S_B - S_A. \qquad (5.42)$$

In the above relation, the equality sign applies if the process is quasi-static. Formula (5.42) is known as the 'Clausius inequality' or 'Clausius–Planck inequality' whereby the integral is referred to as the 'Clausius integral'.

Recall (5.40), that is

$$dS = \frac{dH}{T(\theta)}$$

where S, the specific entropy, is a function of the instantaneous values of $\mathbf{F}$ and $T(\theta)$. This equation is valid, for example, for an infinitesimal step of a homothermal process, whether quasi-static or not, in a perfectly elastic material.

For a material with fading memory, the path in $(\mathbf{F}, T)$ space which may be followed by the non-quasi-static process could also be followed by a quasi-static process. Accordingly, at each point of an arbitrary homothermal process in a material with fading memory, the Clausius inequality (5.42), i.e.

$$\int_A^B \frac{dH}{T(\theta)} \leq S_A - S_B$$

can be replaced by the Clausius–Duhem inequality,

$$\dot{S} \leq \dot{H}/T, \qquad (5.43)$$

where the dot designates material differentiation with respect to time; thus, $\dot{H}$ denotes the rate at which heat enters the body at the instant considered. There is an essential physical difference between the Clausius and Clausius–Duhem inequalities. This may be illustrated by the following comparison given by Rivlin (1975).

Consider a body of material with fading memory to be taken from an equilibrium state A to an equilibrium state B by quasi-static and non-quasi-static, isothermal, homothermal processes which follow the same paths in $(\mathbf{F}, T)$ space. The Clausius inequality states that less heat is fed into the system in the non-quasi-static process than in the quasi-static process. The Clausius–Duhem inequality asserts, however, that the amount of heat fed into the system, in each infinitesimal step of the non-quasi-static process, is no greater than that for the corresponding step of the quasi-static process. Rivlin (1975), however, showed by an example that the Clausius–Duhem inequality may not be valid for all materials and all processes.

(d) Instantaneous response behaviour

In the case of materials with fading memory, instantaneous changes in the deformation gradient tensor $\mathbf{F}$ and the empirical temperature θ result in instantaneous changes in the Piola–Kirchhoff stress $\mathbf{\Sigma}$ and in the specific internal energy e which could be followed by further changes in these quantities.

In order to describe the type of behaviour above, Rivlin (1975), following Green, Rivlin and Spencer (1959), made explicit the dependence of $\mathbf{\Sigma}$ and e, at time t, on the instantaneous values of the deformation gradient tensor $\mathbf{F}$ and of the empirical temperature at time t. Accordingly, the following constitutive equations may be written:

$$\mathbf{\Sigma} = \mathbf{\Sigma}[\mathbf{F}(\tau), \ \theta(\tau); \ \mathbf{F}, \ \theta)] \tag{5.44}$$

and

$$e = e[\mathbf{F}(\tau), \ \theta(\tau); \ \mathbf{F}, \ \theta] \tag{5.45}$$

indicating that $\mathbf{\Sigma}$ and e are functionals of the histories $\mathbf{F}(\tau)$ and $\theta(\tau)$ with support $(-\infty, t)$ and ordinary functions of $\mathbf{F}$ and θ.

For materials possessing instantaneous elasticity, for which the constitutive equations (5.44) and (5.45) are valid, $\mathbf{\Sigma}$ and e are functions of $\mathbf{F}$ and θ only. In this case, we would restrict ourselves to processes for which the histories $\mathbf{F}(\tau)$ and $\theta(\tau)$ are fixed functions of τ, $-\infty < \tau < t$, while only $\mathbf{F}$ and θ may change.

5.4 THERMODYNAMICAL DERIVATION OF THE CONSTITUTIVE RELATIONS

In their derivation of the thermodynamic constitutive equation, Christensen and Naghdi (1967) and Christensen (1971) based their work on the balance of energy equation for the infinitesimal theory and the entropy production postulate (Truesdell and Toupin, 1960). The derivation parallels, in essence, the means of deriving the constitutive equation in the linear isothermal case. However, the situation here is more difficult since the free energy not only depends on the strain history but also depends on the temperature history. At this point, a remark should be cited concerning the free energy relationships. In these relationships, the stress and deformation are conjugate variables. One, therefore, has to make a choice as to which will be the independent variable. If the stress is the independent variable, then the appropriate free energy function is the Gibbs free energy. On the other hand, if the strain is taken as the independent variable, then the corresponding free energy function is the Helmholtz free energy. Hence, considering the latter context, the local balance of energy equation for infinitesimal theory is given (Christensen, 1971) by

$$\rho r - \rho(\dot{A} + \dot{T}S + T\dot{S}) + \sigma_{ij}\dot{\varepsilon}_{ij} - Z_{i,i} = 0. \tag{5.46}$$

In equation (5.46), ρ is the mass density, r is the heat supply function per unit mass, $\dot{A}$ is the time derivative of the Helmholtz free energy per unit mass, T is the absolute temperature, S is the entropy per unit mass and Z_i are the Cartesian components of the heat flux vector measured per unit area per unit time. The related local entropy production inequality (Clausius–Duhem) is given by

$$\rho T\dot{S} - \rho r + Z_{i,i} - Z_i(T_{,i}/T) \geq 0. \tag{5.47}$$

With reference to (5.46), it is usually assumed that ε_{ij} and T are continuous in the interval $-\infty < t < \infty$ and that ε_{ij} tends to zero and T tends to T_0 as t tends to $-\infty$. Based on this assumption, the free energy can be expressed (Christensen, 1971) as a polynomial in a set of real, continuous linear functions of ε_{ij} and T as

$$\rho A = \rho A_0 + \int_{-\infty}^{t} D_{ij}(t-\tau) \frac{\partial \varepsilon_{ij}(\tau)}{\partial \tau} d\tau - \int_{-\infty}^{t} \beta(t-\tau) \frac{\partial \theta(\tau)}{\partial \tau} d\tau$$

$$+ \frac{1}{2} \int_{-\infty}^{t} \int_{-\infty}^{t} R_{ijkl}(t-\tau, t-s) \frac{\partial \varepsilon_{ij}(\tau)}{\partial \tau} \frac{\partial \varepsilon_{kl}(s)}{\partial s} d\tau \, ds$$

$$- \int_{-\infty}^{t} \int_{-\infty}^{t} \phi_{ij}(t-\tau, t-s) \frac{\partial \varepsilon_{ij}(\tau)}{\partial \tau} \frac{\partial \theta(s)}{\partial s} d\tau \, ds$$

$$- \frac{1}{2} \int_{-\infty}^{t} \int_{-\infty}^{t} m(t-\tau, t-s) \frac{\partial \theta(\tau)}{\partial \tau} \frac{\partial \theta(s)}{\partial s} d\tau \, ds \tag{5.48}$$

where $T = T_0 + \theta$ and A_0 is the mean free energy. In (5.48) the integrating material functions are assumed to be continuous for arguments $\tau_i \geq 0$ and vanish identically for $\tau_i < 0$, i.e.

$$\beta(\tau_1) = 0, \qquad D_{ij}(\tau_1) = 0, \qquad R_{ijkl}(\tau_1, \tau_2) = 0,$$

$$\phi_{ij}(\tau_1, \tau_2) = 0 \qquad m(\tau_1, \tau_2) = 0 \qquad \text{for} \quad \tau_1 < 0 \text{ and } \tau_2 < 0. \tag{5.49}$$

For the proposed theory, these integrating functions are necessarily independent of strain and temperature.

Now, if one combines equations (5.46)–(5.48) and, at the same time, carries out the indicated differentiation with respect to time, one obtains (Christensen (1971)

$$
\left[-D_{ij}(0) - \int_{-\infty}^{t} R_{ijkl}(t - \tau, 0) \frac{\partial \varepsilon_{kl}(\tau)}{\partial \tau} \, d\tau \right.
$$

$$
\left. + \int_{-\infty}^{t} \phi_{ij}(0, t - \tau) \frac{\partial \theta(\tau)}{\partial \tau} \, d\tau + \sigma_{ij} \right] \dot{\varepsilon}_{ij}(t)
$$

$$
+ \left[\beta(0) + \int_{-\infty}^{t} m(t - \tau, 0) \frac{\partial \theta(\tau)}{\partial \tau} \, d\tau \right.
$$

$$
+ \int_{-\infty}^{t} \phi_{ij}(t - \tau, 0) \frac{\partial \varepsilon_{ij}(\tau)}{\partial \tau} \, d\tau - \rho S \right] \dot{\theta}(t)
$$

$$
+ \left[-\int_{-\infty}^{t} \frac{\partial}{\partial t} D_{ij}(t - \tau) \frac{\partial \varepsilon_{ij}(\tau)}{\partial \tau} \, d\tau \right.
$$

$$
+ \int_{-\infty}^{t} \frac{\partial}{\partial t} \beta(t - \tau) \frac{\partial \theta(\tau)}{\partial \tau} \, d\tau + \Lambda - Z_i \frac{\theta_{,i}}{T_0} \right] \geq 0 \qquad (5.50)
$$

where

$$
\Lambda = -\frac{1}{2} \int_{-\infty}^{t} \int_{-\infty}^{t} \frac{\partial}{\partial t} R_{ijkl}(t - \tau, t - s) \frac{\partial \varepsilon_{ij}(\tau)}{\partial \tau} \frac{\partial \varepsilon_{kl}(s)}{\partial s} \, d\tau \, ds
$$

$$
+ \frac{1}{2} \int_{-\infty}^{t} \int_{-\infty}^{t} \frac{\partial}{\partial t} \phi_{ij}(t - \tau, t - s) \frac{\partial \varepsilon_{ij}(\tau)}{\partial \tau} \frac{\partial \theta(s)}{\partial s} \, d\tau \, ds
$$

$$
+ \frac{1}{2} \int_{-\infty}^{t} \int_{-\infty}^{t} \frac{\partial}{\partial t} m(t - \tau, t - s) \frac{\partial \theta(\tau)}{\partial \tau} \frac{\partial \theta(s)}{\partial s} \, d\tau \, ds \qquad (5.51)
$$

and the following symmetry properties are implied:

$$
R_{ijkl}(t - \tau, t - s) = R_{klij}(t - s, t - \tau). \qquad (5.52)
$$

The inequality (5.50) must hold for all arbitrary values of $\dot{\varepsilon}_{ij}(t)$ and $\dot{\theta}(t)$; therefore, it is necessary that the coefficients of $\dot{\varepsilon}_{ij}(t)$ and $\dot{\theta}(t)$ in (5.50) vanish. Hence

$$
\sigma_{ij} = D_{ij}(0) + \int_{-\infty}^{t} R_{ijkl}(t - \tau, 0) \frac{\partial \varepsilon_{kl}(\tau)}{\partial \tau} \, d\tau - \int_{-\infty}^{t} \phi_{ij}(0, t - \tau) \frac{\partial \theta(\tau)}{\partial \tau} \, d\tau \qquad (5.53)
$$

and

$$
\rho S = \beta(0) + \int_{-\infty}^{t} \phi_{ij}(t - \tau, 0) \frac{\partial \varepsilon_{ij}(\tau)}{\partial \tau} \, d\tau + \int_{-\infty}^{t} m(t - \tau, 0) \frac{\partial \theta(\tau)}{\partial \tau} \, d\tau. \qquad (5.54)
$$

Relations (5.53) and (5.54) are the constitutive relations for stress and entropy, respectively. From these it is clear that $D_{ij}(0)$ is the initial stress and that $\beta(0)$ is the initial entropy, ρS_0. The integrating functions $R_{ijkl}(t - \tau, 0)$, $\phi_{ij}(0, t - \tau)$, $\phi_{ij}(t - \tau, 0)$ and $m(t - \tau, 0)$ are appropriate relaxation function norms of the material properties. It is the relaxation function $R_{ijkl}(t, 0)$ in this formulation which corresponds to the relation function $R_{ijkl}(t)$ in the isothermal theory.

5.4.1 Reduction to the isotropic theory

For isotropic materials, ϕ_{ij} must be taken as

$$\phi_{ij}(\tau, s) = \delta_{ij} \, \phi(t, s) \tag{5.55}$$

where δ_{ij} is the Kronecker delta. Using the definitions of deviatoric stress and strain, the free energy for isotropic theory can be expressed (Christensen, 1971) with reference to (5.50) as

$$
\begin{aligned}
\rho A = \ &\frac{1}{2} \int_{-\infty}^{t} \int_{-\infty}^{t} G_1(t - \tau, t - s) \frac{\partial \varepsilon_{ij}'(\tau)}{\partial \tau} \frac{\partial \varepsilon_{ij}'(s)}{\partial s} \, d\tau \, ds \\
&+ \frac{1}{6} \int_{-\infty}^{t} \int_{-\infty}^{t} G_2(t - \tau, t - s) \frac{\partial \varepsilon_{kk}(\tau)}{\partial \tau} \frac{\partial \varepsilon_{jj}(s)}{\partial s} \, d\tau \, ds \\
&- \int_{-\infty}^{t} \int_{-\infty}^{t} \phi(t - \tau, t - s) \frac{\partial \varepsilon_{kk}(\tau)}{\partial \tau} \frac{\partial \theta(s)}{\partial s} \, d\tau \, ds \\
&- \frac{1}{2} \int_{-\infty}^{t} \int_{-\infty}^{t} m(t - \tau, t - s) \frac{\partial \theta(\tau)}{\partial \tau} \frac{\partial \theta(s)}{\partial s} \, d\tau \, ds \tag{5.56}
\end{aligned}
$$

where the initial stress and initial entropy effects in (5.50) have been dropped.

Based on the form (5.56) for the free energy, it can be shown that the stress-relaxation equations for isotropic materials are

$$\sigma_{ij}'(t) = \int_{-\infty}^{t} R_1(t - \tau, 0) \frac{\partial \varepsilon_{ij}'(\tau)}{\partial \tau} \, d\tau. \tag{5.57a}$$

and

$$\sigma_{kk}(t) = \int_{-\infty}^{t} R_2(t - \tau, 0) \frac{\partial \varepsilon_{kk}(\tau)}{\partial \tau} \, d\tau - 3 \int_{-\infty}^{t} \phi(0, t - \tau) \frac{\partial \theta(\tau)}{\partial \tau} \, d\tau. \tag{5.57b}$$

If the material functions appearing in the constitutive equations (5.53) and (5.56) are independent of temperature, which in the service life of the material might be true for only small temperature changes, or the temperature is timewise constant, then these constitutive equations will reduce to their counterparts in the isothermal theory. When these two equations do not exist and it is desired to find experimentally the material functions without making any *a priori* assumptions about their temperature dependence, a large number of tests will be needed even for the uniaxial test situation particularly if the temperature varies in a cyclic or a discrete fashion with the time. In this case, one must subject the material specimen to the actual temperature history (Landel and Peng, 1986). The latter approach could prove to be quite impractical. There is, however, experimental evidence (e.g. Schapery, 1974) which implies that viscoelastic characterization for transient temperature applications may be performed by using tests at a set of different constant temperatures. Hence, the phenomenological viscoelastic response description of a large class of polymeric materials and inorganic glasses under nonisothermal conditions is simplified by the adoption of the so-called 'temperature–time equivalence', also known as the 'thermorheologically simple hypothesis'.

5.5 THERMORHEOLOGICALLY SIMPLE MATERIALS

Thermorheologically simple materials (TSMs) are a special class of viscoelastic materials whose temperature dependence of mechanical properties is particularly responsive to analytical description. This group of materials generally constitutes the simplest and most realistic viscoelastic constitutive equation for which response under constant temperatures can be used to predict response under transient temperatures. Two temperature states are studied here, i.e. the constant temperature state and the nonconstant temperature one. For detailed description of the temperature dependent properties of thermorheologically simple materials, reference is made to Leaderman (1943), Schwarzl and Staverman (1952), Morland and Lee (1960), Ferry (1970) and Schapery (1974), among others.

5.5.1 Thermorheologically simple materials under constant temperature states

Following Schapery (1974), the uniaxial creep constitutive relation for thermo-rheologically simple materials can be expressed as

$$\varepsilon(t) = \int_0^t F(\xi - \xi') \frac{d\sigma}{d\xi'} \, d\xi' \tag{5.58}$$

where ε is the uniaxial strain due to stress only, i.e. the total strain less that due to thermal expansion and $F(\xi - \xi')$ is the time- and temperature-dependent creep compliance. In this equation, ξ is called the 'reduced time parameter' and defined by

$$\xi = \xi(t) = \int_0^t \frac{d\tau}{a_T} \tag{5.59}$$

and equivalently

$$\xi' = \xi(t') = \int_0^{t'} \frac{d\tau}{a_T} \tag{5.60}$$

where $a_T = a_T[T(\tau)]$ is the so-called 'temperature shift factor'. The latter is dependent on the absolute temperature T within the time interval $\tau = \xi - \xi'$. In the constitutive equation (5.58), it is assumed that $\sigma = \varepsilon = 0$ when $t \leq 0$.

The inverse of (5.58), i.e. the relaxation constitutive equation, can be determined by using Laplace transform with respect to reduced time. In this context, it can be shown that

$$\sigma(t) = \int_0^t R(\xi - \xi') \frac{d\varepsilon}{d\xi'} \, d\xi' \tag{5.61}$$

in which $R(\xi - \xi')$ is the time- and temperature-dependent relaxation modulus and both the reduced time parameters ξ, ξ' are as expressed previously by (5.59) and (5.60), respectively.

The experimental bases for the constitutive equations (5.58) and (5.61) under constant temperature conditions may be treated by considering isothermal creep and isothermal relaxation tests. That is, for the uniaxial creep test $\sigma(t) = \sigma_0 H(t)$, where σ_0 is the constant stress input and $H(t)$ is the Heaviside step function, the creep constitutive equation (5.58) yields

$$\varepsilon(\xi) = F(\xi)\sigma(t) \tag{5.62}$$

where ε is the resulting uniaxial strain due to the stress only. Similarly, for the relaxation test $\varepsilon(t) = \varepsilon_0 H(t)$, where ε_0 is the constant strain input, the relaxation constitutive equation (5.61) yields

$$\sigma(\xi) = R(\xi)\varepsilon(t) \tag{5.63}$$

with the understanding that for both types of isothermal tests

$$\xi = t/a_T. \tag{5.64}$$

Equation (5.64) indicates that the effect of temperature on the mechanical properties $F(\xi)$ or $R(\xi)$ for a thermorheologically simple material produces only horizontal translations when the property is plotted against log t. Conversely, if it is found that the constant temperature viscoelastic response curves (creep or relaxation) can be superposed so as to form a single curve (master curve) by means of only rigid, horizontal translations then the associated mechanical property (relaxation modulus or creep compliance) would depend only on time and temperature through the one parameter ξ. Such a description is probably a more or less conventional definition of a thermorheologically simple material (Schapery, 1974; Tobolosky and Catsiff, 1956).

An illustration of the time–temperature shift of the stress-relaxation curves at different temperatures to form a master curve associated with a particular reference temperature is given in Fig. 5.1. In Fig. 5.1(a), a series of relaxation moduli curves at different base temperatures are plotted using experimental relaxation data on bisphenol polycarbonate ($M_W = 40\,000$) from Mercier *et al.* (1965). For the purpose of constructing the master curve for these data, the relaxation curve corresponding to a reference temperature $T_R = 141\,°C$ is assigned as the reference curve. The other curves are then shifted along the logarithmic time scale until they superimpose. The relaxation moduli curves corresponding to temperatures above the reference

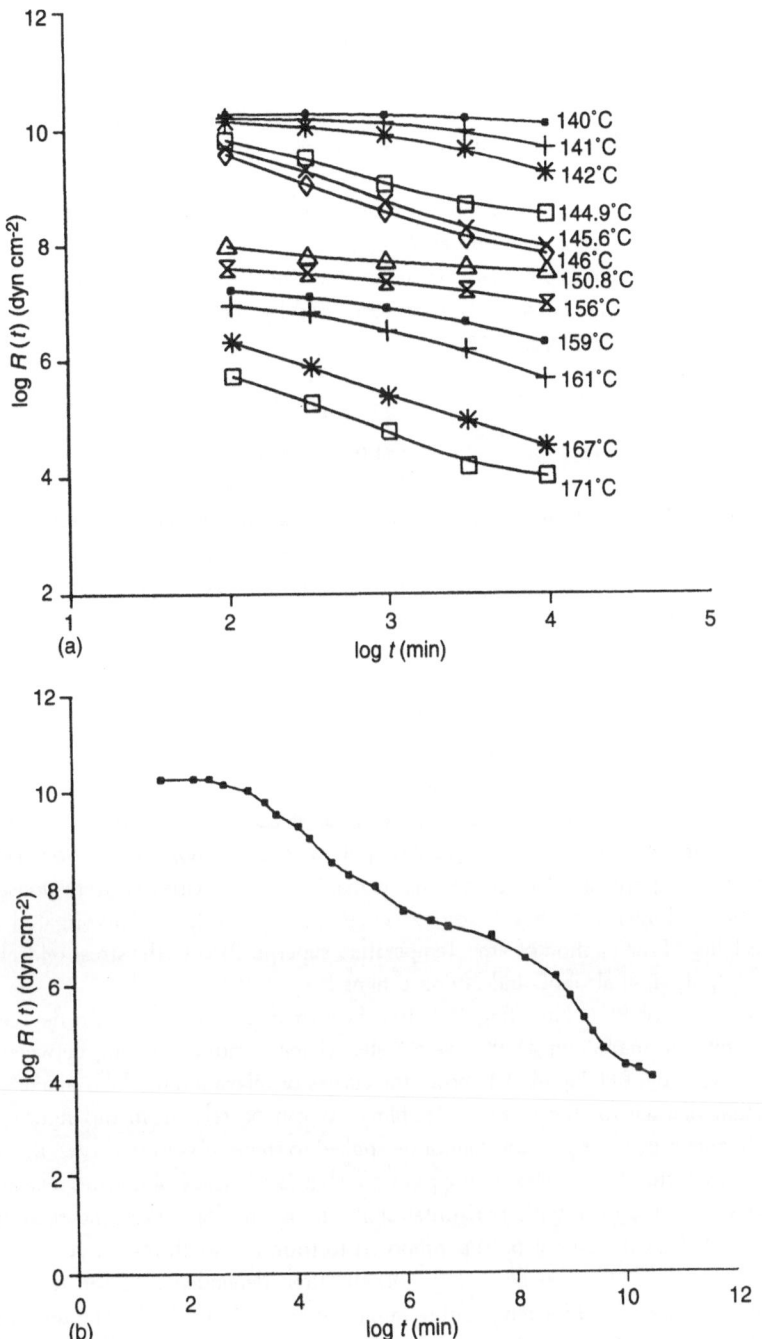

Fig. 5.1 (a) Variation of stress-relaxation moduli with time at different base temperatures for bisphenol polycarbonate ($M_W = 40\,000$); experimental data from Mercier *et al* (1965). (b) Master stress-relaxation curve based on stress-relaxation data presented in (a) with reference temperature $T_R = 141°C$.

temperatures are shifted to the right while those corresponding to temperatures below the reference temperature are shifted to the left of the reference curve. The full master curve is consequently formed as shown in Fig. 5.1(b). It is noticed that the master curve covers a much wider range of time as compared with the original time range covered by the individual relaxation curves. An analogous procedure is followed in Figs. 5.2(a) and 5.2(b) for constructing a master curve using experimental creep compliance data on hot setting epoxy resin from Theocaris (1962). The reference temperature for the data is chosen as $T_R = 25°C$.

For a large class of polymeric systems near their transition temperature, the time–temperature shift factor a_T is often expressed by the following WLF equation (Williams, Landel and Ferry, 1955)

$$\log a_T = -\frac{17.44(T - T_g)}{51.6 + T - T_g} \tag{5.65}$$

where T_g is the glass transition temperature, assumed as the reference temperature for the particular polymer under consideration. Equation (5.65) is considered to be valid for a polymer in the temperature range T_g to $T_g + 100°C$ (e.g. Gittus, 1975). The numerical constants in this equation are established by experiment within the indicated temperature range. If a temperature other than T_g is chosen as the reference temperature, a form analogous to (5.65) may be used to determine a_T, but with the numerical constants corresponding to the chosen reference temperature.

Although the method of time–temperature superposition has been shown to be useful in the characterization of the rheological properties of a large class of amorphous polymers over a wide range of time, it can only be applied to a much smaller range of time for many crystalline polymers (e.g. Onogi et al., 1962; Ferry, 1970). This is primarily due to the predominant nonlinear viscoelastic response of the latter polymers. In this, Onogi et al. (1962), for instance, investigated the applicability of the method of time–temperature superposition to the stress relaxation of PVA (polyvinyl alcohol) and Nylon 6 films.

In the case of PVA films (Fig. 5.3) two heat-treated specimens (with degree of crystallinity of 36.0% and 47.3%) were tested at temperatures varying between 20 and 100°C at 0% RH. Figure 5.3 shows the curves of relaxation modulus against the logarithm of time for the tested PVA films. As can be seen from this figure, the time–temperature superposition cannot be applied to these relaxation curves. In other words, when the shown relaxation curves are shifted vertically along the relaxation modulus axis, together with horizontal shifts along the log t axis, the relaxation curves of PVA films cannot be superimposed to form a smooth master curve.

In the case of Nylon 6 films (Fig. 5.4) the time dependence of the relaxation response was examined for temperature range between 25 and 77°C. The tests were performed at 0% RH. As shown in Fig. 5.4, the curves at temperatures higher than 50°C can be superposed to form a master curve for this temperature range, while those curves corresponding to lower temperatures than 50°C cannot be superposed. According to Onogi et al. (1962), the temperature 50°C conforms closely

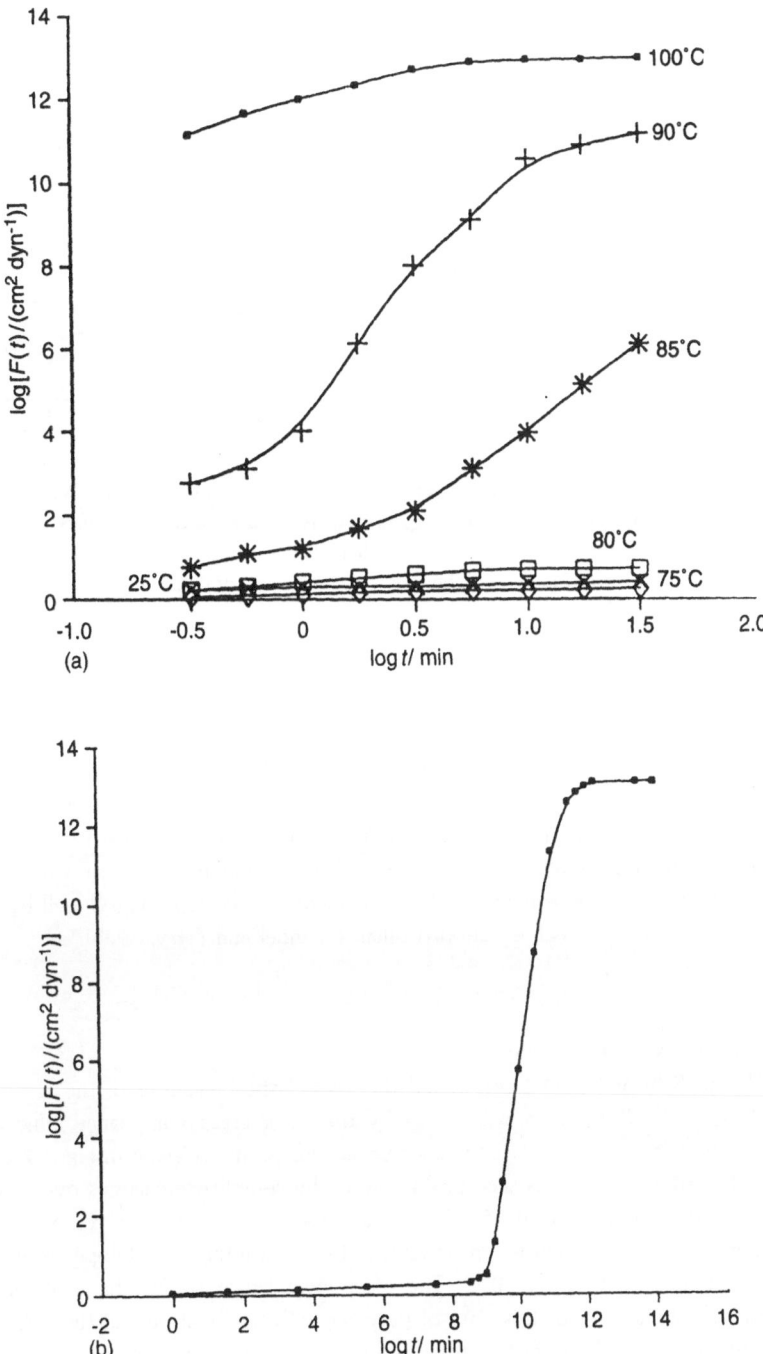

Fig. 5.2 (a) Variation of creep compliance with time at different base temperatures for hot setting epoxy resin (experimental data from Theocaris (1962)). (b) Master creep curve based on creep data presented in (a) with a reference temperature $T_R = 25\,^{\circ}\mathrm{C}$.

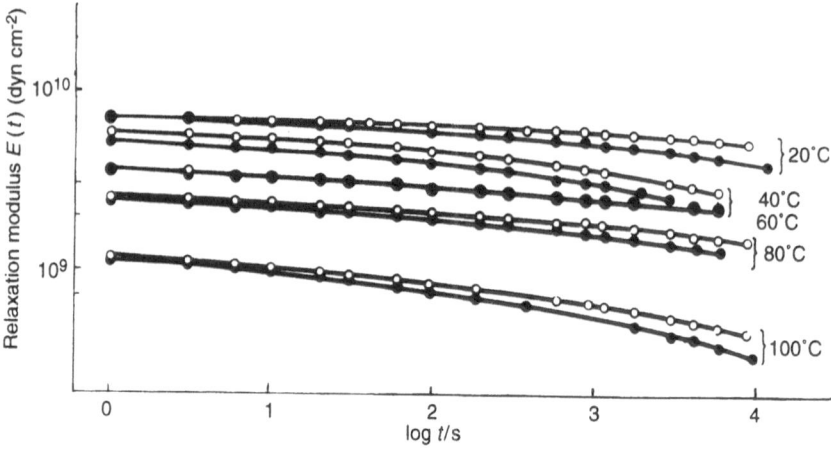

Fig. 5.3 Relaxation modulus versus log(time) at various temperatures for two PVA films of different degrees of crystallinity (0% RH): ○, 47.3%; ●, 36.0%. (Source: Onogi, S., Sasaguri, K., Adachi, T. and Ogihara, S. (1962) Time–humidity superposition in some crystalline polymers. *J. Polymer Sci.*, **58**, 1–17 (John Wiley & Sons Inc. copyright 1962). Reprinted by permission of John Wiley & Sons Inc.)

to the transition temperature of the Nylon 6 film. That is, only the relaxation curves in the transition region can be superposed satisfactorily. The master curve obtained by Onogi *et al.* (1962) with a reference temperature of 50°C is shown in Fig. 5.5. The shift factor $\log a_T$ versus the reciprocal absolute temperature is shown in Fig. 5.6. The temperature dependence of the shift factor can be represented well by the following form of the WLF equation (Williams, Landel and Ferry, 1955), i.e.

$$\log a_T = 25.6(\theta - 50°C)/(85.2 + \theta - 50°C)$$

where θ (°C) is the temperature.

Link and Schwarzl (1987) considered the viscoelastic behaviour of the technical polystyrene PS N7000 in a wide range of the shear creep compliance, time and temperature. The shear creep compliance versus creep time is given in Fig. 5.7 on a double-logarithmic scale. As seen in the figure, the compliance changes over seven orders in magnitude from $10^{-8}\,Pa^{-1}$ to $10^{-1}\,Pa^{-1}$. The course of compliance is determined over more than seven decades in time. Meantime, the temperature was varied between 95°C and 170°C. The creep compliance shows the well-known characteristic behaviour of this class of polymer with temperature and time. At the lower temperatures, the transition region is seen with a steep rise of nearly constant slope. At the temperature of 100°C, the beginning of the rubbery plateau may first be seen at longer times. Increasing the temperature further, the rubbery plateau becomes shorter and the viscous contribution would start to dominate even

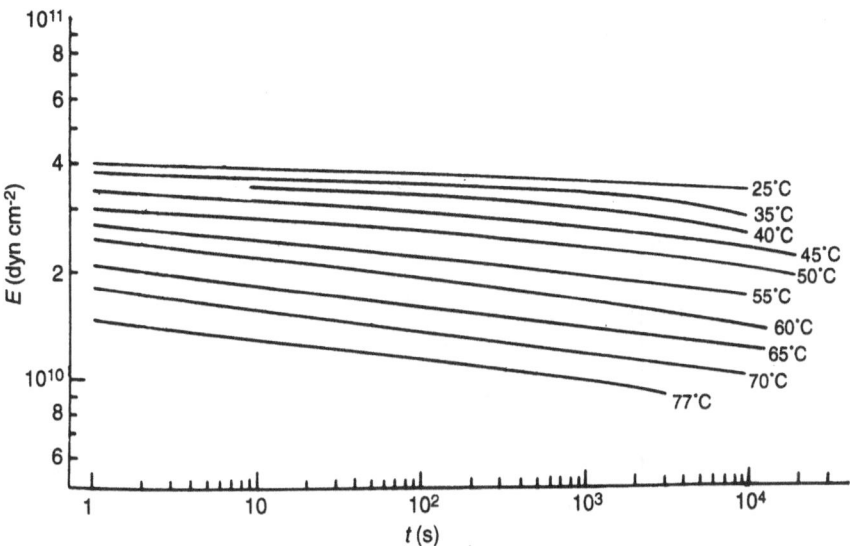

Fig. 5.4 Relaxation modulus versus log(time) at various temperatures for heat-treated nylon 6 film (0% RH). (Source: Onogi, S., Sasaguri, K., Adachi, T. and Ogihara, S. (1962) Time–humidity superposition in some crystalline polymers. *J. Polym. Sci.*, **58**, 1–17 (John Wiley & Sons Inc. copyright 1962). Reprinted by permission of John Wiley & Sons Inc.)

at shorter times. In the flow region, the creep compliance increases with time with a slope which, as seen in the figure, approaches unity on a double-logarithmic scale.

The recoverable creep compliance of polystyrene PS N7000 is shown at the same temperatures in Fig. 5.8 from Link and Schwarzl (1987). As indicated by these authors, two transitions can be seen in the course of the recoverable creep compliance. In addition to the glass–rubber transition, a second pronounced transition becomes

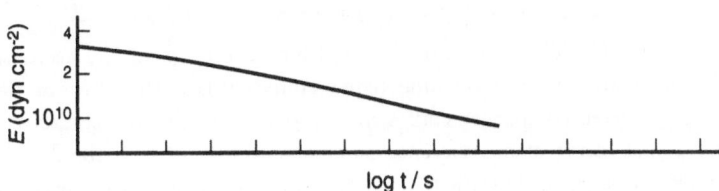

Fig. 5.5 Master relaxation curve for heat-treated nylon 6 film (50°C–77°C) (0% RH) as obtained from relaxation moduli curves of Fig. 5.4. (Source: Onogi, S., Sasaguri, K., Adachi, T. and Ogihara, S. (1962) Time–humidity superposition in some crystalline polymers. *J. Polym. Sci.*, **58**, 1–17 (John Wiley & Sons Inc. copyright 1962). Reprinted by permission of John Wiley & Sons Inc.)

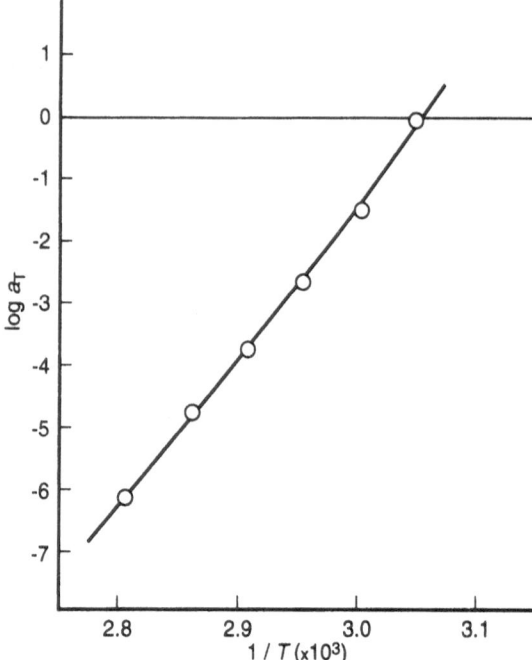

Fig. 5.6 Shift factor a_T against reciprocal of absolute temperature (relaxation moduli data for heat-treated nylon 6 film at 0% RH, Fig. 5.4). (Source: Onogi, S., Sasaguri, K., Adachi, T. and Ogihara, S. (1962) Time–humidity superposition in some crystalline polymers. *J. Polym. Sci.*, **58**, 1–17 (John Wiley & Sons Inc. copyright 1962). Reprinted by permission of John Wiley & Sons Inc.)

evident whereby the recoverable compliance rises from the rubber level of about $5 \times 10^{-6}\,\mathrm{Pa}^{-1}$ up to a long-time-limiting value of $4.7 \times 10^{-4}\,\mathrm{Pa}^{-1}$ which is the steady recoverable compliance. This transition is often referred to as a 'network transition'.

As shown in Fig. 5.9 (Link and Schwarzl, 1987), the reference temperature was chosen as 126.7°C. All the measured compliance curves (Fig. 5.7) were shifted along the time scale with the same time–temperature shift law. The latter was derived by shifting the creep compliance curves from 140°C to 170°C to coincide with the creep compliance curve at the reference temperature of 126.7°C in the flow region at a compliance level of $4 \times 10^{-3}\,\mathrm{Pa}^{-1}$. The corresponding master curve for the recoverable creep compliance is shown in Fig. 5.10 at the same reference temperature of 126.7°C.

Blatz (1956) considered the rheological behaviour of a typical composite solid propellant that is based on a cross-linked polymeric binder. Each of the propellant

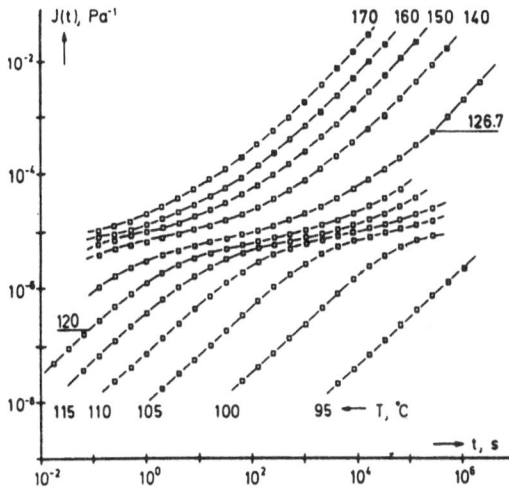

Fig. 5.7 Double-logarithmic plot of the creep compliance versus time for polystyrene (PSN N 7000). (Source: Link, G. and Schwarzl, F. R. (1987) Shear creep and recovery of a technical polystyrene. *Rheol. Acta*, **26**(4), 375–84 (Steinkopff Verlag Darmstadt). Reprinted with permission of Steinkopff Verlag Darmstadt.)

formulations studied includes the following constituents: a linear polymer 'R', a trifunctional cross-linking agent 'X', a low molecular weight plasticizer 'P', an inorganic oxidizer 'F' and a polymeric binder. The latter holds all the above-mentioned constituents except the oxidizer filler 'F'. Figure 5.11 represents the temperature dependence of the creep performance of a propellant of composition (by weight) R–8X–0.6P–60F over a temperature range from −40 to 150°F. The creep

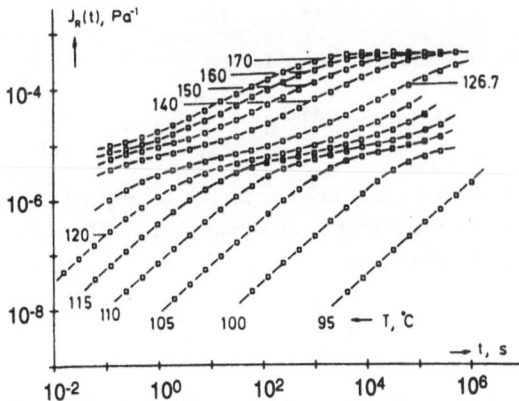

Fig. 5.8 Recoverable creep compliance versus creep time for polystyrene (PSN N 7000). (Source: Link, G. and Schwarzl, F. R. (1987) Shear creep recovery of a technical polystyrene. *Rheol. Acta*, **26**(4), 375–84 (Steinkopff Verlag Darmstadt). Reprinted with permission of Steinkopff Verlag Darmstadt.)

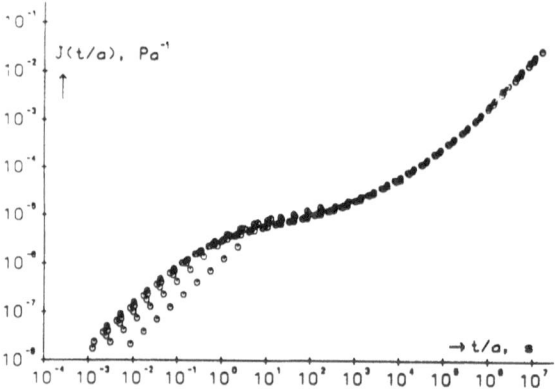

Fig. 5.9 Creep compliance versus reduced time t/a for polystyrene (PSN N 7000); reference temperature, 126.7°C; applied shift function determined in the flow region. (Source: Link, G. and Schwarzl, F. R. (1987) Shear creep and recovery of a technical polystyrene. *Rheol. Acta,* **26**(4), 375–84 (Steinkopff Verlag Darmstadt). Reprinted with permission of Steinkopff Verlag Darmstadt.)

compliance is plotted as a function of temperature on a double-logarithmic scale in Fig. 5.12 for a propellant formulation R–6X–0.6P–60F. The master curve for the creep compliance data of Fig. 5.12 is given in Fig. 5.13. In the latter figure, the master curve is adjusted so that its inflection point is at unity on the reduced time scale (Blatz, 1956).

In the three-dimensional case, the creep equation for the anisotropic thermo-rheologically simple material is given by (Schapery, 1974).

$$\varepsilon_{ij}(t) = \int_0^t F_{ijkl}(\xi - \xi') \frac{\partial \sigma_{kl}}{\partial t'} \, dt' + \int_0^t \alpha_{ij}(\xi - \xi') \frac{\partial \Delta T}{\partial t'} \, dt'. \qquad (5.66)$$

The corresponding stress-relaxation equation is

$$\sigma_{ij}(t) = \int_0^t R_{ijkl}(\xi - \xi') \frac{\partial \varepsilon_{kl}}{\partial t'} \, dt' - \int_0^t \beta_{ij}(\xi - \xi') \frac{\partial \Delta T}{\partial t'} \, dt' \qquad (5.67)$$

where the material functions are identical to the corresponding functions in the isothermal case except for the change in argument from physical time to reduced time where the latter is defined by

$$\xi = \xi(t) = \int_0^t \frac{d\tau}{a_T}, \quad \xi' = \xi(t') = \int_0^{t'} \frac{d\tau}{a_T} \qquad (5.68)$$

as represented earlier by equations (5.59) and (5.60) respectively. The second-order tensors α_{ij} and β_{ij} in (5.66) and (5.67), respectively, are associated with the thermal expansion characteristics of the thermorheologically simple material and define, respectively, thermal strains in the absence of applied stress and thermal stresses in a completely constrained body.

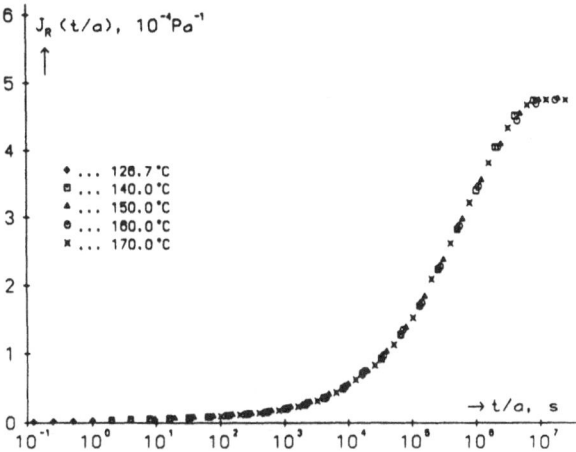

Fig. 5.10 Semilogarithmic plot of the recoverable creep compliance versus reduced creep time t/a for polystyrene at various temperatures (reference temperature, 126.7°C). (Source: Link, G. and Schwarzl, F. R. (1987) Shear creep and recovery of a technical polystyrene. *Rheol. Acta*, **26**(4), 375–84 (Steinkopff Verlag Darmstadt). Reprinted with permission of Steinkopff Verlag Darmstadt.)

The constitutive equations (5.66) and (5.67) may be seen as results of the linear hereditary theory where the input variables (stresses or strains) are combined with the temperature change which is not applied until time $t = 0$ on a non-aging type of material of a reduced time scale ξ. These equations have been derived from thermodynamics theory which predicts complete symmetry of the material functions involved (Schapery, 1974).

5.5.2 Thermorheologically simple materials under nonconstant temperature states

The effects to be studied here are outside the scope of the first-order linear theory. Consequently, a coupled thermoviscoelastic theory which includes the temperature dependence of mechanical properties is necessarily nonlinear. Guided by the work of Crochet and Naghdi (1969), Christensen (1971) presented a nonlinear theory of thermoviscoelasticity which, on the usual linearization of stress and strain, still retains a nonlinear dependence on temperature. Concerning this, Christensen attempted to derive the special results appropriate to the stress–strain constitutive relation without consideration of the other field variables such as energy, entropy, and the heat flux vector which necessarily are involved in the general theory. For this purpose, Christensen extended the uncoupled theory of linear thermoviscoelasticity to account for the temperature dependence of the relevant mechanical properties. The non-constant, nonuniform temperature history is considered to be known.

In Christensen's (1971) work, the starting point is the statement of a general nonlinear function which expresses the dependence of the current value of strain on

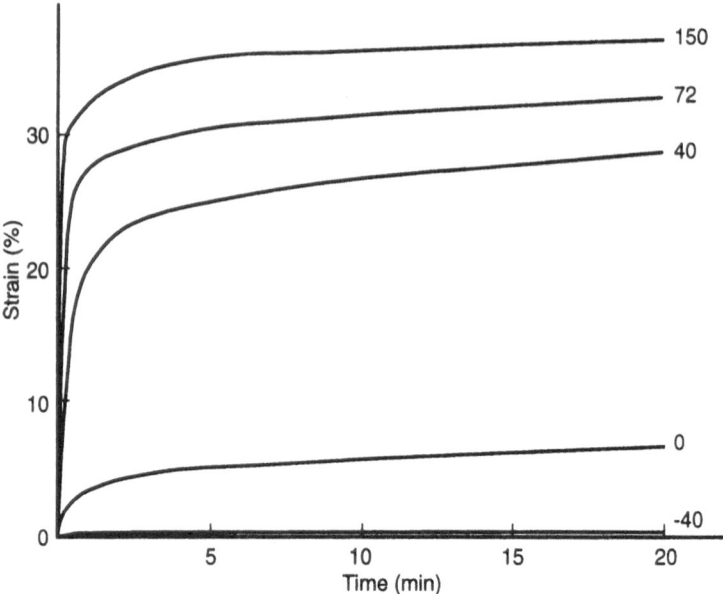

Fig. 5.11 Effect of temperature (°F) on creep of a class of composite solid propellants. (Source: Blatz, P. J. (1956) Rheology of composite solid propellants. *J. Ind. Eng. Chem.*, **48**(4), 727–9 (American Chemical Society). Reprinted with permission of American Chemical Society.)

the histories of stress and temperature, together with their current values. That is,

$$\mathbf{E}(t) = \underset{\tau=0}{\overset{\infty}{\psi}} \ [\mathbf{\Sigma}(t - \tau), \ T(t - \tau), \ \mathbf{\Sigma}(t), \ T(t)] \qquad (5.69)$$

In (5.69), $\mathbf{E}(t)$ is the nonlinear strain measure at time t, $\mathbf{\Sigma}$ is the Piola–Kirchhoff stress tensor and T is the absolute temperature.

For the material in a stress-free state, but with nonconstant temperature history, equation (5.69) may be expressed as a separate functional of temperatures only, i.e.

$$\mathbf{E}(t)|_{\mathbf{\Sigma}} = \underset{\tau=0}{\overset{\infty}{\psi'}} \ [T(t - \tau), \ T(t)]. \qquad (5.70)$$

The functional (5.70) could be restricted (Crochet and Naghdi, 1969, 1979) to express strain at zero stress as a function of current temperature, i.e. (5.70) reduces to

$$\underset{\tau=0}{\overset{\infty}{\psi'}} \ [T(t - \tau), \ T(\tau)] = \alpha T(t) \qquad (5.71)$$

where α is the coefficient of linear thermal expansion. Equation (5.71) may be considered as a special case of the type of behaviour allowed in the infinitesimal theory.

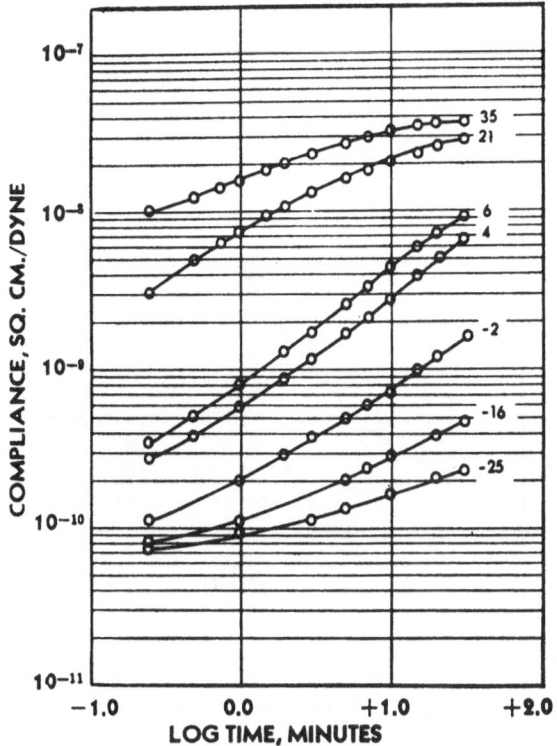

Fig. 5.12 Flexural creep of a class (weighted formulation, R–6X–0.6P–60F) of composite solid propellants as function of temperature. (Source: Blatz, P. J. (1956) Rheology of composite solid propellants. *J. Ind. Eng. Chem.*, **48**(4), 727–9 (American Chemical Society). Reprinted with permission of American Chemical Society.)

Thus, by decomposing $E(t)$ of (5.69) into two parts such that

$$\overset{\infty}{\underset{\tau=0}{\psi}} [\Sigma(t-\tau),\ T(t-\tau),\ \Sigma(t),\ T(t)] = \overset{\infty}{\underset{\tau=0}{\psi'}} [T(t-\tau),\ T(t)]$$

$$+ \overset{\infty}{\underset{\tau=0}{\psi''}} [\Sigma(t-\tau),\ T(t-\tau);\ \Sigma(t),\ T(t)] \quad (5.72)$$

where, by adopting (5.70), one must take

$$\overset{\infty}{\underset{\tau=0}{\psi''}} [0,\ T(t-\tau),\ 0,\ T(t)] = 0 \quad (5.73)$$

Thus, using (5.70) and (5.71) in (5.72), gives the form

$$E(t) - \alpha T(t) = \overset{\infty}{\underset{\tau=0}{\psi''}} [\Sigma(t-\tau),\ T(t-\tau).\ \Sigma(t),\ T(t)] \quad (5.74)$$

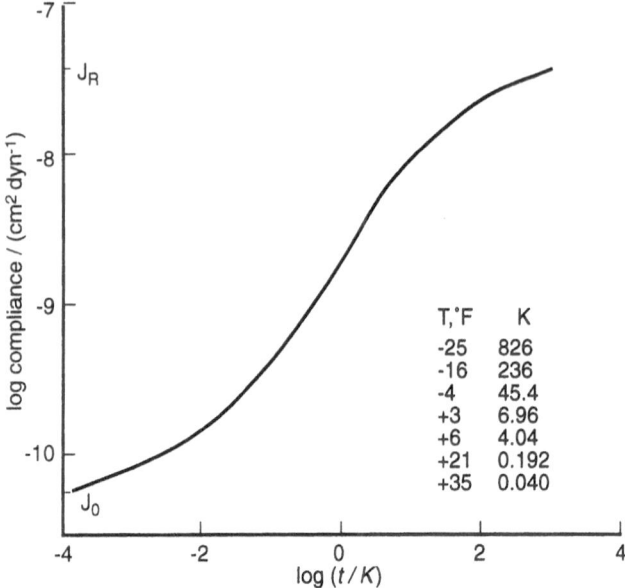

Fig. 5.13 Master curve for creep of a class of composite solid propellants ($-25°$F to $+35°$F) (Fig. 5.12). (Source: Blatz, P. J. (1956) Rheology of composite solid propellants. *J. Ind. Eng. Chem.*, **48**(4), 727–9 (American Chemical Society). Reprinted with permission of American Chemical Society.)

with the understanding from previous definitions that $\mathbf{E}(t)$ is the total strain including both the stress and temperature effects and $\alpha T(t)$ is the thermal strain in the absence of stress.

In a similar manner, on the assumption that the inverse of (5.69) exists, the stress-relaxation equation can be written as

$$\Sigma(t) = \overset{\infty}{\underset{\tau=0}{\mathbf{Y}}} \; [\mathbf{E}(t-\tau), \; T(t-\tau), \; \mathbf{E}(\tau), \; T(t)]. \tag{5.75}$$

Now, if one assumes, following Christensen (1971), that the nonisothermal stress constitutive relation is determined by the corresponding isothermal function with $\mathbf{E}$ replaced by $\mathbf{E} - \alpha$ and with a modified time scale ξ_τ to account for the history of the temperature, the nonisothermal stress relaxation may be expressed as

$$\Sigma(t) = \overset{\infty}{\underset{\tau=0}{\mathbf{Y}}} \; [\mathbf{E}(t-\xi_\tau) - \alpha(t-\xi_\tau), \; \mathbf{E}(t) - \alpha(t)] \tag{5.76}$$

In equation (5.76), the modified time scale ξ_τ is given by

$$\xi_\tau = \overset{\infty}{\underset{s=0}{\gamma}} \; [T(t-s), \; \tau] \tag{5.77}$$

where $\overset{\infty}{\underset{s=0}{\gamma}} (\cdot)$ has properties such as

$$\xi_\tau|_{\tau=0} = 0, \tag{5.78a}$$

and

$$\frac{\partial \xi_\tau}{\partial \tau} \geq 0 \tag{5.78b}$$

and

$$\xi_\tau|_{T=T_0} = \tau \tag{5.78c}$$

with the understanding that T_0 designates some constant base temperature.

It has been shown by Christensen (1971) that the appropriate infinitesimal theory form of (5.76) is

$$\sigma_{ij}(t) = R_{ijkl}(0)[\varepsilon_{kl}(t) - \alpha_{kl}(t)]$$

$$+ \int_0^\infty [\varepsilon_{kl}(t - \xi_\tau) - \alpha_{kl}(t - \xi_\tau)] \frac{\partial R_{ijkl}(\tau)\,d\tau}{d\tau}. \tag{5.79}$$

In (5.79), the usual infinitesimal theory definitions of stress and strain are taken; however, the general nonlinear dependence on temperature is retained.

5.6 THERMORHEOLOGICALLY COMPLEX MATERIALS

This is the class of viscoelastic materials whose temperature dependence of mechanical properties is not particularly responsive to analytical description through the time–temperature shift phenomenon. Following Schapery (1974), two classifications of such materials are defined, namely TCM-1 and TCM-2.

An example of the class of TCM-1 materials would be a composite material system consisting of two or more TSM phases. The mechanical behaviour of different composite systems belonging to the class TCM-1 has been considered by different researchers under isothermal conditions at different temperatures. In this context, Halpin (1969) dealt with a composite system consisting of two types of elastomers with different glass transition temperatures (-29 and $-75°$C). Fesco and Tschoegl (1971) considered the behaviour of two-phase block copolymers at various constant temperatures. The case of TCM-1 under transient temperatures is particularly complicated and further research effort is needed in this area (e.g. Schapery, 1966, 1968, 1969, 1974).

The TCM-2 class of viscoelastic materials is defined (Schapery, 1974) by the following uniaxial constitutive relationship for the cases of constant or transient temperatures

$$\varepsilon(t, T) = F_0(T)\sigma_0 + \int_0^t \Delta F(\xi - \xi') \frac{d}{dt'} \left[\frac{\sigma}{a_G(T)} \right] dt' \tag{5.80}$$

where $F_0(T)$ is the initial value of creep compliance, $a_G(T)$ is a new shift factor for the class of material considered and ξ, ξ' are reduced time parameters introduced earlier by (5.68). Applying (5.80) to an isothermal creep test, the creep compliance $F_T = \varepsilon/\sigma$, is expected to be of the form

$$F_T = F_0(T) + \Delta F(\xi)/a_G(T) \tag{5.81}$$

where $\xi = t/a_T$. This is with the understanding that, since $F_0(T)$ is the initial compliance, $\Delta F(0) = 0$. However, in order to relate the two shift factors a_T and a_G to experimental data, one writes (5.81) in the following logarithmic form:

$$\log[F_T - F_0(T)] = \log \Delta F(\xi) - \log a_G(T). \tag{5.82}$$

One also recalls from (5.68) that

$$\log \xi = \log t - \log a_T \tag{5.83}$$

Equations (5.82) and (5.83) imply that a plot of $\log[F_T - F_0(T)]$ versus $\log t$ at a test temperature T will be identical to that at an arbitrary selected reference temperature T_R, apart from rigid horizontal and vertical translations of $|\log t|$ and $|\log a_T|$ respectively.

A case of particular interest of the constitutive equation (5.80) is when

$$F_0 = F_0(T_R)/a_G(T). \tag{5.84}$$

In this particular case, the creep compliance (5.81) reduces to

$$F_T = F(\xi)/a_G(T) \tag{5.85}$$

in which

$$F(\xi) + F_0(T_R)/\Delta F(\xi). \tag{5.86}$$

On the other hand, assuming as a normalization case that $a_G = a_T = 1$ at T_R, one can write, with reference to (5.81), the following equation:

$$\Delta F(t) = F_T(t, T_R) - F_0(T_R). \tag{5.87}$$

Meantime, by using (5.87) with the argument t replaced by ξ, it can be shown that the constitutive equation (5.80) becomes

$$\varepsilon(t, T) = \int_0^t F(\xi - \xi') \frac{\mathrm{d}}{\mathrm{d}t'} \left(\frac{\sigma}{a_G} \right) \mathrm{d}t'. \tag{5.88}$$

The inversion of (5.88) may be expressed as

$$\sigma = a_G \int_0^t R(t - t') \frac{\mathrm{d}\varepsilon}{\mathrm{d}t} \mathrm{d}t'. \tag{5.89}$$

Accordingly, the relaxation modulus, R_T, for a constant strain input, is

$$R_T = a_G(T)R(\xi). \tag{5.90}$$

Equations (5.85) and (5.90), within their applicability, indicate that master curves of $F = F(\xi)$ and $R = R(\xi)$ can be plotted by making horizontal (log a_T) and vertical (log a_G) shifts of the experimentally derived creep compliance F_T and relaxation modulus R_T.

For the experimental verification of the constitutive equations (5.80) and (5.88), reference is made to Schapery, Beckwith and Conrad (1973), Schapery (1974) and Watkins (1973). The reader is also referred to McCrum and Pogany (1970) for different procedures by which experimental data can be represented by master curves. At this point, it should be mentioned that the strain response expressed by (5.88) may be converted to total strain by using the procedure presented earlier for the case of thermorheologically simple materials (section 5.5).

For a three-dimensional representation of TCMs, we restrict our analysis, following Schapery (1974), to the isotropic case. For this purpose, two independent material parameters are used. In the creep case, one can use both the uniaxial creep compliance as presented earlier by (5.81), i.e.

$$F_T = F_0(T) + \Delta F(\xi)/a_G(T) \tag{5.91a}$$

together with the shear creep compliance

$$J_T = J_0(T) + \Delta J(\xi)/a_G(T). \tag{5.91b}$$

Thus, with reference to the one-dimensional constitutive creep equation (5.80), the components for the three-dimensional (isotropic) case are

$$\varepsilon_{11}(t, T) = F_1\sigma_{11} - \left(\frac{J_0}{2} - F_0\right)(\sigma_{22} + \sigma_{33}) + \int_0^t \Delta F(\xi - \xi') \frac{\partial}{\partial t'}\left(\frac{\sigma_{11}}{a_G}\right) dt'$$
$$- \int_0^t \left[\frac{\Delta J(\xi - \xi')}{2} - \Delta F(\xi - \xi')\right] \frac{\partial}{\partial t'}\left(\frac{\sigma_{22} + \sigma_{33}}{a_G}\right) dt' \tag{5.92a}$$

and two additional equations for ε_{22} and ε_{33}. Similarly for the shear components of the strain tensor, one has

$$2\varepsilon_{23}(t, T) = J_0\sigma_{23} + \int_0^t \Delta F(\xi - \xi') \frac{\partial}{\partial t'}\left(\frac{\sigma_{23}}{a_G}\right) dt' \tag{5.92b}$$

and two additional equations for ε_{12} and ε_{13}.

From (5.92), it can be shown that the bulk creep compliance $K_T(t, T) = 3\varepsilon_{ii}/\sigma_{ii}$, where σ_{ii} is constant, is expressed by

$$K_T(t, T) = 3(3F_T - J_T). \tag{5.93}$$

Meantime, the isothermal value of Poisson's ratio is determined at each temperature by

$$v = (J_T/2F_T) - 1. \tag{5.94}$$

The constitutive equations for thermorheologically simple materials can be immediately deduced from (5.92) by setting $a_G = 1$ and assuming that F_1, J_1 and α are constants.

REFERENCES

Biot, M. A. (1958) Linear thermodynamics and the mechanics of solids, in Proc. 3rd US Natl Congr. Appl. Mech., pp. 1–18.

Biot, M. A. (1973) Nonlinear thermoelasticity, irreversible thermodynamics and elastic instability. *Indiana Univ. Math. J.*, **23**, 309–35.

Blatz, P. J. (1956) Rheology of composite solid propellants. *Ind. Eng. Chem.*, **48**(4), 727–9.

Breuer, S. (1969) Lower bounds on work in linear viscoelasticity. *Q. Appl. Math.*, **27**(2), 139–46.

Breuer, S. and Onat, E. T. (1964) On the determination of free energy in linear viscoelastic solids. *Z. Angew. Math. Phys.*, **15**, 184–91.

Christensen, R. M. (1971) *Theory of Viscoelasticity*, Academic Press, New York.

Christensen, R. M. and Naghdi, P. M. (1967) Linear non-isothermal viscoelastic solids. *Acta Mech.*, **3**, 1–12.

Coleman B. D. (1963) The thermodynamics of elastic materials with heat conduction and viscosity. *Arch. Ration. Mech. Anal.*, **13**, 167–78.

Coleman, B. D. (1964a) Thermodynamics of materials with memory. *Arch. Ration. Mech. Anal.*, **17**, 1–46.

Coleman B. D. (1964b) On thermodynamics, strain impulses, and viscoelasticity. *Arch. Ration. Mech. Anal.*, **17**, 230–54.

Coleman, B. D. and Gurtin, M. E. (1967a) Thermodynamics with internal state variables. *J. Chem. Phys.*, **47**, 597–613.

Coleman, B. D. and Gurtin, M. E. (1967b) Equipresence and constitutive equations for rigid heat conductors. *Z. Angew. Math. Phys.*, **18**, 199–208.

Coleman, B. D. and Mizel, V. J. (1963) Thermodynamics and departures from Fourier's law of heat conduction. *Arch. Ration. Mech. Anal.*, **13**, 245–61.

Coleman, B. D. and Mizel, V. J. (1964) Existence of coloric equations of state in thermodynamics. *J. Chem. Phys.*, **40**, 1116–25.

Coleman, B. D. and Mizel, V. J. (1967) A general theory of dissipation in materials with memory. *Arch. Ration. Mech. Anal.*, **27**, 255–74.

Coleman, B. D. and Mizel, V. J. (1968) On the general theory of fading memory. *Arch. Ration. Mech. Anal.*, **29**, 18–31.

Coleman, B. D. and Noll, W. (1960) An approximation theorem for functionals with applications in continuum mechanics. *Arch. Ration. Mech. Anal.*, **6**, 355–70.

Coleman, B. D. and Noll. W. (1961) Foundations of linear viscoelasticity. *Rev. Mod. Phys.*, **33**, 239–49.

Coleman, B. D. and Noll, W. (1963) The thermodynamics of elastic materials with heat conduction and viscosity. *Arch. Ration. Mech. Anal.*, **13**, 167–78.

Coleman, B. D. and Noll, W. (1964) Simple fluids with fading memory, in Proc. Int. Symp., Second Order Effects, Heifa, 1962, Macmillan, New York, pp. 530–52.

Coleman B. D. and Owen, D. R. (1970) On the thermodynamics of materials with memory. *Arch. Ration. Mech. Anal.*, **36**, 245–69.

Crochet, M. J. (1975) A non-isothermal theory of viscoelastic materials, in *Theoretical Rheology* (eds J. F. Hutton, J. R. A. Pearson and K. Walters), Applied Science, London, pp. 111–22.

Crochet, M. J. and Naghdi, P. M. (1969) A class of simple solids with fading memory. *Int. J. Eng. Sci.*, **7**, 1173–98.

Crochet, M. J. and Naghdi, P. M. (1974) On a restricted non-isothermal theory of simple materials. *J. Méc.*, **13**, 97–114.

Crochet, M. J. and Naghdi, P. M. (1979) On 'thermo-rheologically simple' solids, in Proc. IUTAM Symp., Thermoelasticity, Springer, New York, pp. 59–86.

Day, W. A. (1970) Reversibility, recoverable work and free energy in linear viscoelasticity. *Q. J. Mech. Appl. Math.*, **23**(1), 1–15.

Day, W. A. (1972) *The Thermodynamics of Simple Materials with Fading Memory*, Springer, New York.

Ferry, J. D. (1970) *Viscoelastic Properties of Polymers*, 2nd edn, Wiley, New York.

Fesco, D. G. and Tschoegl, N. W. (1971) Time–temperature superposition in thermo-rheologically complex materials. *J. Polym. Sci. C*, **35**, 51–69.

Fried, N. (1970) In *Mechanics of Composite Materials* (eds F. W. Wendt, H. Liebowitz and N. Perrone), Pergamon, Oxford, pp. 813–37.

Fung, Y. C. (1965) *Foundations of Solid Mechanics*, Prentice-Hall, Englewood Cliffs, NJ, pp. 377–446.

Gittus, J. (1975) *Creep, Viscoelasticity and Creep Fracture in Solids*, Wiley, New York.

Green, A. E. and Laws, N. (1967) On the formulation of constitutive equations in thermo-mechanical theories of continua. *Q. J. Mech. Appl. Math.*, **20**, 265–75.

Green, A. E. and Naghdi, P. M. (1965) A general theory of an elastic–plastic continuum. *Arch. Ration. Mech. Anal.*, **18**, 251–81.

Green, A. E., Rivlin, R. S. and Spencer, A. J. M. (1959) The mechanics of nonlinear materials with memory, Part II. *Arch. Ration. Mech. Anal.*, **3**, 82–90.

Gurtin, M. E. (1965) Thermodynamics and the possibility of spatial interaction in elastic materials. *Arch. Ration. Mech. Anal.*, **19**, 339–52.

Gurtin, M. E. and Williams, W. O. (1966) On the inclusion of the complete symmetry group in unimodular group. *Arch. Ration. Mech. Anal.*, **23**, 163–72.

Halpin, J. C. (1969) Characterization of orthotropic (fiber-reinforced) polymeric solids. Doctoral Dissertation, University of Akron, OH.

Kestin, J. (1966) *A Course in Thermodynamics*, Vol. 1, Blaisdell, Waltham, MA.

Landel, R. F. and Peng, S. T. T. (1986) Equations of state and constitutive equations. *J. Rheol.*, **30**(4), 741–65.

Laws, N. (1967) On the thermodynamics of certain materials with memory. *Int. J. Eng. Sci.*, **5**, 427–34.

Leaderman, H. (1943) *Elastic and Creep Properties of Filamentous Materials and Other Polymers*, Textile Foundation, Washington, DC, pp. 175–85.

Link, G. and Schwarzl, F. R. (1987) Shear creep and recovery of a technical polystyrene. *Rheol. Acta*, **26**, 375–84.

McCrum, N. G. and Pogany, G. A. (1970) Time–temperature superposition in the α-region of an epoxy resin. *J. Macromol. Sci. Phys. B*, **4**(1), 109–25.

Meixner, J. (1969) Processes in simple thermodynamic materials. *Arch. Ration. Mech. Anal.*, **33**, 33–53.

Mercier, J. P., Aklonis, J. J., Litt, M. and Tobolsky, A. V. (1965) Viscoelastic behaviour of the polycarbonate of bisphenol A. *J. Appl. Polym. Sci.*, **9**, 447–59.

Morland, L. W. and Lee, E. H. (1960) Stress analysis for linear viscoelastic materials with temperature variation. *Trans. Soc. Rheol.*, **4**, 233–63.

Müller, I. (1967) On the entropy inequality. *Arch. Ration. Mech. Anal.*, **26**, 118–41.

Noll, W. (1958) A mathematical theory of the mechanical behaviour of continuous media. *Arch. Ration. Mech. Anal.*, **2**, 197–226.

Onogi, S., Sasaguri, K., Adachi, T. and Ogihara, S. (1962) Time–humidity superposition in some crystalline polymers. *J. Polym. Sci.*, **58**, 1–17.

Owen, D. R. (1968) Thermodynamics of materials with elastic range. *Arch. Ration. Mech. Anal.*, **31**, 91–112.

Owen, D. R. (1970) A mechanical theory of materials with elastic range. *Arch. Ration. Mech. Anal.*, **37**, 85–110.

Rivlin, R. S. (1975) The thermodynamics of materials with fading memory, in *Theoretical Rheology* (Eds J. F. Hutton, J. R. A. Pearson and K. Walters), Applied Science, London, pp. 83–103.

Schapery, R. A. (1964) Application of thermodynamics to thermomechanical, fracture, and birefringent phenomena in viscoelastic media. *J. Appl. Phys.*, **35**(5), 1451–65.

Schapery, R. A. (1966) A theory of nonlinear thermoviscoelasticity based on irreversible thermodynamics, in Proc. 5th US Natl Congr. of Appl. Mech., ASME, pp. 511–30.

Schapery, R. A. (1968) On a thermodynamic constitutive theory and its application to various nonlinear materials, in Proc. IUTAM Symp., East Kilbride, pp. 259–85.

Schapery, R. A. (1969) On a thermodynamic constitutive theory and its application to various nonlinear materials, in Proc. IUTAM Symp. on Thermoinelasticity, Springer, Berlin.

Schapery, R. A. (1974) Viscoelastic behaviour and analysis of composite materials, in *Mechanics of Composite Materials*, Vol. 2 (Ed. G. Sandeskj), Academic Press, new York, pp. 86–168.

Schapery, R. A., Beckwith, S. W. and Conrad, N. (1973) Studies on the viscoelastic behaviour of fiber-reinforced plastic. *Mech. Mater. Res. Center Rep. MM 2702-73-3 (AFML-TR-73-179)*, Texas A & M University.

Schwarzl, F. and Staverman, A. J. (1952) Time–temperature dependence of linear viscoelastic behaviour. *J. Appl. Phys.*, **23**(8), 838–43.

Steel, D. J. (1965) The creep and stress-rupture of reinforced plastics. *Trans. J. Plast. Inst.*, **33**, 161–7.

Theocaris, P. S. (1962) Viscoelastic properties of epoxy resins derived from creep and relaxation tests at different temperatures. *Rheol. Acta*, **2**(2), 92–6.

Tobolosky, A. F. and Catsiff, E. (1956) Elastoviscous properties of polyisobutylene (and other amorphous polymers) from stress–relaxation studies, IX. A summary of results. *J. Polym. Sci.*, **19**, 111–21.

Truesdell, C. (1951) A new definition of a fluid. II. The Maxwellian fluid. *J. Math. Pures Appl.*, **30**, 111–58.

Truesdell, C. and Toupin, R. A. (1960) Classical field theories, in *Handbuch der Physik*, Vol. III/1 (Ed. S. Flügge), Springer, Berlin, pp. 226–790.

Tsai, S. W. (1970) In *Mechanics of Composite Materials* (eds F. W. Wendt, H. Liebowitz and N. Peronne), Pergamon, Oxford, pp. 749–67.

Wang, C. C. and Bowen, R. M. (1966) On the thermodynamics of nonlinear materials with quasi-elastic response. *Arch. Ration. Mech. Anal.*, **22**, 79–99.

Watkins, L. A. (1973) Creep of an epoxy resin under transient temperatures. M.S. Thesis, Civil Engineering, Texas A & M. University.

Williams, M. L., Landel, R. F. and Ferry, J. D. (1955) the temperature dependence of relaxation mechanisms in amorphous polymers and other glass-forming liquids. *J. Am. Chem. Soc.*, **77**, 3701–7.

Wilson, A. H. (1957) *Thermodynamics and Statistical Mechanics*, Cambridge University Press, Cambridge.

FURTHER READING

Bataille, J. and Kestin, J. (1979) Irreversible processes and physical interpretation of rational thermodynamics. *J. Non-Equilib. Thermodyn.*, **4**, 229–58.

Biot, M. A. (1954) Theory of stress–strain relationship in anisotropic viscoelasticity and relaxation phenomena. *J. Appl. Phys.*, **25**(11), 1385–91.

Eringen, A. C. (1960) Irreversible thermodynamics and continuum mechanics. *Phys. Rev.*, **117**, 1174–83.

Freeman, J. W. and Voorhees, H. R. (1956) Relaxation properties of steels and super-strength alloys at elevated temperatures. *ASTM, Spec. Tech. 187*, August. (also, *ASTM Publ. DS-114*, August 1961).

Freudenthal, A. M. (1954) Effect of rheological behaviour on thermal stresses. *J. Appl. Phys.*, **25**(9), 1110–7.

Hunter, S. C. (1961) Tentative equations for the propagation of stress, strain and temperature fields in viscoelastic solids. *J. Mech. Phys. Solids*, **9**, 39–51.

Koh, S. L. and Eringen, A. C. (1963) On the foundation of nonlinear thermoviscoelasticity. *Int. J. Eng. Sci.*, **1**, 199–229.

Lee, E. H. (1955) Stress analysis in viscoelastic bodies. *Q. Appl. Math.*, **13**(2), 183–90.

Manjoine, M. J. and Voorhees, H. R. (1982) Compilation of stress–relaxation data for engineering alloys. *ASTM Data Ser. Publ. DS60.*

Pindera, J. T. and Straka, P. (1974) On physical measures of rheological responses of some materials in wide ranges of temperature and spectral frequency. *Rheol. Acta,* **13**, 338–51.

Prager, W. (1956) Thermal stresses in viscoelastic structures. *J. Appl. Math. Phys.,* **7**, 230–8.

Rivlin, R. S. (1972) On the principles of equipresence and unification. *Q. Appl. Math.,* **30**, 227–8.

Wolosewick, R. M. and Gratch, S. (1965) Transient response in a viscoelastic material with temperature-dependent properties and thermomechanical coupling. *J. Appl. Mech.,* **32**(3), 620–2.

6

Transition to nonlinear viscoelasticity

6.1 INTRODUCTION

Material systems are often used under conditions which do not comply with the infinitesimal deformation postulates of the linear theory. The existence of imperfections and discontinuities in the material is particularly cited as a likely source of much of the nonlinearity of the behaviour. In the case of two-phase materials such as particulate and fibre composites, for instance, microstructural damage is often considered to be the most predominant cause of the nonlinear behaviour of such materials. Hence, these materials usually exhibit nonlinearity and strain rate dependent hysteresis over a wide range of temperature and stress or strain rate (e.g. Schapery, 1974).

For a nonlinear viscoelastic material, one or both of criteria (2.64) and (2.65) mentioned in Chapter 2 concerning, respectively, homogeneity and superposition of input histories are not met. However, the memory hypothesis previously introduced in linear viscoelasticity is still valid for the case of the nonlinear theory. This means, for the two theories, that the current value of the output is determined by the complete past history of the input. As we noticed in Chapter 2, this memory hypothesis was the starting point in the development of the linear theory of viscoelasticity and it is also the starting point for the treatment of nonlinear viscoelasticity (e.g. Christensen, 1971).

Linear viscoelastic behaviour of materials has been the subject of extensive studies for over a century, but it is only in the last few decades that researchers have started to pay particular attention to the more complex subject of nonlinear viscoelasticity. This was primarily motivated by the observation of certain nonlinear effects during the course of the study of the performance of a class of viscoelastic materials (Weissenberg, 1948) and, hence, by the failure of the linear theory to predict reasonably the viscoelastic behaviour of such materials. Considerable research efforts characterizing the nonlinear viscoelastic nature of materials have been recorded since then in the literature. A number of nonlinear theories of viscoelasticity have been proposed and many practical problems have been analysed, particularly those pertaining to the steady flow of viscoelastic media. The dynamical aspects of the nonlinear viscoelastic theory have also been dealt with by a number of researchers, but interest in the study of wave propagation in nonlinear viscoelastic materials did not develop until recently (Chapter 8). Further research efforts have also been directed towards incorporating the coupled mechanical and nonmechanical effects in a nonlinear viscoelastic medium (e.g. Christensen, 1971). In this, some results, for instance, have been obtained on the behaviour of heat-conducting viscoelastic materials, but no significant problems have been solved to study other coupled effects. Theorems on the uniqueness of the existence of solutions for initial and boundary value problems in nonlinear viscoelasticity are still at an early stage.

Although the rational approach to nonlinear viscoelasticity may be considered as has been developed, experimental work concerning the validity of the theory is still lagging behind significantly. This is mainly due to the complexity of the experimental programme required to determine the pertaining material functions even in the one-dimensional case (Lockett, 1965, 1972).

6.2 ILLUSTRATIONS OF NONLINEAR VISCOELASTIC PERFORMANCE OF MATERIALS

We provide below some illustrations of the nonlinear behaviour of a class of viscoelastic materials. The examples cited are taken from the work of Ward and Onat (1963), on oriented polypropylene monofilaments. As will be seen from the discussions below, the performance of this class of nonfilaments is nonlinear. For a nonlinear viscoelastic material the following should be noted.

1. Creep compliance is dependent on the level of loading. Let $\varepsilon(\sigma_0, t)$ denote the strain response in a creep experiment performed under the loading

$$\sigma(t) = 0, \quad t < 0; \quad \sigma(t) = \sigma_0 = \text{constant}, \quad t > 0.$$

   If the material is linear, the constitutive creep response is expressed by equation (2.67), that is

$$\varepsilon(t) = F(t)\sigma_0 \tag{6.1}$$

   where $F(t)$ designates the creep function of the linear material; it is independent of the loading σ_0. On the other hand, for a nonlinear viscoelastic material, the corresponding form of the constitutive relation (6.1) may be written as

$$\varepsilon(\sigma_0, t) = F(\sigma_0, t)\sigma_0 \tag{6.2}$$

   where the dependence of the creep function $F(\sigma_0, t)$ on the loading σ_0 is indicated.

   Figure 6.1 shows the creep function (compliance) versus time from creep experiments conducted by Ward and Onat (1963) on oriented polypropylene monofilaments under five different load levels. It is seen from the figure that except for short times ($t \approx 10^2$ s) and only for low intensities of loading (67.6 gf and 129.6 gf) the creep compliances do not coincide. This is in contradiction to the response predicted by (6.1), but in agreement with (6.2). This indicates that the

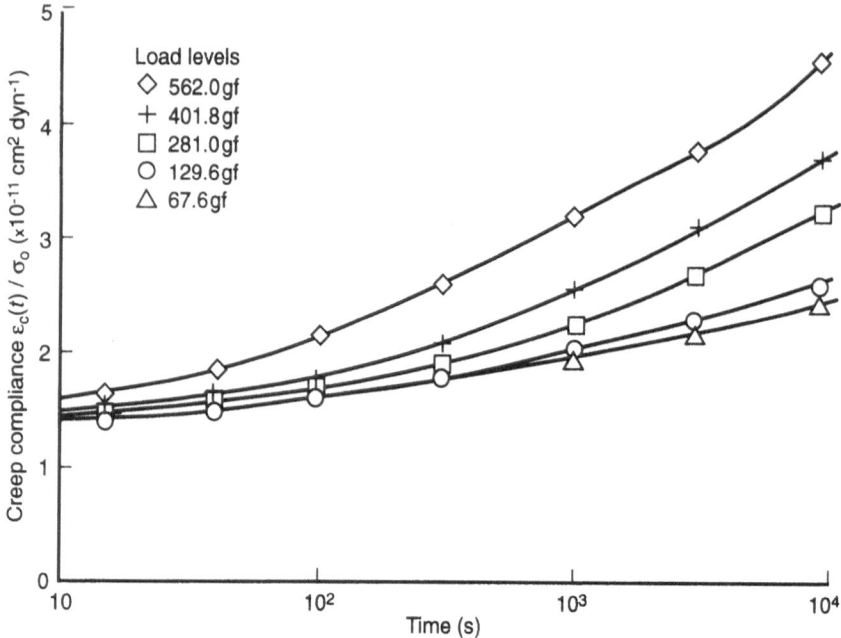

Fig. 6.1 Creep compliance curves of oriented polypropylene under different levels of load. (Reprinted with permission from *J. Mech. Phys. Solids,* **11**, Ward, I. M. and Onat, E. T., Nonlinear mechanical behaviour of oriented polypropylene, copyright (1963), Pergamon Press Ltd.)

material tested is nonlinear in behaviour. To affirm the above characteristics of the nonlinear behaviour of oriented polypropylene monofilaments, Fig. 6.2 (Ward and Onat, 1963) shows the creep compliance against the level of loading for various fixed values of the time t. For a linear material, the creep compliance, as indicated by (6.1), is independent of the loading σ_0 and these curves would be accordingly horizontal. However, in Fig. 6.2, the curves resemble more parabolas, thus indicating the dependence of the creep compliance on the loading level. It is only for small times and low load levels that the tested oriented polypropylene monofilaments may be considered as linear.

2. Recovery compliance is dependent on the level of loading. Denoting by $\varepsilon_R(t, t_1)$ the strain response in a recovery test, then the linear constitutive equation in case of recovery is written in correspondence to (6.1) as

$$\varepsilon_R(t, t_1) = F_R(t - t_1)\sigma_0. \tag{6.3}$$

In (6.3), $F_R(t - t_1)$ is the linear recovery compliance; it is independent of the loading level. The corresponding recovery equation to (6.3) in the nonlinear case may be expressed as

$$\varepsilon_R(\sigma_0, t, t_1) = F_R(\sigma_0, t - t_1)\sigma_0. \tag{6.4}$$

indicating that the recovery compliance, in the nonlinear case, is dependent on

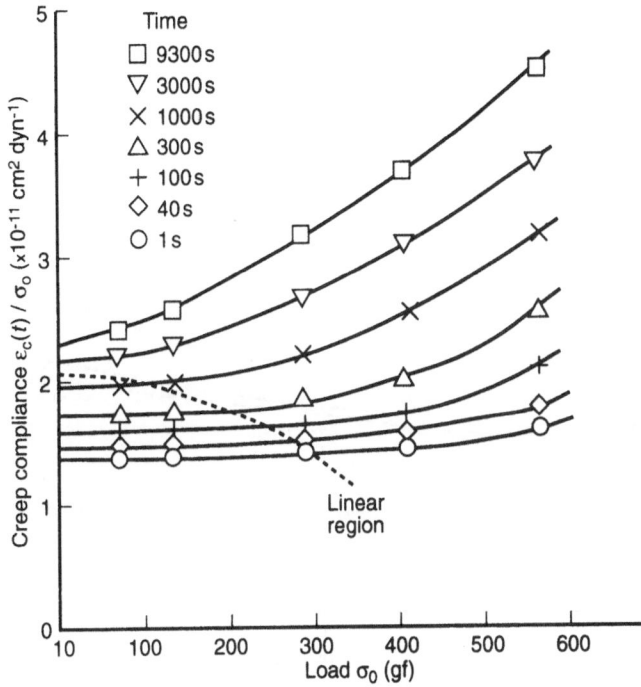

Fig. 6.2 Creep compliance $\varepsilon_c(t)/\sigma_0$ versus load σ_0 for various times (oriented polypropylene monofilaments). (Reprinted with permission from *J. Mech. Phys. Solids*, **11**, Ward, I. M. and Onat, E. T., Nonlinear mechanical behaviour of oriented polypropylene, copyright (1963), Pergamon Press Ltd.)

the level of loading. This is illustrated in Figs. 6.3 and 6.4 (after Ward and Onat, 1963) showing, respectively, the creep compliance versus the time and loading level for oriented polypropylene monofilaments. Thus, the prediction of the linear theory concerning the recovery of the tested oriented polypropylene monofilaments could be contested in view of the observed behaviour of the recovery compliance in Figs. 6.3 and 6.4, similar to what was discussed earlier concerning the creep compliance.

3. With regard to successive creep and recovery, for a nonlinear viscoelastic material, creep and recovery curves do not coincide for a given level of loading. The above remarks concerning the dependence of both the creep and recovery compliances, of a nonlinear material, on the level of loading may be further supported by observing the behaviour of the two compliances at different levels of loading. This is demonstrated in Fig. 6.5, taken from Ward and Onat (1963), concerning the creep and recovery of oriented polypropylene monofilaments. Time on the abscissa of this figure refers, in view of (6.1), to time t for creep curves and, according to (6.3), to time $(t - t_1)$ for recovery curves whereby t_1 is taken as 9.3×10^3 s for the data presented in the figure. According to the linear theory,

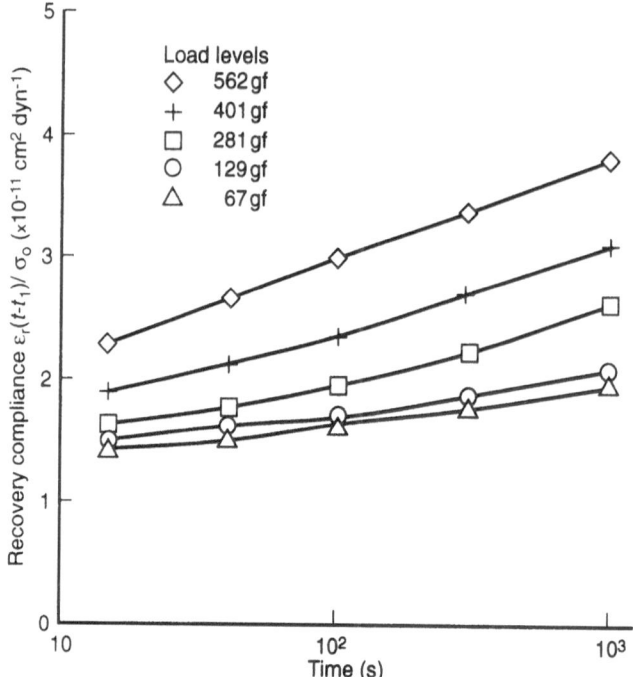

Fig. 6.3 Recovery curves of oriented polypropylene monofilaments for different levels of load (time of loading, 9.3×10^3 s). (Reprinted with permission from *J. Mech. Phys. Solids*, **11**, Ward, I. M. and Onat, E. T., Nonlinear mechanical behaviour of oriented polypropylene, copyright (1963), Pergamon Press Ltd.)

constitutive equations (6.1) and (6.3), creep and recovery curves of Fig. 6.5 would coincide for a given level of loading. As shown in this figure, however, these curves do not coincide for the tested monofilaments, except for low load levels, indicating thereby the nonlinearity of these materials. In Fig. 6.5, it is interesting to observe that the instantaneous recovery, as well as the short time recovery, is larger than the initial creep response and such difference increases with increasing level of loading (Ward and Onat, 1963). For a review of the viscoelastic properties of polymers in general, the reader is referred to Ferry (1960) and Turner (1971, 1973), amongst others.

6.3 OBJECTIVITY PRINCIPLE

In the treatment of a nonlinear theory of viscoelasticity, one must deal with a nonlinear analysis of the deformation process and, hence, nonlinear kinematical quantities need to be introduced. In this context, it is necessary that the particular measures of deformation, or strain quantities, involved in the nonlinear theory have the proper coordinate invariance characteristics. This would also apply to all other

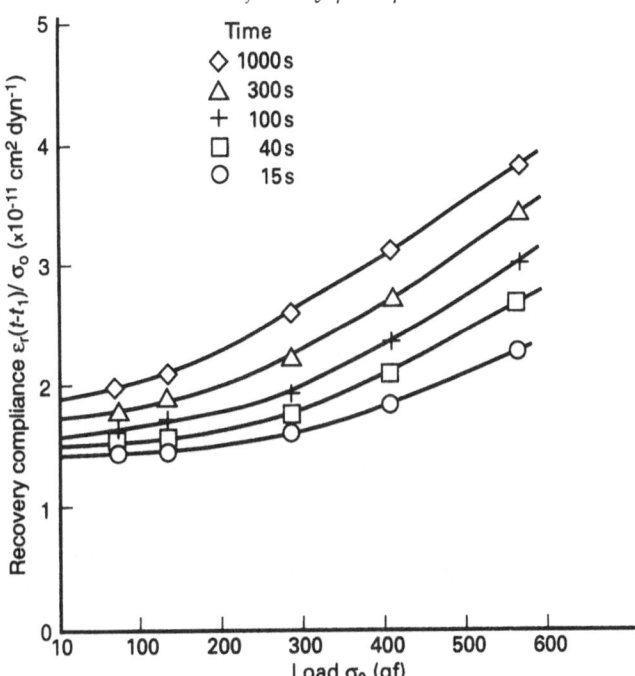

Fig. 6.4 Recovery compliance $\varepsilon_r(t - t_1)/\sigma_0$ against load σ_0 (time of loading 9.3×10^3 s) for various times for oriented polypropylene monofilaments. (Reprinted with permission from J. *Mech. Phys. Solids*, **11**, Ward, I. M. and Onat, E. T., Nonlinear mechanical behaviour of oriented polypropylene, copyright (1963), Pergamon Press Ltd.)

physical quantities and their analysis in the sense that they must be independent of the particular frame of reference with respect to which they are described (e.g. Christensen, 1971). This requirement is often referred to as the principle of objectivity, or material frame indifference (Truesdell and Noll, 1965). This principle is a reflection of the fact that material deformation occurs independently of the observer, i.e. frame independent as we discussed earlier in the introduction to Chapter 1.

As dealt with in Chapters 2 and 4, a viscoelastic material is generally defined as one for which the stress tensor depends on the entire history of the deformation measures involved. A 'simple viscoelastic material', on the other hand, could be characterized (Noll, 1958), by a dependence of the stress tensor σ_{ij} at time t on the deformation gradient $F_{rs}(t - \tau)$ only, describing the motion of the body from time τ up to the time t. Accordingly, a simple material may be defined by the constitutive equation

$$\sigma_{ij}(t) = \pi_{ij} \underset{\tau = -\infty}{\overset{t}{}} [F_{rs}(t - \tau)] \tag{6.5}$$

where π_{ij} is a particular functional of the deformation gradient tensor function $F_{rs}(t - \tau)$ taken with respect to the configuration at time t where F_{rs} is defined by

$$F_{rs}(t - \tau) = x_{m,r}x_{m,s} \tag{6.6}$$

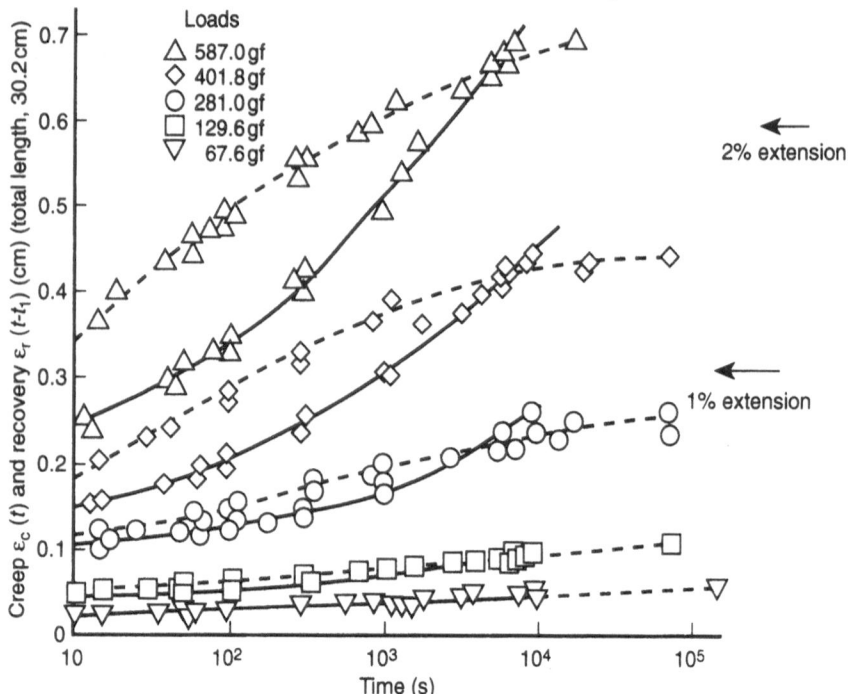

Fig. 6.5 Successive creep (———) and recovery (- - -) of oriented polypropylene conditioned fibres. (Reprinted with permission from *J. Mech. Phys. Solids*, **11**, Ward, I. M. and Onat, E. T., Nonlinear mechanical behaviour of oriented polypropylene, copyright (1963), Pergamon Press Ltd.)

where $x_{m,r}$ is the deformation gradient at time t at the spatial point x_m taken with respect to the configuration at time t.

The deformation gradient tensor $\mathbf{F}$, i.e. $F_{rs}(t - \tau)$, appearing in (6.5) may be decomposed (Truesdell and Noll, 1965) as follows (also Lockett, 1972):

$$\mathbf{F} = \mathbf{RU} = \mathbf{VR} \tag{6.7}$$

where $\mathbf{R}$ is an orthogonal matrix and $\mathbf{U}$ and $\mathbf{V}$ designate positive-definite symmetric matrices. In other words, the deformation measure $\mathbf{F}$ may be regarded, from a physical point of view, as a rigid rotation $\mathbf{R}$ of the continuous body that is either preceded by a pure stretch $\mathbf{U}$ or followed by a pure rotation $\mathbf{V}$.

A restriction on the form of the functional $\boldsymbol{\pi}$ of (6.5) follows from the application of the principle of objectivity introduced earlier. Consequently, it can be shown (Truesdell and Noll, 1965) that the functional relation (6.5) may be expressed in the form

$$\sigma_{ti}(t) = F_{rs}(t) \overset{t}{\underset{-\infty}{G_{ij}}} [U_{rs}(t - \tau)] F_{rs}^{\mathsf{T}}(t) \tag{6.8}$$

where $\mathbf{F}^{\mathsf{T}}$ is the transpose of $\mathbf{F}$ and $\mathbf{G}$ is a new arbitrary functional (Lockett, 1972).

Alternatively, by combining (6.7) and (6.8) it can be shown that

$$\boldsymbol{\sigma}(t) = \mathbf{R}(t) \overset{t}{\underset{-\infty}{\mathbf{H}}} [\boldsymbol{U}(t - \tau)]\mathbf{R}^{\mathrm{T}}(t) \tag{6.9}$$

where again **H** is a new arbitrary functional of the stretch tensor $\mathbf{U}(t - \tau)$. The constitutive relation (6.9) implies, through the principle of objectivity, that the current stress depends only on the current value of the rotation tensor **R**, but, meantime, it depends on the entire past history of the stretch **U**. Expression (6.9) can be simplified further by considering the relation

$$\mathbf{R}^{\mathrm{T}}\mathbf{R} = \mathbf{R}\mathbf{R}^{\mathrm{T}} = \mathbf{I}. \tag{6.10}$$

since **R** is an orthogonal matrix and where **I** is the identity matrix. Accordingly, the constitutive relation (6.9) may be written in the following form:

$$\boldsymbol{\Sigma}(t) = \mathbf{R}^{\mathrm{T}}\boldsymbol{\sigma}\mathbf{R} = \overset{t}{\underset{-\infty}{\mathbf{H}}} [\mathbf{U}(t - \tau)] \tag{6.11}$$

where $\boldsymbol{\Sigma}$ is the stress matrix evaluated in a coordinate system which is related to the rigid-body-rotation component of the deformation. Equation (6.11) implies, through the principle of objectivity, that a permissible form of the constitutive relation is one in which the stress tensor $\boldsymbol{\Sigma}$ is expressed as an arbitrary functional of the deformation measure **U**. Accordingly, **U** is an 'objective' deformation measure. However, it is not the only objective measure of deformation as any simple function of **U** would also be objective. For instance, the right Cauchy–Green strain measure **C** and the finite strain measure **E** defined, respectively by

$$\mathbf{C} = \mathbf{F}^{\mathrm{T}}\mathbf{F} = \mathbf{U}^{\mathrm{T}}\mathbf{R}^{\mathrm{T}}\mathbf{R}\mathbf{U} = \mathbf{U}^{\mathrm{T}}\mathbf{U} = \mathbf{U}^2 \tag{6.12}$$

and

$$2\mathbf{E} = \mathbf{C} - \mathbf{I} = \mathbf{F}^{\mathrm{T}}\mathbf{F} - \mathbf{I} \tag{6.13}$$

are both objective.

In view of the relations (6.12) and (6.13) between **C, F** and **U**, the following alternative forms of the constitutive equation (6.11) can be expressed

$$\boldsymbol{\Sigma}(t) = \overset{t}{\underset{\tau = -\infty}{\mathbf{f}}} [\mathbf{C}(t - \tau)] \tag{6.14}$$

and

$$\boldsymbol{\Sigma}(t) = \overset{t}{\underset{\tau = -\infty}{\mathbf{J}}} [\mathbf{E}(t - \tau)] \tag{6.15}$$

where **f** and **J** are constitutive functionals. Their detailed forms need to be established for the particular viscoelastic material under consideration. For an isotropic material, **f** and **J** are supposed to be isotropic functionals (Lockett, 1972).

Under the conditions that (6.15) is invertible, its corresponding equation in the creep case may be expressed as

$$\mathbf{E}(t) = \mathop{\mathbf{Y}}_{\tau = -\infty}^{t} [\Sigma(t - \tau)],\tag{6.16}$$

which expresses the current strain as a functional of stress history.

Following Lockett (1965, 1972), it is convenient to consider (6.15) and (6.16) at the same time by utilizing

$$\mathbf{Q}(t) = \mathop{\mathbf{F}}_{\tau = -\infty}^{t} [\mathbf{P}(t - \tau)]\tag{6.17}$$

where $\mathbf{Q}$ and $\mathbf{P}$ can be interpreted as either

$$\mathbf{Q} = \Sigma = \mathbf{R}^{\mathsf{T}}\boldsymbol{\sigma}\mathbf{R}, \quad \mathbf{P} = \mathbf{E}\tag{6.18}$$

or

$$\mathbf{Q} = \mathbf{E}, \quad \mathbf{P} = \Sigma = \mathbf{R}^{\mathsf{T}}\boldsymbol{\sigma}\mathbf{R}\tag{6.19}$$

6.4 CHARACTERIZATION OF NONLINEAR VISCOELASTIC BEHAVIOUR

The purely mechanical, one-dimensional response of a nonlinear material with memory may be characterized by the following constitutive assumptions (Schuler, Nunziato and Walsh, 1973).

1. At a given material point, the stress at time t is determined by the entire history of the strain. This may be expressed in the following manner:

$$\boldsymbol{\sigma}(t) = \mathbf{L}[\boldsymbol{\varepsilon}(t, \tau)], \quad 0 \leq (\tau, t) < \infty\tag{6.20}$$

   where $\mathbf{L}$ is a constitutive functional of strain.

2. The material may exhibit 'fading memory'. This implies (Coleman and Noll, 1960, 1961) that the functional $\mathbf{L}$ in (6.20) has certain smoothness properties. In this context, Schuler, Nunziato and Walsh (1973) defined the norm of strain histories by

$$\| \boldsymbol{\varepsilon}(t, \tau) \| = \| \boldsymbol{\varepsilon}(t - \tau) \| = \left[|\boldsymbol{\varepsilon}(t)|^2 + \int_0^\infty h^2(\tau)|\boldsymbol{\varepsilon}(t, \tau)|^2 \, d\tau \right]^{1/2}\tag{6.21}$$

where $h(t)$ is a continuous monotonically decreasing, square-integrable function. Thus, the material behaviour would be more influenced by the events occurring in the recent past than those occurring in the distant past. With reference to (6.21), the set of all histories with finite norm forms a Hilbert space. The constitutive functional $\mathbf{L}[\boldsymbol{\varepsilon}(t, \tau)]$, in (6.20), is assumed to be defined in this space and to be twice continuously differentiable with respect to the present value of the strain (Schuler, Nunziato and Walsh, 1973).

Based on the assumed smoothness of the functional **L**, Coleman and Noll (1961) have shown that the response behaviour of nonlinear material with memory may be approximated in the case of small relative strain histories $\Delta \boldsymbol{\varepsilon}(t, \tau)$ by the linear theory (Chapter 2); also, Coleman, Gurtin and Herrera (1965) should be referred to.

6.5 CONSTITUTIVE EQUATIONS OF NONLINEAR VISCOELASTIC MATERIALS

Based on the thermomechanical theory of materials with memory due to Coleman (1964) (Chapter 5), Christensen (1971) presented a general formulation of the nonlinear viscoelastic response of materials. Although Coleman's approach is not restricted to isothermal conditions, Christensen, for simplification of the presentation, limited his derivation to the isothermal case.

Christensen's derivation of the stress constitutive equation is based on the combination of the local balance of energy expression and the local entropy production. This is, then, joined with the definition of the norm in the collection of past histories within the associated fading memory hypothesis.

As previously discussed in the context of the linear theory (Chapters 2 and 5), the concept of memory in viscoelastic materials provides an expression of the postulate that the current value of a field variable, such as the stress and stored energy, depends not only on the current value of the stimulus but also on the past history of such stimulus. Accordingly, the stress constitutive relation may be expressed through the following functional expression:

$$\sigma_{ij}(t) = f_{ij}[x_{i,L}(t - \tau), \quad x_{j,L}(t)], \quad 0 \le (\tau, t) \le \infty, \tag{6.22}$$

in which σ_{ij} is the Cauchy stress tensor and $x_{i,L}$ is the deformation gradient defined by

$$x_{i,L}(X_K, t) = \frac{\partial x_i(X_K, t)}{\partial X_L}. \tag{6.23}$$

Similarly, the stored energy per unit mass, A, may be expressed as functional of the past history of deformation

$$A(t) = \mathop{\phi}_{\tau=0}^{\infty} [E_{KL}(t - \tau), E_{KL}(t)], \quad 0 \le (\tau, t) \le \infty. \tag{6.24}$$

The function (6.24) is assumed to be a continuous function of the nonlinear strain history $E_{KL}(t, \tau)$. Further, both the strain tensor E_{KL} and its first derivative are assumed to be continuous. This is in contrast to the requirements of the linear derivations where only the input histories were needed to be continuous. In view of the objectivity principle, it is necessary that all mathematical descriptions of physical quantities and events be independent of the particular frame of reference employed.

Combining the local balance of energy equation and the local entropy production inequality, both under isothermal conditions, one obtains

$$-\rho \dot{A}(t) + \sigma_{ij}(t)d_{ij}(t) \ge 0. \tag{6.25}$$

In equation (6.25), ρ is the mass density and $d_{ij}(t)$ is the deformation rate tensor defined by

$$d_{ij}(t) = v_{i,j}(t) + v_{j,i}(t) \tag{6.26}$$

where

$$v_i(t) = \dot{x}_i(t, X_K). \tag{6.27}$$

To be able to use the stored energy per unit mass, A, from (6.24) in (6.25), it is necessary to obtain its time derivative $\dot{A}$. In order to accomplish this, Christensen (1971) proposed that the nonlinear strain function $E_{KL}(s)$, $s = t - \tau$, constituted a history. In this case, the norm in the collection of histories is defined by

$$\|E\| = \left[\int_0^{\infty} E_{KL}(t - \tau) E_{KL}(t - \tau) h^2(\tau) \, d\tau \right]^{1/2} \tag{6.28}$$

where $h(\tau)$ is, as mentioned earlier in connection with (6.21), a monotonically decreasing function of the time parameter τ.

The collection of histories with a finite norm, such as (6.28), constitutes a Hilbert space. It is assumed that the stored energy function $\underset{\tau=0}{\overset{\infty}{\phi}} (\cdot)$ (equation (6.24)), is the Fréchet differential in the Hilbert space corresponding to $h(\tau)$. Accordingly, the derivative of the stored energy per unit mass can be defined as

$$\dot{A}(t) = \frac{\partial}{\partial E_{KL}(t)} \underset{\tau=0}{\overset{\infty}{\phi}} [E_{KL}(t - \tau), E_{KL}(t)] \dot{E}_{KL}$$

$$+ \delta \underset{\tau=0}{\overset{\infty}{\phi}} \left[E_{KL}(t - \tau), E_{KL}(t) \Big| \frac{dE_{KL}(t - \tau)}{d\tau} \right], \quad 0 \leq (\tau, t) \leq \infty. \tag{6.29}$$

This is on the understanding that the strain rate history $\dot{E}_{KL}(t), -\infty \leq t \leq \infty$, is continuous. Combining (6.25) and (6.29), the former equation can be written as

$$\left\{ \sigma_{ij}(t) - \rho \frac{\partial}{\partial E_{KL}(t)} \underset{\tau=0}{\overset{\infty}{\phi}} [E_{KL}(t - \tau), E_{KL}(t)] x_{i,K}(t) x_{j,K}(t) \right\} d_{ij}(t) \rho \underset{\tau=0}{\overset{\infty}{\Lambda}} (\cdot) \geq 0,$$

$$0 \leq (\tau, t) \leq \infty, \tag{6.30}$$

where

$$\underset{\tau=0}{\overset{\infty}{\Lambda}} (\cdot) = \delta \underset{\tau=0}{\overset{\infty}{\phi}} \left[E_{KL}(t - \tau), E_{KL}(t) \Big| \frac{dE_{KL}(t - \tau)}{dt} \right], \quad 0 \leq (\tau, t) \leq \infty. \tag{6.31}$$

is a measure of the rate of dissipation of energy in the viscoelastic material during the deformation process.

In order for (6.30) to be satisfied for a given deformation history, it is necessary that the coefficient of d_{ij} vanishes. This leads to

$$\sigma_{ij}(t) = \rho(t) \frac{\partial}{\partial E_{KL}(t)} \underset{\tau=0}{\overset{\infty}{\phi}} [E_{KL}(t - \tau), E_{KL}(t)] x_{i,K}(t) x_{j,K}(t). \tag{6.32}$$

This leaves (6.30) as

$$\rho(t) \bigwedge_{\tau=0}^{\infty} (\cdot) \geq 0 \qquad (6.33)$$

which states that the rate of dissipation of energy in the viscoelastic material, during deformation, is nonnegative.

Equation (6.33) is the stress constitutive equation in isothermal nonlinear viscoelasticity which can be expressed, with reference to (6.24), as

$$\sigma_{ij} = \rho \frac{\partial A}{\partial E_{EL}} x_{i,K} x_{J,L}. \qquad (6.34)$$

Similar formulations to the above have been followed by Coleman and Mizel (1967).

Christensen (1971) presented further reduction of the above formulation to the case of infinitesimal deformation. For this purpose, the following expression for the free energy was used:

$$\rho_0 A(t) = \frac{1}{2} \int_{-\infty}^{t} \int_{-\infty}^{t} \lambda(2t - \tau - \eta) \frac{\partial E_{KK}(\tau)}{\partial \tau} d\tau \, d\eta$$
$$+ \int_{-\infty}^{t} \int_{-\infty}^{t} \mu(2t - \tau - \eta) \frac{\partial E_{KL}(\tau)}{\partial \tau} \frac{\partial E_{KL}(\eta)}{\partial \eta} d\tau \, d\eta \qquad (6.35)$$

in which ρ_0 is the mass density in the reference configuration and $\lambda(\cdot)$, $\mu(\cdot)$ are time-dependent relaxation functions. Hence, based on equation (6.35) and by employing the general nonlinear stress constitutive equation (6.32), one obtains

$$\sigma_{ij} = \frac{\rho}{\rho_0} \left[\delta_{KL} \int_{-\infty}^{t} \lambda(t - \tau) \frac{\partial E_{mm}(\tau)}{\partial \tau} d\tau \right.$$
$$\left. + 2 \int_{-\infty}^{t} \mu(t - \tau) \frac{\partial E_{KL}(\tau)}{\partial \tau} \right] x_{i,K} x_{j,L} \, d\tau. \qquad (6.36)$$

This constitutive expression can be rewritten, within the infinitesimal deformation theory and with the assumption that the mass density is constant, as

$$\sigma_{ij}(t) = \delta_{ij} \int_{-\infty}^{t} \lambda(t - \tau) \frac{\partial \varepsilon_{kk}(\tau)}{\partial \tau} d\tau + 2 \int_{-\infty}^{t} \mu(t - \tau) \frac{\partial \varepsilon_{ij}(\tau)}{\partial \tau} d\tau \qquad (6.37)$$

in which ε_{ij} is the small strain tensor. Thus, expression (6.37) represents the reduced form of the general nonlinear viscoelastic stress constitutive equation (6.32) within the context of the infinitesimal deformation theory.

Coleman and Noll (1961) derived a special nonlinear theory for solids known as finite linear viscoelasticity. This theory, however, restricts the deformation to be slowly changing in the recent past. In this manner, the current value of stress can be determined by linearly integrating the deformation history with reference to the current configuration. However, in contrast to the infinitesimal theory, the integrating functions in the constitutive integrals are nonlinear functions of the current state of deformation. Green and Rivlin (1957), based on an earlier work by Rivlin and Ericksen (1955), discussed the form of the constitutive equations governing the deformation of a class of materials with memory. It was assumed that the stress in an element of the material depends not only on the deformation gradients in the element at the instant of time considered but also on those of previous instants of time. The limitations imposed on the constitutive formulation by the principle of objectivity (frame indifference) were examined. This was carried out by first considering that the stress depends on the deformation gradients at a number of discrete times up to the instant of measurement. Then, the number of instants of time was considered to increase indefinitely so that the expression for the stress becomes a functional of the deformation gradients. In this analysis, it was found that the form invariance of the constitutive equation under a relation of the physical system leads naturally to a particular form of dependence of the stress on the deformation gradients at the instant of measurement. In this work, Green and Rivlin (1957) assumed that the expression for the stress as a functional of the deformation gradients at times up to and including the instant of measurement is continuous. Green, Rivlin and Spencer (1959) do not, however, make this assumption, but assume that the stress has arbitrary polynomial dependence on the deformation gradients at the instant of measurement, while its functional dependence on the deformation gradients at times preceding the instant of measurement is continuous. Under these assumptions, the limitations imposed by the isotropy of the material in its undeformed state on the form the constitutive relations have been dealt with by these authors.

Green and Rivlin (1960) assumed a constitutive equation in the form of implicit relations between the stress, its time derivatives and gradients of displacement, velocity, acceleration, etc., at a number of instants of time within a specified time interval. In this work, the limitations on the form of the constitutive equation resulting from the requirement that it is unaltered by the imposition on the body of an additional arbitrary rigid rotation, etc., are discussed.

Green and Rivlin (1957, 1960) and Green, Rivlin and Spencer (1959) presented a three-dimensional theory for the nonlinear viscoelastic material based on a multiple integral formulation whereby the output is expressed as a polynomial expansion in linear function of the input histories. In this context, the most general form for an initially isotropic material to third order (Meixner and König, 1958; Lockett, 1965, 1972) is

$$\mathbf{Q}(t) = \int_{-\infty}^{t} (\mathbf{I}\psi_1 Y_1 + \psi_2 \mathbf{M}_1) \, d\tau_{(1)}$$

$$+ \iint_{-\infty}^{t} (\mathbf{I}\psi_3 Y_1 Y_2 + \mathbf{I}\psi_4 Y_{12} + \psi_5 Y_1 \mathbf{M}_2 + \psi_6 \mathbf{M}_1 \mathbf{M}_2) \, d\tau_{(2)}$$

$$+ \iiint_{-\infty}^{t} (\mathbf{I}\psi_7 Y_{123} + \mathbf{I}\psi_8 Y_1 Y_{23} + \psi_9 Y_1 Y_2 \mathbf{M}_3$$

$$+ \psi_{10} Y_{12} \mathbf{M}_3 + \psi_{11} Y_1 \mathbf{M}_2 \mathbf{M}_3 + \psi_{12} \mathbf{M}_1 \mathbf{M}_2 \mathbf{M}_3) \, d\tau_{(3)} \qquad (6.38)$$

where

$$\mathbf{M}_\alpha = \dot{\mathbf{P}}(\tau_\alpha),$$

$$Y_\alpha = \mathrm{tr}(\mathbf{M}_\alpha),$$

$$Y_{\alpha\beta} = \mathrm{tr}(\mathbf{M}_\alpha \mathbf{M}_\beta), \quad Y_{\alpha b\gamma} = \mathrm{tr}(\mathbf{M}_\alpha \mathbf{M}_\beta \mathbf{M}_\gamma),$$

$$d\tau_{(N)} = d\tau_1 \, d\tau_2 \ldots d\tau_N.$$

In the constitutive equation (6.38), matrices $\mathbf{Q}$ and $\mathbf{P}$ represent, interchangeably, nonlinear measures of input and output (stress and strain) and the superposed dot denotes a material time derivative. Material functions ψ_1 and ψ_2 are functions of one variable s_1, $\psi_3, \ldots, \psi_6$ are functions of s_1 and s_2 and $\psi_7, \ldots, \psi_{12}$ are functions of s_1, s_2 and s_3, where $s_\alpha = t - \tau_\alpha$. These functions characterize the mechanical behaviour of given material. It has been shown by Lockett (1965) that the following symmetries may be assumed without loss of generality:

$$\psi_3, \psi_4, \psi_6, \psi_9, \psi_{10} = s(1, 2), \psi_7, = s(1, 2, 3),$$

$$\psi_8, \psi_{11} = s(2, 3) \text{ and } \psi_{12} = s(1, 3) \tag{6.39}$$

where $s(i, j)$ denotes symmetry in the ith and jth arguments. A creep response formulation is given by defining $\mathbf{P} = \Sigma$ and $\mathbf{Q} = \mathbf{E}$ and a stress relaxation formulation is given by setting $\mathbf{P} = \mathbf{E}$ and $\mathbf{Q} = \Sigma$ where stress measure Σ and strain $\mathbf{E}$ are defined by

$$\Sigma = \sigma = \begin{cases} \mathbf{R}^\mathsf{T}\sigma\mathbf{R} \\ \mathbf{F}^\mathsf{T}\sigma\mathbf{F} \end{cases}, \quad \mathbf{F} = \mathbf{R}\mathbf{U}, \quad 2\mathbf{E} = \mathbf{U}^2 - \mathbf{I} = \mathbf{F}^\mathsf{T}\mathbf{F} - \mathbf{I}. \tag{6.40}$$

In (6.40), the decomposition of the deformation gradient $\mathbf{F}$ expresses the idea that the deformation may be considered to be a pure stretch $\mathbf{U}$ followed by a rigid rotation $\mathbf{R}$, as discussed previously in section 6.3 (equation (6.7)).

In the general theory, the integral polynomial (6.38) is of infinite order. It is, however, assumed that the terms of the first three orders, written explicitly in (6.38), are the dominant ones. The constitutive equation (6.38) may also be written in terms of the classical strain ε since

$$E = \varepsilon + \tfrac{1}{2}\varepsilon^2. \tag{6.41}$$

This is, of course, with the understanding that the associated material functions will be different in this case.

The one-dimensional constitutive relation corresponding to (6.38) can be written as

$$Q(t) = \int_{-\infty}^{t} \phi_1(t - \tau_1)\dot{P}(\tau_1) \, d\tau_{(1)}$$

$$+ \iint_{-\infty}^{t} \phi_2(t - \tau_1, t - \tau_2)\dot{P}(\tau_1)\dot{P}(\tau_2) \, d\tau_{(2)}$$

$$+ \iiint_{-\infty}^{t} \phi_3(t - \tau_1, t - \tau_2, t - \tau_3)\dot{P}(\tau_1)\dot{P}(\tau_2)\dot{P}(\tau_3) \, d\tau_{(3)} \tag{6.42}$$

where $d\tau_{(n)} = d\tau_1, d\tau_2, \ldots, d\tau_n$ and $\phi_n(\cdot)$, $n = 1, 2, \ldots$, are material functions which can be assumed without loss of generality to be symmetric with respect to their indicated arguments. In the constitutive equation (6.42), $P(t)$ and $Q(t)$ denote, alternatively, stress and strain components in a one-dimensional relation that may be referred to as input and output functions. Thus, if $P(t)$ designates a strain measure, then ϕ_n are stress-relaxation functions of order n for the mode of deformation relevant to the definitions of P and Q. Alternatively, if $P(t)$ denotes stress, then ϕ_n represent creep functions.

The first-order term in (6.42) represents the usual term in the constitutive relation of linear viscoelasticity in which the function $\phi_1(\cdot)$ may be determined from a creep experiment. $\phi_1(\cdot)$ is, then, the response to a homogeneous unit stimulus applied at time $t = 0$.

The set of loading programmes required for the determination of the material functions of the constitutive equation (6.42) is given by Lockett (1965, 1972). For the general three-dimensional case, equation (6.38), Lockett considered the simplification when $P_{ij} = 0$ for $i \neq j$ and it was noted that all of the material functions except ψ_{12} may be determined by experiment. The corresponding set of test programmes required to determine $\psi_1, \ldots, \psi_{11}$ is summarized in Lockett's paper. Lifshitz and Kolsky (1967) have confirmed, however, that these tests are exceedingly difficult to perform accurately. On the other hand, an expression of the form (6.42) was employed by Ward and Onat (1963) to interpret the results of uniaxial experiments for the case of oriented polypropylene. In this context, it was indicated by the latter authors that the experimental results may be described adequately by considering only the first- and third-order terms in the constitutive equation (6.42), and some information was concluded about the material functions involved. Turner (1966), on the other hand, was unable to model creep data conveniently for a polypropylene using the constitutive equation (6.42). Meantime, Gittus (1975) presented an argument that the constitutive equation (6.42) could be invalid for the case of cyclic loading. In this regard, Gittus (1975) shows that, for the same creep load, equation (6.42) can result in creep rate in compression greater than that in tension. Pipkin and Rogers (1968), however, describe the application of the multiple-integral representation of the constitutive equation (6.42), to be effective if the viscoelastic response is weakly nonlinear. Strong nonlinearity would, then, require the inclusion of higher-order multiple integrals.

In order to avoid the enormous number of tests required to determine the material functions in (6.38) and (6.42), as described in Lockett's programme (Lockett, 1965), Pipkin and Rogers (1968) give an integral series representation of the constitutive equation. The first term in this series is a single integral with a nonlinear integrand, and may be determined by a single-step creep or relaxation test. The second term generalizes the exact representation to two-step inputs and so forth. The nth term of the series is obtained by the difference, if any is observed, between the experimental data for n-step tests and the prediction based on $(n - 1)$-step data. The series has the characteristic that it terminates at the nth term whenever the input (stress or strain) is an n-step history. In a study to demonstrate the consistency of the Green–Rivlin (Green and Rivlin, 1957, 1960) representation, Ward and Wolfe (1966) report the results of four different two-stage uniaxial load programmes. An analysis of some of the data presented in their paper could be used to demonstrate the predictive capability of Pipkin and Rogers' one-step extrapolation (Pipkin and Rogers, 1968). Stafford (1969), however, criticizes Pipkin and Rogers' model in that their second function $C_2(\sigma_1, t - t_1, \sigma_2, t - t_2)$ requires the evaluation of a function of four variables which may appear impractical in view of the inability of Lifshitz and Kolsky (1967) to evaluate a function of two variables $K(t_1, t_2)$. Further, Lifshitz and Kolsky's results indicate that the two-step response is a separable function of time and strain. In this context, Stafford (1969) argues that, if such separability were to hold for all step tests, Pipkin and Roger's constitutive equation would reduce to a superposition form. As an alternative simplification of (6.38), Pipkin (1964) developed the following four-function stress-relaxation relation for incompressible materials:

$$\left.\begin{matrix} \mathbf{R}^T\boldsymbol{\sigma}\mathbf{R} \\ \mathbf{F}^T\boldsymbol{\sigma}\mathbf{R} \end{matrix}\right\} = \int_{-\infty}^{t} \psi_2 \mathbf{M}_1 \, d\tau_{(1)} + \iint_{-\infty}^{t} \psi_6 \mathbf{M}_1 \mathbf{M}_2 \, d\tau_{(2)}$$

$$+ \iiint_{-\infty}^{t} (\psi_{10} Y_{12} \mathbf{M}_3 + \psi_{12} \mathbf{M}_1 \mathbf{M}_2 \mathbf{M}_3) \, d\tau_{(3)} \tag{6.43}$$

Lockett and Stafford (1969) have indicated that a similar equation to (6.43) can be derived for the creep case. It has been demonstrated in the latter reference that the resulting formulation can be further reduced to a relationship involving only three material functions. The experimental programmes for evaluating the material functions involved are described by Lockett and Stafford (1969). The number of tests required indicates, however, that such a constitutive relationship may still be too complex for routine practical applications. Stafford (1969) presented a further step toward the simplification of the constitutive equation in nonlinear viscoelasticity. In his paper, Stafford directed particular attention to reducing the evaluation of material functions to simple (one-step) creep or stress-relaxation tests. For this purpose, an isothermal theory of (initially) isotropic materials was primarily considered in the explicit form of a third-order integral polynomial (6.42), taking into consideration the following three simplifications.

1. *Physical linearity.* In this case, Stafford (1969) considers that the material functions would be expressed as a sum of single-argument functions. This approximation produces a response which is linearly dependent on the input history but is nonlinearly dependent on the present value of the input. The application of this approximation to (6.42) yields

$$Q = \int_0^t [\psi_1(t - \tau) + f_2(t - \tau)P(t) + f_3(t - \tau)P^2(t)]\dot{P}(t) \, d\tau. \tag{6.44}$$

The constitutive equation (6.44) is interpreted by Stafford (1969) to be directly related to the theory of finite viscoelasticity due to Coleman and Noll (1961).

2. *Superposition.* In this type of simplification, Stafford (1969) defines the multiple-argument material function in terms of a single function of one argument. This produces a response in which each term has a linear dependence on the history of an input measure; however, each measure is a nonlinear function of the input. Hence, the resulting constitutive equation is in the form of a convolution integral and may be regarded to be of the Boltzmann superposition type for large deformations. The application of the superposition postulate to (6.42) yields the constitutive equation

$$Q = \int_0^t [\psi_1(t - \tau) + f_2(t - \tau)2P(\tau) + f_3(t - \tau)3P^2(\tau)\dot{P}(\tau)] \, d\tau. \qquad (6.45)$$

Findley and Lai (1966) and Lai and Findley (1968) have also considered a type of nonlinear superposition with equivalent resulting equations.

3. *Product nonlinearity.* Here, in view of the one-step test requirement, Stafford (1969) postulates that the second- and higher-order material functions are separable and may be expressed as the product

$$\psi_i(x_1, x_2, \ldots, x_n) = f_i^{(1)}(x_1) f_i^{(2)}(x_2) \ldots f_i^{(n)}(x_n) \qquad (6.46)$$

where functions $f_i^{(j)}$ are related by the symmetry conditions (6.39). Lai and Findley (1968) and Findley and Onaran (1968) have also proposed forms similar to (6.46).

The application of the postulate of product nonlinearity to (6.42) produces the result

$$Q = \int_0^t \psi_1(t - \tau)\dot{P}(\tau) \, d\tau + \left[\int_0^t f_2^{1/2}(t - \tau)\dot{P}(\tau) \, d\tau \right]^2$$

$$+ \left[\int_0^t f_3^{1/3}(t - \tau)\dot{P}(\tau) \, d\tau \right]^3 \qquad (6.47)$$

where the square and cube roots of f_2 and f_3 are introduced to make the equations identical for one input tests. Thus, for an input $P = P_0 H(t)$,

$$Q = P_0 \psi_1(t) + P_0^2 f_2(t) + P_0^3 f_3(t). \qquad (6.48)$$

However, the response of the three equations (6.44), (6.45) and (6.47) to general loads is quite different. Further details concerning the above constitutive equations and the validity of their application can be found in Stafford's (1969) paper. In general, the selection of a constitutive equation must be guided by long-range information such as that obtained from a multi-step test. In this regard, Pipkin and Rogers (1968) observed that multistep inputs constitute a more critical test of the applicability of a constitutive equation. The tests required for two- and three-dimensional equations offer, however, additional complications. In this regard, it has been indicated by Lockett and Stafford (1969) that, when the material functions are completely symmetric, uniaxial tests may be sufficient. Thus, only the superposition form (6.45) can be determined from uniaxial tests, while the linear and product forms, (6.44) and (6.27) respectively, would require biaxial tests (Stafford, 1969).

An alternative approach to the multiple-integral formulation of the constitutive equation is to adopt the reduced-time constitutive equations such as the one developed by Schapery (1966),

$$\varepsilon = g(\sigma) + f(\sigma) \int_0^t D[Z(t) - Z(\tau)] \frac{\mathrm{d}}{\mathrm{d}\tau} \left[\frac{\sigma}{C(\sigma)} \right] \mathrm{d}\tau, \quad Z(t) = \int_o^t \frac{B(\sigma)}{C(\sigma)} \, \mathrm{d}\tau, \text{ (6.49)}$$

where $D(\cdot)$ is the linear creep compliance. In equation (6.49), the material functions $B(\sigma)$, $C(\sigma)$ and $f(\sigma)$ as well as $g(\sigma)$ and $D(\cdot)$ are determinable from creep and recovery data for several stress levels and times. Here, the evaluation of $B(\sigma)$, for instance, would require only one two-step test, one three-step test, etc., while an integral polynomial theory would require at least n two-step tests, $\frac{1}{2}n^2$ three-step tests, etc. In this regard, Stafford (1969) noted that choosing $B(\sigma)$ to be linear in σ gives excellent results for PVC. However, as indicated in this reference, the reduced-time equations will only follow data in which the functional form of the response is unchanged by additional steps; if it does change, however, one must resort to the more general integral polynomial theories.

6.6 COUPLED EFFECTS IN NONLINEAR VISCOELASTICITY

As we indicated in Chapter 5, in the course of introducing the thermomechanical theory of viscoelasticity, viscoelastic materials are sensitive to thermal variations as well as other environmental effects. Such effects should be coupled with the mechanical effects if a comprehensive constitutive theory were to be developed. Unfortunately, it is only recently that researchers in the field started to incorporate such coupling of effects in the prediction of the response behaviour of nonlinear viscoelastic materials. As we discussed earlier in section 6.3, concerning the objectivity principle, coupled thermomechanical behaviour of nonlinear materials should be treated in the light of the invariance principle of constitutive theory.

Koh and Eringen (1963) proposed a thermomechanical theory of viscoelasticity that includes nonlinear effects. The theory introduces the formalism for the analysis of the response behaviour of nonlinear heat-conducting viscoelastic materials. It postulates a constitutive formulation based essentially on the two sets of equations

$$\boldsymbol{\sigma} = \boldsymbol{\sigma}(\mathbf{c}^{-1}, \mathbf{d}, \mathbf{b})$$

and   (6.50)

$$\mathbf{h} = \mathbf{h}(\mathbf{c}^{-1}, \mathbf{d}, \mathbf{b})$$

where $\boldsymbol{\sigma}$, the stress tensor, is considered a function of the three variables: $\mathbf{c}^{-1}$, the deformation tensor defined by

$$\mathbf{c}^{-1} = C_{kl}^{-1} = x_{k,M} x_{M,l}; \tag{6.51}$$

d, the deformation rate tensor defined by

$$\mathbf{d} = 2d_{kl} = v_{k,l} + v_{l,k};$$ (6.52)

b, the temperature gradient bi-vector defined by

$$\mathbf{b} = b_{kl} = \varepsilon_{klm}\theta_{,m}$$ (6.53)

where θ is the temperature and ε_{klm} is the skew-symmetric alternating tensor. Meantime, the second equation of (6.50) relates the heat flux bi-vector **h** to the same independent variables $\mathbf{c}^{-1}$, **d**, **b**. Accordingly, **h** is defined by

$$\mathbf{h} = h_{kl} = \varepsilon_{klm}q_m.$$ (6.54)

In addition to the above constitutive equations (6.50)–(6.54), the following field equations would be required in the course of the development of the theory:

● continuity equation

$$\dot{\rho} + \rho v_{k,k} = 0;$$ (6.55)

● equations of motion

$$\sigma_{ij,i} + \rho\chi_j = \rho a_j,$$
$$\sigma_{ij} = \sigma_{ji}.$$ (6.56)

In (6.55) and (6.56), v_k are the velocity components, ρ is the density, χ_j are the body forces components and a_j are the components of the acceleration.

Further, the conservation of energy must also be satisfied. The latter principle is expressed by

$$\rho\dot{e} = \sigma_{ij}d_{ji} - q_{j,j} + \rho r$$ (6.57)

where e is the specific internal energy and r is the supply of energy.

With the application of a caloric equation of state, it was concluded by Koh and Eringen (1963) that the principle of conservation of energy (6.57) reduces to a heat conduction equation which for the incompressible case simplifies to

$$\rho K\dot{\theta} = \sigma_{ij}d_{ji} - q_{j,j} + \rho r$$ (6.58)

where K is the specific heat.

The theory of Koh and Eringen (1963) demonstrates the mechanism by which thermomechanical coupling could be introduced to other purely mechanical theories. These authors considered further the modification of their theory to situations involving hygrosteric materials (e.g. Noll, 1955) and to a particular subclass of Rivlin–Ericksen viscoelastic fluids (for instance, Rivlin and Ericksen, 1955).

6.7 EXPERIMENTAL CHARACTERIZATION OF NONLINEAR VISCOELASTIC MATERIALS

6.7.1 One-dimensional nonlinear behaviour

The basis for the experimental work discussed below is the Green–Rivlin multiple integral response relation (6.42),

$$
Q(t) = \int_{-\infty}^{t} \phi_1(t - \tau_1)\dot{P}(\tau_1)\, d\tau_{(1)}
$$

$$
+ \iint_{-\infty}^{t} \phi_2(t - \tau_1, t - \tau_2)\dot{P}(\tau_1)\dot{P}(\tau_2)\, d\tau_{(2)}
$$

$$
+ \iiint_{-\infty}^{t} \phi_3(t - \tau_1, t - \tau_2, t - \tau_3)\dot{P}(\tau_1)\dot{P}(\tau_2)\dot{P}(\tau_3)\, d\tau_{(3)} \qquad (6.59)
$$

The identification of the terms appearing in the above equation is mentioned in section 6.5 where we dealt with this equation. If sufficient data are available to determine the nonlinear material functions ϕ_1, ϕ_2 and ϕ_3, then equation (6.59) can be used to predict the response of the material to other input histories.

The experimental plan that would be required to determine the values of these material functions has been discussed by Ward and Onat (1963), Hadley and Ward (1965), Lockett (1965, 1972) and Lifshitz and Kolsky (1967).

Lockett (1972) dealt comprehensively with the development of an experimental programme to determine the material functions in (6.59) for both creep and stress relaxation. In this context, with reference to (6.59) and depending on whether the input function $P(\tau)$ is a stress or strain, the corresponding experiment is of creep or stress-relaxation type respectively.

Following Lockett (1972), one may consider the simple test

$$
P(t) = pH(t), \qquad (6.60)
$$

where p is a positive constant input and $H(t)$ is the Heaviside step function. In view of (6.59), the response corresponding to (6.59) is

$$
Q(t) = p\phi_1(t) + p^2\phi_2(t, t) + p^3\phi_3(t, t, t). \qquad (6.61)
$$

That is, the measured response $Q(t)$ from one experiment alone is insufficient to determine any of the three material functions ϕ_1, ϕ_2 and ϕ_3. However, if the simple test (6.60) is performed for three different values of p, this will provide three output equations, similar to (6.61), for determining $\phi_1(t)$, $\phi_2(t, t)$ and $\phi_3(t, t, t)$. Accordingly, the material function ϕ_1 is known over the time scale of the experiment, but ϕ_2 and ϕ_3 are known only when their arguments are all equal. To establish further values of ϕ_2 and ϕ_3, Lockett (1972) proposed the additional test

$$P(t) = p_1 H(t) + p_2 H(t - k) \qquad (6.62)$$

in which p_1, p_2 and k are positive constants. This produces, in view of (6.59), the following response:

$$\begin{aligned}
Q(t) = {} & 2 p_1 p_2 \phi_2(t, \, t - k) \\
& + 3 p_1 p_2 [\, p_1 \phi_3(t, \, t, \, t - k) + p_2 \phi_3(t, \, t - k, \, t - k)\,] \\
& + \cdots
\end{aligned} \qquad (6.63)$$

with the assumption that ϕ_α are symmetric functions with respect to their arguments. Thus, if three experiments are carried out corresponding to three independent combinations of the parameters p_1 and p_2, then, according to (6.63), the measured responses would allow the determination of the following values of the material functions:

$$\phi_2(t, \, t - k), \quad \phi_3(t, \, t, \, t - k), \quad \phi_3(t, \, t - k, \, t - k). \qquad (6.64)$$

The number N of the values of k which are required will be determined by the experimental data (on an *ad hoc* basis) during the course of the experiment. Lockett (1972) estimated, however, that N could be of the order of 10 for reasonable characterization.

Similar remarks may apply concerning the value of ϕ_3 obtained from (6.64). In this case, ϕ_3 may be determined along lines in its three-dimensional argument space. The two values of ϕ_3 mentioned in (6.64) lead to the determination of this function when two of its arguments are equal. Accordingly, as a result of $3N$ experiments of the type (6.62), the function ϕ_2 is determined completely and ϕ_3 is established except when its arguments are all unequal. In order to obtain these unknown values of ϕ_3, one may need to apply a triple-step input of the form

$$P(t) = p_1 H(t) + p_2 H(t - k) + p_3 H(t - l) \qquad (6.65)$$

where p_1, p_2, p_3, k and l are positive constants. Corresponding to the input (6.65), equation (6.59) gives the following output:

$$Q(t) = 6 p_1 p_2 p_3 \phi_3(t, \, t - k, \, t - l) + \cdots. \qquad (6.66)$$

Accordingly, (6.65) leads to the determination of the material function $\phi_3(t, \, t - k, \, t - l)$ and repetition of the experiment for other values of k and l would establish ϕ_3 completely. In this regard, because of symmetry of ϕ_3, one may not wish to repeat the experiment for N values of both k and l. It could be sufficient to consider $0 < k < l$ (both k and l positive). Thus, the total number of experiments of the form (6.66) is $\frac{1}{2} N(N - 1)$.

The complete set of experiments that may be required for the determination of ϕ_α, as discussed above, is given by Lockett (1965, 1972).

6.7.2 Three-dimensional nonlinear behaviour

All of the experiments that are required for three-dimensional characterization of the nonlinear viscoelastic behaviour would generally involve multiple-step inputs and are, in essence, similar to the experiments introduced in the foregoing concerning the one-dimensional behaviour. In the three-dimensional characterization, however, it would be necessary (Lockett, 1965, 1972) to apply different input steps of the input matrix and to measure more than one component of the output matrix.

The basis for the experimental programme here is the Green–Rivlin three-dimensional constitutive relation (6.38). Further to the previous discussion, equation (6.38) is an implicit relation as the matrix $\mathbf{P}$ appears in the definitions (6.18) and (6.19) of $\mathbf{P}$ and $\mathbf{Q}$. In this, following Lockett (1965, 1972), it would be natural to consider first situations in which $\mathbf{R} = \mathbf{I}$. Such situations would arise when $P_{ij} = 0$ for $i \neq j$. In the experimental programme of Lockett (1965), introduced briefly below, it is shown that all the material functions $\psi_1, \cdots, \psi_{11}$ can be determined from experiments of this type.

Lockett (1965, 1972) introduced an experimental programme for the determination of the material parameters appearing in the Green and Rivlins' three-dimensional constitutive equation (6.38). In this constitutive equation, terms up to and including those of third order in the input matrix are retained. Since $\mathbf{P}$ and $\mathbf{Q}$ appearing in (6.38) may be given either of the interpretations (6.18) and (6.19), the experiments referred to below may be of either the creep or the stress-relaxation type. In (6.38), the following material functions are identified:

- ψ_1 and ψ_2 are functions of $t - \tau_1$;
- $\psi_3, \cdots, \psi_6$ are functions of $t - \tau_1$ and $t - \tau_2$;
- $\psi_7, \cdots, \psi_{12}$ are functions of $t - \tau_1$, $t - \tau_2$ and $t - \tau_3$.

That is, an experimental programme would be required to determine

- two functions of a single variable,
- four functions of two variables and
- six functions of three variables.

The set of experimental inputs required to determine the functions $\psi_1, \cdots, \psi_{12}$ is given by Lockett (1965, 1972). The reader is also referred to McGuirt and Lianis (1967).

REFERENCES

Christensen, R. M. (1971) *Theory of Viscoelasticity*, Academic Press, New York.

Coleman, B. D. (1964) Thermodynamics of materials with memory. *Arch. Ration. Mech. Anal.*, **17**, 1–46.

Coleman, B. D. and Mizel, V. (1967) A general theory of dissipation in materials with memory. *Arch. Ration. Mech. Anal.*, **27**, 255–74.

Coleman, B. D. and Noll, W. (1960) An approximation theorem for functionals with applications in continuum mechanics. *Arch. Ration. Mech. Anal.*, **6**, 355–70.

Coleman, B. D. and Noll, W. (1961) Foundations of linear viscoelasticity. *Rev. Mod. Phys.*, **33**, 239–49.

Coleman, B. D., Gurtin, M. E. and Herrera, R. I. (1965) Waves in materials with memory, I. The velocity of one-dimensional shock and acceleration waves. *Arch. Ration. Mech. Anal.*, **19**, 1–19.

Ferry, J. D. (1960) *Viscoelastic Properties of Polymers*, Wiley, New York.

Findley, W. N. and Lai, J. S. Y. (1966) *Brown University Rep. EMRL-27.*

Findley, W. N. and Onaran, K. (1968) Product form of kernel functions for nonlinear viscoelasticity of PVC under constant rate stressing. *Trans. Soc. Rheol.*, **12**(2), 217–42.

Gittus, J. (1975) *Creep, Viscoelasticity and Creep Fracture in Solids*, Wiley, New York.

Green, A. E. and Rivlin, R. S. (1957) The mechanics of non-linear materials with memory, part I. *Arch. Ration. Mech. Anal.*, **1**, 1–21.

Green, A. E. and Rivlin, R. S. (1960) The mechanics of non-linear materials with memory, III. *Arch. Ration. Mech. Anal.*, **4**, 387–404.

Green, A. E., Rivlin, R. S. and Spencer, A. J. M. (1959) The mechanics of non-linear materials with memory, II. *Arch. Ration. Mech. Anal.*, **3**, 82–90.

Hadley, D. W. and Ward, I. M. (1965) Nonlinear creep and recovery behaviour of polypropylene fibres. *J. Mech. Phys. Solids*, **13**, 397–411.

Koh, S. L. and Eringen, A. C. (1963) On the foundations of nonlinear thermoviscoelasticity. *Int. J. Eng. Sci.*, **1**, 199–229.

Lai, J. S. Y. and Findley, W. N. (1968) Stress relaxation of nonlinear viscoelastic material under uniaxial strain. *Trans. Soc. Rheol.*, **12**(2), 259–80.

Lifshitz, J. M. and Kolsky, H. (1967) Nonlinear viscoelastic behaviour of polyethylene. *Int. J. Solid Struct.*, **3**, 383–97.

Lockett, F. J. (1965) Creep and stress-relaxation experiments for non-linear materials. *Int. J. Eng. Sci.*, **3**, 59–75.

Lockett, F. J. (1972) *Nonlinear Viscoelastic Solids*, Academic Press, New York.

Lockett, F. J. and Stafford, R. O. (1969) On special constitutive relations in nonlinear viscoelasticity. *Int. J. Eng. Sci.*, **7**, 917–30.

McGuirt, C. W. and Lianis, G. (1967) Constitutive equations for viscoelastic solids under finite uniaxial and biaxial deformations. *Trans. Soc. Rheol.*, **14**, 117–34.

Meixner, J. and König, H. (1958) Zür Theorie der Linearen dissipativen Systeme. *Rheol. Acta*, **1**(2–3), 190–3.

Noll, W. (1955) On the continuity of the solid and the fluid states. *J. Ration. Mech. Anal.*, **4**, 3–81.

Noll, W. (1958) A mathematical theory of the mechanical behaviour of continuous media. *Arch. Ration. Mech. Anal.*, **2**, 197–226.

Pipkin, A. C. (1964) Small finite deformations of viscoelastic solids. *Rev. Mod. Phys. Solids*, **36**, 1034–41.

Pipkin, A. C. and Rogers, T. G. (1968) A nonlinear integral representation of viscoelastic behaviour. *J. Mech. Phys. Solids*, **16**, 59–72.

Rivlin, R. S. and Ericksen, J. L. (1955) Stress–deformation relations for isotropic materials. *J. Ration. Mech. Anal.*, **4**, 323–425.

Schapery, R. A. (1966) A theory of nonlinear thermoviscoelasticity based on irreversible thermodynamics, in Proc. 5th US Natl. Congr. on Applied Mechanics, pp. 511–30.

Schapery, R. A. (1974) Viscoelastic behaviour and analysis of composite materials, in *Mechanics of Composite Materials*, Vol. 2 (ed. G. Sendickj), Academic Press, New York, Chap. 4, pp. 85–168.

Schuler, K. W., Nunziato, J. W. and Walsh, E. K. (1973) Recent results in non-linear viscoelastic wave propagation. *Int. J. Solid Struct.*, **9**, 1237–81.

Stafford, R. O. (1969) On mathematical forms for the material functions in nonlinear viscoelasticity. *J. Mech. Phys. Solids*, **17**, 339–58.

Truesdell, C. and Noll, W. (1965) The nonlinear field theories of mechanics, in *Encyclopedia of Physics*, Vol. III/3 (ed. S. Flügge), Springer, Berlin.

Turner, S. (1966) The strain response of plastics to complex stress histories. *Polym. Eng. Sci.*, **6**, 306–16.

Turner, S. (1971) Creep studies on plastics, in *Applied Polymer Symposium 17*, Wiley, New York, pp. 25–43.

Turner, S. (1973) *Mechanical Testing of Plastics*, Iliffe, London.

Ward, I. M. and Onat, E. T. (1963) Nonlinear mechanical behaviour of oriented polypropylene. *J. Mech. Phys. Solids*, **11**, 217–29.

Ward, I. M. and Wolfe, J. M. (1966) The nonlinear mechanical behaviour of polypropylene fibres under complex loading programmes. *J. Mech. Phys. Solids*, **14**, 131–40.

Weissenberg, K. (1948) Abnormal substances and abnormal phenomena of flow, in *Proc. Int. Congr. on Rheology*, pp. I-29–I-46.

FURTHER READING

Beham, P. P. and Hutchinson, S. J. (1971) A comparison of constant and complex creep loading programs for several thermoplastics. *Polym. Eng. Sci.*, **11**(4), 335–43.

Bernstein, B., Kearsley, E. A. and Zapas, L. J. (1963) A study of stress relaxation with finite strain. *Trans. Soc. Rheol.*, **7**, 391–410.

Bychawski, Z. and Fox, A. (1967) Generalized creep function and the problem of inversion in the theory of nonlinear viscoelasticity. *Bull. Acad. Pol. Sci.*, **15**, 297–304.

Christensen, R. M. (1968) On obtaining solutions in nonlinear viscoelasticity. *J. Appl. Mech.*, **35**, 129–33.

Distéfano, N. and Todeschini, R. (1973a) Modeling, identification and prediction of a class of nonlinear viscoelastic materials, part I. *Int. J. Solids Struct.*, **9**, 805–18.

Distéfano, N. and Todeschini, R. (1973b) Modeling, identification and prediction of a class of nonlinear viscoelastic materials, part II. *Int. J. Solids Struct.*, **9**, 1431–8.

Drescher, A. and Kwaszczynska, K. (1970) An approximate description of nonlinear viscoelastic materials. *Int. J. Nonlinear Mech.*, **5**, 11–22.

Eringen, A. C. (1962) *Nonlinear Mechanics of Continua*, McGraw-Hill, New York.

Findley, W. N. (1976) *Creep and Relaxation of Nonlinear Viscoelastic Media*, North-Holland, Amsterdam.

Findley, W. N. and Lai, J. S. Y. (1967) A modified superposition principle applied to creep of nonlinear viscoelastic material under abrupt changes in state of combined stress. *Trans. Soc. Rheol.*, **11**(3), 361–80.

Findley, W. N. and Stanely, C. A. (1968) Non-linear combined stress creep experiments on rigid polyurethane foam with application to multiple integral and modified superposition theory. *ASTM J. Mater.*, **3**, 916–49.

Findley, W. N., Lai, J. S. and Onaran, K. (1976) *Creep and Relaxation of Non-linear Viscoelastic Materials*, North-Holland, Amsterdam.

Gottenberg, N. G., Bird, J. O. and Agrawal, G. L. (1969) An experimental study of a nonlinear viscoelastic solid in uniaxial tension. *J. Appl. Mech.*, **36**, 558–64.

Gradowczyk, M. H. (1969) On the accuracy of the Green–Rivlin representation for viscoelastic materials. *Int. J. Solid Struct.*, **5**, 873–7.

Haddad, Y. M. (1987) Un modèle de réponse en viscoelasticité non-linéaire de matériaux. *Res Mech.*, **20**, 235–53.

Haddad, Y. M. (1988) On the theory of the viscoelastic solid. *Res Mech.*, **25**, 225–59.

Hlavacek, B., Seyer, F. A. and Stanislav, J. (1973) Quantitative analogies between the linear and the nonlinear viscoelastic functions. *Can. J. Chem. Eng.*, **51**, 412–7.

Huang, N. C. and Lee, E. H. (1966) Nonlinear viscoelasticity for short time ranges. *J. Appl. Mech.*, **33**, 313–21.

Lai, J. S. Y. and Findley, W. N. (1968) Prediction of uniaxial stress relaxation from creep of nonlinear viscoelastic material. *Trans. Soc. Rheol.*, **12**(2), 243–57.

Lai, J. S. Y. and Findley, W. N. (1969) Behaviour of nonlinear viscoelastic material under simultaneous stress relaxation in tension and creep in torsion. *J. Appl. Mech.*, **36**, 22–7.

Lockett, F. J. and Turner, S. (1971) Nonlinear creep of plastics. *J. Mech. Phys. Solids*, **19**, 201–14.

Lubliner, J. (1967) Short-time approximation in nonlinear viscoelasticity. *Int. J. Solid Struct.*, **3**, 513–20.

McGuirt, C. W. and Lianis, G. (1969) Experimental investigation of nonlinear, non-isothermal viscoelasticity. *Int. J. Eng. Sci.*, **7**, 579–99.

Molinari, A. (1973) Relation between creep and relaxation in nonlinear viscoelasticity. *C.R. Acad. Sci., Sci. Math.*, **277**, 621–3.

Nakada, O. (1960) Theory of nonlinear responses. *J. Phys. Soc. Jpn*, **15**, 2280–8.

Neis, V. V. and Sackman, J. L. (1967) An experimental study of a nonlinear material with memory. *Trans. Soc. Rheol.*, **11**(3), 307–33.

Ng, T. H. and Williams, H. L. (1986) Stress–strain properties of linear aromatic polyesters in the nonlinear viscoelastic range. *J. Appl. Polym. Sci.*, **32**, 4883–96.

Nolte, K. G. and Findley, W. N. (1970) A linear compressibility assumption for the multiple integral representation of nonlinear creep of polyurethane. *J. Appl. Mech.*, **37**, 441–8.

Nolte, K. G. and Findley, W. N. (1971) Multiple step, nonlinear creep of polyurethane predicted from constant stress creep by three integral representations. *Trans. Soc. Rheol.*, **15**(1), 111–33.

Nolte, K. G. and Findley, W. N. (1974) Approximation of irregular loading by intervals of constant stress rates to predict creep and relaxation of polyurethane by three integral representation. *Trans. Soc. Rheol.*, **18**(1), 123–43.

Onaran, K. and Findley, W. N. (1965) Combined stress–creep experiments on a nonlinear viscoelastic material to determine the kernel functions for a multiple integral representation of creep. *Trans. Soc. Rheol.*, **9**(2), 299–327.

Onaran, K. and Findley, W. N. (1971) Experimental determination of some kernel functions in the multiple integral method for nonlinear creep of polyvinylchloride. *J. Appl. Mech.*, **38**, 30–8.

Peters, L. (1955) A note on nonlinear viscoelasticity. *Textile Res. J.*, **25** (March), 262–5.

Pipkin, A. C. and Rivlin, R. S. (1961) Small deformations superposed on large deformations in materials with fading memory. *Arch. Ration. Mech. Anal.*, **8**, 297–308.

Pipkin, A. C. and Rogers, T. G. (1968) A nonlinear integral representation for viscoelastic behaviour. *J. Mech. Phys. Solids*, **16**, 59–72.

Robotonov, Y. N., Papernik, L. K. and Stepanychev, E. I. (1973). Application of nonlinear theory of heredity to the description of time effects in polymeric materials. *Polym. Mech.*, **7**, 63–73.

Schapery, R. A. (1969) On the characterization of nonlinear viscoelastic materials. *Polym. Eng. Sci.*, **9**(4), 295–310.

Scholtens, B. J. R. and Bodit, H. C. (1986) Nonlinear viscoelastic analysis of uniaxial stress–strain measurements of elastomers at constant stretching rates. *J. Rheol.*, **30**(2), 301–12.

Smart, J. and Williams, J. G. (1972) A comparison of single-integral nonlinear viscoelasticity theories. *J. Mech. Phys. Solids*, **20**, 313–24.

Stouffer, D. C. and Wineman, A. S. (1972) Constitutive representation for nonlinear aging, environmental-dependent viscoelastic materials. *Acta Mech.*, **13**, 31–53.

Ting, E. C. (1971) Approximations in nonlinear viscoelasticity. *Int. J. Eng. Sci.*, **9**, 995–1006.

Turner, S. (1966) The strain response of plastics to complex stress histories. *Polym. Eng. Sci.*, **6**, 306–16.

Yanas, I. V. and Haskell, V. C. (1971) Utility of the Green–Rivlin theory in polymer mechanics. *J. Appl. Phys.*, **42**, 610–13.

Yuan, H. and Lianis, G. (1972) Experimental investigation of nonlinear viscoelasticity in combined finite torsion–tension. *Trans. Soc. Rheol.*, **16**, 615–33.

7

Numerical analysis in viscoelasticity

7.1 INTRODUCTION

As discussed earlier, in Chapters 2 and 4, two functions are significant in the description of the linear viscoelastic behaviour of materials, i.e. the creep function and the relaxation function. Quite generally, they are expressed, in view of Boltzmann's superposition principle, by relations (4.4) and (4.9). Recalling the latter two relations in the same order, one may write that

$$\varepsilon(t) = \mathbf{F}(t)\sigma(0) + \int_0^t \mathbf{F}(t - \tau)\dot{\sigma}(\tau) \, d\tau \qquad (7.1)$$

and

$$\sigma(t) = \mathbf{R}(t)\varepsilon(0) + \int_0^t \mathbf{R}(t - \tau)\dot{\varepsilon}(\tau) \, d\tau \qquad (7.2)$$

where $\varepsilon(t)$ and $\sigma(t)$ are the time-dependent strain and stress tensors, respectively. In (7.1), $\mathbf{F}(t - \tau)$ is the creep function; it is a monotonically increasing function of time. In (7.2), the function $\mathbf{R}(t - \tau)$ is the relaxation function, which is usually a decreasing function of time.

In order to account for a nonlinearity in the viscoelastic response behaviour, Boltzmann's response relations above may be modified, following Distéfano (1970, 1971) and Distéfano and Todeschini (1973a, b), to read for the uniaxial case as follows:

$$\varepsilon(t) = g[\sigma(t), l_1, l_2, \cdots] + \int_0^t h[\sigma(\tau), b_1, b_2, \cdots]F(t - \tau) \, d\tau \qquad (7.3)$$

and

$$\sigma(t) = g_1[\varepsilon(t), k_1, k_2, \cdots] + \int_0^t h_1[\varepsilon(\tau), c_1, c_2, \cdots]R(t - \tau) \, d\tau \qquad (7.4)$$

in which $\varepsilon(t)$ and $\sigma(\tau)$ are the scalar components of strain and stress, respectively, at any time t in the sample subjected to uniaxial loading. In the above relations, the functions $g(\cdot)$ and $g_1(\cdot)$ correspond to the instantaneous elastic response in a parametric form, whereas the functions $h(\cdot)$ and $h_1(\cdot)$ account for the nonlinear hereditary effects and are also given in a parametric form. In both of these relations, unknown constants $l_1, l_2, \cdots, k_1, k_2, \cdots, b_1, b_2, \cdots$ and $c_1, c_2, \cdots$ are involved. The choice of the functions $g(\cdot)$, $g_1(\cdot)$, $h(\cdot)$ and $h_1(\cdot)$ is guided by a qualitative knowledge of the viscoelastic behaviour of the material. The functions $F(\cdot)$ and $R(\cdot)$ are, respectively, the creep and relaxation functions of the material for the uniaxial case. These functions may be assumed (Haddad and Tanary, 1987) to satisfy an Nth-order differential equation with constant coefficients of the following form respectively:

$$a_0 F + a_1 F^{(1)} + a_2 F^{(2)} + \cdots + a_{N-1} F^{(N-1)} + F^{(N)} = 0 \tag{7.5a}$$

and

$$q_0 R + q_1 R^{(1)} + q_2 R^{(2)} + \cdots + q_{N-1} R^{(N-1)} + R^{(N)} = 0 \tag{7.5b}$$

where $a_0, a_1, \cdots, a_{N-1}$ and $q_0, q_1, \cdots, q_{N-1}$ are unknown coefficients to be determined.

In the following section, we continue with the identification problem of the creep response. A similar development could, however, be carried out for the characterization of the stress relaxation behaviour of the material as pointed out below in section 7.3.

7.2 CHARACTERIZATION OF THE CREEP RESPONSE

For the simplification of the analysis, we introduce for the instantaneous elastic response a single elastic constant, E, only. Let the constant stress being applied on the material specimen during a creep experiment i be denoted by $\breve{\sigma}_i$ $(i = 1, 2, \cdots, n)$· where n is the number of creep experiments. Hence, the creep response equation (7.3) can be written as

$$\varepsilon_i(t) = E^{-1} \breve{\sigma}_i + h(\breve{\sigma}_i, b_1, b_2, \cdots) \int_0^t F(\tau) \, d\tau \qquad (i = 1, 2, \cdots, n). \tag{7.6}$$

In order, however, to find the unknown constants $b_1, b_2, \cdots$ and to solve for the creep function $F(\tau)$, it is convenient to adopt differential approximation and minimization procedures as outlined below. In this context, the adopted method will be presented first while an actual numerical evaluation for the nonlinear viscoelastic response of a particular viscoelastic material is shown in section 7.4.1.

7.2.1 Approximate solution of the creep function $F(\tau)$

While equation (7.6) has been shown in a general form for the creep of a nonlinear viscoelastic material, a solution of the kernel $F(\tau)$ can only be given in an approximate form. For this purpose, equation (7.7) will be used which is based on experimental observations of the creep behaviour of a class of materials (for instance, Figs. 2.3–2.5).

Thus, denoting the experimental stress or strain by ' '', one may express analytically the experimental creep curves as a function of the applied stress and time, for a number of experiments $i = 1, 2, \cdots, n$, in the following manner:

$$\check{\varepsilon}_i(t) = \check{\varepsilon}_i + G_i \theta_i(t, m_{i1}, m_{i2}, \cdots) \qquad (i = 1, 2, \cdots, n) \tag{7.7}$$

in which $\check{\varepsilon}_i = E^{-1}\check{\sigma}_i$ is the instantaneous experimental strain at $t = 0$, G_i are constants and $\theta_i(t, m_{i1}, m_{i2}, \cdots)$ are functions of time. The shape of the experimental creep curves will suggest the forms of the functions $\theta_i(\cdot)$ of (7.7) for the type of material under consideration. In this equation, the constants G_i and $m_{i1}, m_{i2}, \cdots$ are required to be determined, for each creep experiment i, by a fitting procedure.

In order to use the experimentally available data of the creep of the material by means of equation (7.7), it is evident that the coefficients in (7.5a) have to be found by an optimization procedure such that the kernel $\int_0^t F(\tau)$ in (7.6) is approximated by functions $\theta_i(t, m_{i1}, m_{i2}, \cdots)$ in equation (7.7). Hence, one can write

$$F(\tau) \approx \frac{d}{dt} \theta_i(t, m_{i1}, m_{i2}, \cdots)\Bigg]_0^T \tag{7.8}$$

where the time T represents the total time for each creep experiment. Thus, using this approximation, it is convenient for the subsequent analytical development to identify

$$\Gamma_i = \frac{d}{dt} \theta_i(t, m_{i1}, m_{i2}, \cdots). \tag{7.9}$$

By substituting the values of Γ_i corresponding to the above equation into (7.5a) and using the method of differential approximation (Bellman, 1970), one requires that the functional

$$\sum_{i=1}^{n} \int_0^T (a_0 \Gamma_i + a_1 \Gamma_i^{(1)} + a_2 \Gamma_i^{(2)} + \cdots + a_{N-1} \Gamma_i^{(N-1)} + \Gamma_i^{(N)})^2 \, dt \tag{7.10}$$

be a minimum with respect to all possible choices of the coefficients $a_0, a_1, \cdots, a_{N-1}$.

It is evident that the minimization of expression (7.10) with respect to the coefficients a_j ($j = 0, 1, \cdots, N - 1$), i.e.

$$\sum_{i=0}^{n} \int_0^T \frac{\partial}{\partial a_j} (a_0 \Gamma_i + a_1 \Gamma_i^{(1)} + a_2 \Gamma_i^{(2)} + \cdots + a_{N-1} \Gamma_i^{(N-1)} + \Gamma_i^{(N)})^2 \, dt = 0$$

$$(i = 1, 2, \cdots, n \text{ and } j = 0, 1, \cdots, N - 1), \tag{7.11}$$

leads to a system of N simultaneous linear algebraic equations in the following manner:

$$\sum_{i=1}^{n} \int_{0}^{T} \Gamma_i(a_0 \Gamma_i^{(1)} + a_2 \Gamma_i^{(2)} + \cdots + a_{N-1} \Gamma_i^{(N-1)} + \Gamma_i^{(N)}) \, dt = 0,$$

$$\sum_{i=1}^{n} \int_{0}^{T} \Gamma_i^{(1)}(a_0 \Gamma_i^{(1)} + a_2 \Gamma_i^{(2)} + \cdots + a_{N-1} \Gamma_i^{(N-1)} + \Gamma_i^{(N)}) \, dt = 0,$$

$$\sum_{i=1}^{n} \int_{0}^{T} \Gamma_i^{(2)}(a_0 \Gamma_i^{(1)} + a_2 \Gamma_i^{(2)} + \cdots + a_{N-1} \Gamma_i^{(N-1)} + \Gamma_i^{(N)}) \, dt = 0, \qquad (7.12)$$

$$\vdots$$

$$\sum_{i=1}^{n} \int_{0}^{T} \Gamma_i^{(N-1)}(a_0 \Gamma_i^{(1)} + a_2 \Gamma_i^{(2)} + \cdots + a_{N-1} \Gamma_i^{(N-1)} + \Gamma_i^{(N)}) \, dt = 0,$$

$$(i = 1, 2, \cdots, n).$$

This can be written in a more compact form as

$$\sum_{i=1}^{n} \int_{0}^{T} \Gamma_i^{(j)}(a_0 \Gamma_i + a_1 \Gamma_i^{(1)} + a_2 \Gamma_i^{(2)} + \cdots + a_{N-1} \Gamma_i^{(N-1)}) \, dt = - \sum_{i=1}^{n} \int_{0}^{T} \Gamma_i^{(j)} \Gamma_i^{(N)} \, dt$$

$$(i = 1, 2, \cdots, n \text{ and } j = 0, 1, 2, \cdots, N - 1). \qquad (7.13)$$

The set of equations (7.13) may be expressed in matrix form as

$$[a_j] [x_{jk}] = [x_k]$$

$$(j, k = 0, 1, 2, \cdots, N - 1) \qquad (7.14)$$

where

$$[a_j] = \begin{bmatrix} a_0 \\ a_1 \\ a_2 \\ \vdots \\ a_{N-1} \end{bmatrix}, \qquad (7.15a)$$

$$[x_{jk}] = \begin{bmatrix} \sum_{i=1}^{n} \int_{0}^{T} \Gamma_i \Gamma_i \, dt & \sum_{i=1}^{n} \int_{0}^{T} \Gamma_i \Gamma_i^{(1)} \, dt & \cdots & \sum_{i=1}^{n} \int_{0}^{T} \Gamma_i \Gamma_i^{(N-1)} \, dt \\ \sum_{i=1}^{n} \int_{0}^{T} \Gamma_i^{(1)} \Gamma_i \, dt & \sum_{i=1}^{n} \int_{0}^{T} \Gamma_i^{(1)} \Gamma_i^{(1)} \, dt & \cdots & \sum_{i=1}^{n} \int_{0}^{T} \Gamma_i^{(1)} \Gamma_i^{(N-1)} \, dt \\ \sum_{i=1}^{n} \int_{0}^{T} \Gamma_i^{(2)} \Gamma_i \, dt & \sum_{i=1}^{n} \int_{0}^{T} \Gamma_i^{(2)} \Gamma_i^{(1)} \, dt & \cdots & \sum_{i=1}^{n} \int_{0}^{T} \Gamma_i^{(2)} \Gamma_i^{(N-1)} \, dt \\ \vdots & \vdots & & \vdots \\ \sum_{i=1}^{n} \int_{0}^{T} \Gamma_i^{(N-1)} \Gamma_i \, dt & \sum_{i=1}^{n} \int_{0}^{T} \Gamma_i^{(N-1)} \Gamma_i^{(1)} \, dt & \cdots & \sum_{i=1}^{n} \int_{0}^{T} \Gamma_i^{(N-1)} \Gamma_i^{(N-1)} \, dt \end{bmatrix}$$

$$(7.15b)$$

and

$$[x_k] = - \begin{bmatrix} \sum\limits_{i=1}^{n} \int_{0}^{T} \Gamma_i \Gamma_i^{(N)} \, dt \\[2mm] \sum\limits_{i=1}^{n} \int_{0}^{T} \Gamma_i^{(1)} \Gamma_i^{(N)} \, dt \\[2mm] \sum\limits_{i=1}^{n} \int_{0}^{T} \Gamma_i^{(2)} \Gamma_i^{(N)} \, dt \\[2mm] \vdots \\[2mm] \sum\limits_{i=1}^{n} \int_{0}^{T} \Gamma_i^{(N-1)} \Gamma_i^{(N)} \, dt \end{bmatrix} \tag{7.15c}$$

In principle, the solution of equation (7.14) can be found by a straightforward matrix inversion method, i.e.

$$[a_j] = [x_{jk}]^{-1} [x_k] \tag{7.16}$$

where $[x_{jk}]^{-1}$ is the inverse of the matrix $[x_{jk}]$ as given by equation (7.15b) with the understanding that before the inverse exists the determinant of $[x_{jk}]$ must be nonzero.

Having determined the coefficients of the linear differential equation (7.5a), one solution of this equation may be assumed to be of the form $F(\tau) = \exp(A\tau)$. In order that this solution satisfies the linear differential equation (7.5a), identically, one must have the characteristic equation

$$a_0 + a_1 A + a_2 A^2 + \cdots + a_{N-1} A^{N-1} + A^N = 0. \tag{7.17}$$

This expression yields N roots A_I $(I = 1, 2, \cdots, N)$ which correspond to the N solutions of the original linear differential equation (7.5a). Hence, the general solution can be written as a linear combination of the N solutions in the following form:

$$F(\tau) = \sum_{I=1}^{N} C_I \exp(A_I \tau) \qquad (I = 1, 2, \cdots, N) \tag{7.18}$$

where C_I $(I = 1, 2, \cdots, N)$ are constants which can be determined by an optimization procedure as outlined below.

Integration of both sides of (7.18) with respect to the time t gives

$$\int_{0}^{t} F(\tau) \, d\tau = \sum_{I=1}^{N} D_I [\exp(A_I t) - 1] \qquad (I = 1, 2, \cdots, N) \tag{7.19}$$

in which

$$D_I = \frac{C_I}{A_I} \qquad (I = 1, 2, \cdots, N).$$

By substituting the value of the integrand corresponding to (7.19) into the analytical expression (7.1), it follows that

$$\varepsilon_i(t) = E^{-1}\breve{\sigma}_i + h(\breve{\sigma}_i, b_1, b_2, \cdots) \sum_{I=1}^{N} D_I[\exp(A_I t) - 1]$$

$$(i = 1, 2, \cdots, n \quad \text{and} \quad I = 1, 2, \cdots, N). \tag{7.20}$$

The identification problem of the nonlinear viscoelastic response of the material can now be formalized in the following manner.

Given n independent experimental functionals $\breve{\varepsilon}_i$ ($i = 1, 2, \cdots, n$), equation (7.7), corresponding to applied stresses $\breve{\sigma}_i$, one wishes to find the constants $b_1, b_2, \cdots$ and D_I ($I = 1, 2, \cdots, N$) in the constitutive equation (7.20) such that the functional

$$\text{II}(b_1, b_2, \cdots; D_I) = \sum_{i=1}^{n} \gamma_i \int_0^T [\varepsilon_i(t) - \breve{\varepsilon}_i(t)]^2 \, dt$$

$$(i = 1, 2, \cdots, n \quad \text{and} \quad I = 1, 2, \cdots, N) \tag{7.21}$$

is minimized. For the purpose of minimizing this functional, a quadratic expression (Mikhlin, 1965) of the difference between the theoretical strain $\breve{\varepsilon}_i(t)$ ($i = 1, 2, \cdots, n$), which is given by (7.20), and the experimental strain $\breve{\varepsilon}_i(t)$ as given by (7.7) is used and where γ_i ($i = 1, 2, \cdots, n$) represent suitable positive weighting factors.

Hence, by substituting for $\varepsilon_i(t)$ and $\breve{\varepsilon}_i(t)$ their corresponding expressions from (7.20) and (7.7), respectively, into (7.21) it follows that

$$\text{II}(b_1, b_2, \cdots; D_I) = \sum_{i=1}^{n} \gamma_i \int_0^T \left\{ h(\breve{\sigma}_i, b_1, b_2, \cdots) \sum_{I=1}^{N} D_I[\exp(A_I t) - 1] \right.$$

$$\left. - G_i \theta_i(t, m_{i1}, m_{i2}, \cdots) \right\}^2 dt$$

$$(i = 1, 2, \cdots, n \quad \text{and} \quad I = 1, 2, \cdots, N). \tag{7.22}$$

The minimization of expression (7.22) can be carried out by using a number of numerical optimization techniques as demonstrated in section 7.4.1.

Once the values of the unknown constants of equation (7.22) have been determined by the above procedure, this expression can be used to represent a 'model equation' for the creep response of a real viscoelastic material.

7.3 CHARACTERIZATION OF THE RELAXATION RESPONSE

We propose in this case an analytical development similar to that followed above for the characterization of the creep response. Let the constant strain being applied on the specimen during a relaxation experiment i be denoted by $\breve{\varepsilon}_i$ ($i = 1, 2, \cdots, n$) where n denotes the number of relaxation experiments. Thus, the stress relaxation equation, analogous to (7.6), is written as

$$\sigma_i(t) = E\breve{\varepsilon}_i + h_1(\breve{\varepsilon}_i, c_1, c_2, \cdots) \int_0^t R(\tau) \, d\tau \tag{7.23}$$

whereby the relaxation function $R(\tau)$ is assumed to satisfy an Nth-order differential equation of the form given by (7.5b), namely

$$q_0 R + q_1 R^{(1)} + q_2 R^{(2)} + \cdots + q_{N-1} R^{(N-1)} + R^{(N)} = 0$$

in which $q_0, q_1, \cdots, q_{N-1}$ are coefficients to be determined.

Further, in a manner corresponding to (7.7), the experimental relaxation curves may be expressed analytically as a function of the applied strain and time for a number of relaxation experiments $i = 1, 2, \cdots, n$ as

$$\check{\sigma}_i(t) = \check{\sigma}_i + g_i \phi_i(t, z_{i1}, z_{i2}, \cdots) \qquad (i = 1, 2, \cdots, n) \tag{7.24}$$

where $\check{\sigma}_i = E\check{\varepsilon}_i$ is the instantaneous experimental stress at $t = 0$, g_i are constants and $\phi_i(t, z_{i1}, z_{i2}, \cdots)$ are functions of time. The shape of the experimental relaxation curve will suggest the forms of the functions $\phi_i(\cdot)$ for the type of material under consideration. In the above equation, the constants g_i and $z_{i1}, z_{i2}, \cdots$ will be required to be determined for each relaxation experiment i. Meantime, the model response equation in the relaxation phase can be written, with reference to (7.23), in an analogous manner to (7.20), as

$$\sigma_i(t) = E\check{\varepsilon}_i + h_1(\check{\varepsilon}_i, c_1, c_2, \cdots) \sum_{I=1}^{N} \hat{D}_I[\exp(B_I t) - 1]$$

$$(i = 1, 2, \cdots, n \quad \text{and} \quad I = 1, 2, \cdots, N) \tag{7.25}$$

in which the functions $h_1(\cdot)$ and the constants $\hat{D}_I$ and B_I $(I = 1, 2, \cdots, N)$ are to be identified for the considered range of applied strain and the extent of time covered during the relaxation experiments $i = 1, 2, \cdots, n$.

7.4 NUMERICAL ILLUSTRATION

7.4.1 Creep response

Figure 7.1 illustrates the experimental creep data (Hill, 1967) for a class of wood pulp fibres. In this figure, the first creep strain is presented against the logarithm of time for dry summerwood fibers of a longleaf pine holocellulose pulp after conditioning at 50% RH and 23 °C. Three creep experimental curves are shown, i.e. for initial stress inputs of 29.1, 38.4 and 49.1 dyn μm^{-2}.

Thus, by referring to expression (7.7) and the shape of the creep curves of Fig. 7.1, one may express the experimental creep strain, for the above fibres, as

$$\check{\varepsilon}_i(t) = \check{\varepsilon}_i + G_i t^{m_i} \qquad (i = 1, 2, \cdots, n), \quad n = 3, \tag{7.26}$$

where $\check{\varepsilon}_i$ is the instantaneous strain at $t = 0$ and G_i and m_i are constants. The values of the parameters G_i and m_i in the above equation can be determined from the corresponding experimental data (Fig. 7.1) by a standard fitting procedure. In this context, the outcome data of each creep experiment can be derived from the

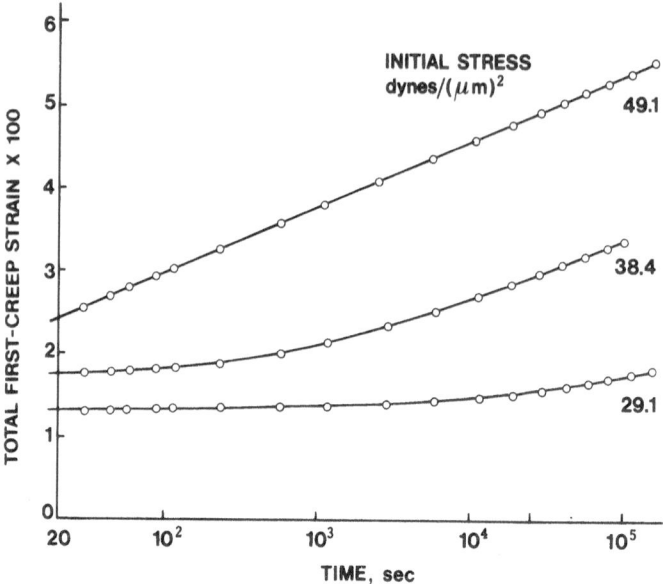

Fig. 7.1 Experimental first creep curves of dry summerwood fibres of a longleaf pine holocellulose pulp after conditioning at 50% RH and 23 °C. (Source: Hill, R. L. (1967) The creep behaviour of individual pulp fibres under tensile stress. *Tappi*, **50**(8), 432–40. Reprinted by permission of Tappi.)

creep curves of Fig. 7.1 and expressed as follows:

$$[t_1, \check{\varepsilon}_i(t_1)], [t_2, \check{\varepsilon}_i(t_2)], \cdots, [t_l, \check{\varepsilon}_i(t_l)], \cdots, [t_\chi, \check{\varepsilon}_i(t_\chi)]$$

where $t_1 < t_2 < \cdots < t_l < t_\chi$ within the domain of T and the time T represents, as pointed out in section 7.2, the total time for each relaxation experiment.

It is now required to fit for each creep experiment i the parameters G_i and m_i equation (7.26) such that the resulting strain $\check{\varepsilon}_i(t)$ gives a reasonable representation of the outcome data of the creep experiment i. One fits for each creep experiment i by minimizing a functional

$$Y_i(G_i, m_i) = \sum_{l=1}^{\chi} [\check{\varepsilon}_i(t_l) - (\check{\varepsilon}_i + G_i t_l^{m_i})]^2. \tag{7.27}$$

By carrying out the minimization of expression (7.27) for each creep experiment i with respect to all possible choices of the parameter G_i, i.e.

$$\frac{\partial}{\partial G_i} Y_i(G_i, m_i) = \frac{\partial}{\partial G_i} \sum_{l=1}^{\chi} [\check{\varepsilon}_i(t_l) - (\check{\varepsilon}_i + G_i t_l^{m_i})]^2$$

$$(l = 1, 2, \cdots, \chi \quad \text{and} \quad i = 1, 2, \cdots, n; \quad n = 3), \tag{7.28}$$

it follows that

$$G_i = \frac{\sum\limits_{l=1}^{\chi} [\check{\varepsilon}_i(t_l) - \check{\varepsilon}_i]t_l^{m_i}}{\sum\limits_{l=1}^{\chi} t_l^{2m_i}}. \tag{7.29}$$

Similarly, by carrying out the minimization of expression (7.27), for each creep experiment i, with respect to all possible choices of the parameter m_i, i.e.

$$\frac{\partial}{\partial m_i} Y_i(G_i, m_i) = \frac{\partial}{\partial m_i} \sum_{l=1}^{\chi} [\check{\varepsilon}_i(t_l) - (\check{\varepsilon}_i + G_i t_l^{m_i})]^2 = 0$$

$$(l = 1, 2, \cdots, \chi \quad \text{and} \quad i = 1, 2, \cdots, n; \quad n = 3), \tag{7.30}$$

one obtains, by combining (7.29) and (7.30), the following relation:

$$\sum_{l=1}^{\chi} [\check{\varepsilon}_i(t_l) - \check{\varepsilon}_i]t_l^{m_i} \ln(t_l) - \frac{\sum\limits_{l=1}^{\chi} [\check{\varepsilon}_i(t_l) - \check{\varepsilon}_i]}{\sum\limits_{l=1}^{\chi} t_l^{2m_i}} t_l^{m_i} \sum_{l=1}^{\chi} t_l^{2m_i} \ln(t) = 0. \tag{7.31}$$

Thus, by using the available experimental data, one can solve numerically equation (7.31) for each experiment i to find the parameter m_i, and then, by substituting the resulting value of m_i into (7.29), the value of the corresponding parameter G_i can be determined.

The secant method (for instance, Jones, Smith and Welford, 1970) has been used in the present work to solve the equation (7.31). The numerical results of the fitting procedure, as applied to the creep curves of Fig. 7.1, are presented in Table 7.1.

(a) Differential approximation of the creep kernel

With the foregoing obtained information from the experimental creep data by means of equation (7.26), it is now possible to proceed using the method of differential approximation indicated in section 7.2 to determine the coefficients of the linear

Table 7.1 Material parameters pertaining to the suggested, experimental creep expression (7.26)

i	$\check{\sigma}_i(\text{dyn } \mu m^{-2})$	$\check{\varepsilon}_i\ (\mu m\ \mu m^{-1})$	$G_i(h^{-m_i})$	m_i
1	29.1	0.012	0.164×10^{-2}	0.192
2	38.4	0.016	0.816×10^{-2}	0.388
3	49.1	0.021	0.205×10^{-2}	0.224

Dry summerwood fibres of longleaf pine holocellulose pulp after conditioning at 50% RH and 23 °C.

differential equation (7.5a). Having determined these coefficients, one may continue to find the roots of the characteristic equation (7.17).

Comparing (7.26) and (7.7), it is evident that, for the present case of wood pulp fibres, the functions $\theta_i(\cdot)$ take the form

$$\theta_i(\cdot) = t^{m_i}. \tag{7.32}$$

Consequently, one can write (7.9) as

$$\Gamma_i = \frac{d}{dt}(t^{m_i}) \tag{7.33}$$

from which the jth derivative required for (7.13) is expressed as

$$\Gamma_i^{(j)} = \frac{d^{j+1}}{dt^{j+1}}(t^{m_i}) = m_i(m_i - 1)(m_i - 2) \cdots (m_i - j)t^{m_i - j - 1}$$

$$(j = 0, 1, 2, \cdots, N - 1). \tag{7.34}$$

Hence, by substituting the expression of $\Gamma_i^{(j)}$ corresponding to (7.34) into (7.13) and carrying out the integration of the elements of the matrices $[x_{jk}]$ and $[x_k]$ as indicated in (7.15), then

$$a_j = \begin{bmatrix} a_0 \\ a_1 \\ a_2 \\ \vdots \\ a_{N-1} \end{bmatrix}, \tag{7.35a}$$

and

$$[x_k] = - \begin{bmatrix} \sum_{i=1}^{n} \frac{m_i^2(m_i - 1)(m_i - 2) \cdots (m_i - N)t^{2m_i - N - 1}}{2m_i - N - 1} \Big]_0^T \\ \sum_{i=1}^{n} \frac{m_i^2(m_i - 1)^2(m_i - 2) \cdots (m_i - N)t^{2m_i - N - 2}}{2m_i - N - 2} \Big]_0^T \\ \sum_{i=1}^{n} \frac{m_i^2(m_i - 1)^2(m_i - 2)^2 \cdots (m_i - N)t^{2m_i - N - 3}}{2m_i - N - 3} \Big]_0^T \\ \vdots \\ \sum_{i=1}^{n} \frac{m_i^2(m_i - 1)^2(m_i - 2)^2 \cdots (m_i - N + 1)^2(m_i - N)t^{2m_i - 2N}}{2m_i - 2N} \Big]_0^T \end{bmatrix}$$

$$(k = 0, 1, \cdots, N - 1 \quad \text{and} \quad i = 1, 2, \cdots, n). \tag{7.35c}$$

In (7.35b) and (7.35c) it is apparent that the elements of the matrices $[x_{jk}]$ and $[x_k]$ contain terms such as $t^{2m_i - 1}$, $t^{2m_i - 2}$, $\cdots$, etc. Thus, in view of Table 7.1, since m_i has values ranging from 0.192 to 0.388, we expect the exponent of each of the

$$|x_{j,k}| = \begin{vmatrix} \sum_{i=1}^{n} m_i^2\frac{t^{2m_i-1}}{2m_i-1}\Big]_0^T & \sum_{i=1}^{n} m_i^2(m_i-1)\frac{t^{2m_i-2}}{2m_i-2}\Big]_0^T & \cdots & \sum_{i=1}^{n} m_i^2(m_i-1)(m_i-2)\cdots(m_i-N+1)\frac{t^{2m_i-N}}{2m_i-N}\Big]_0^T \\[2ex] \sum_{i=1}^{n} m_i^2(m_i-1)\frac{t^{2m_i-2}}{2m_i-2}\Big]_0^T & \sum_{i=1}^{n} m_i^2(m_i-1)^2\frac{t^{2m_i-3}}{2m_i-3}\Big]_0^T & \cdots & \sum_{i=1}^{n} m_i^2(m_i-1)^2(m_i-2)\cdots(m_i-N+1)\frac{t^{2m_i-N-1}}{2m_i-N-1}\Big]_0^T \\[2ex] \sum_{i=1}^{n} m_i^2(m_i-1)(m_i-2)\frac{t^{2m_i-3}}{2m_i-3}\Big]_0^T & \sum_{i=1}^{n} m_i^2(m_i-1)^2(m_i-2)\frac{t^{2m_i-4}}{2m_i-4}\Big]_0^T & \cdots & \sum_{i=1}^{n} m_i^2(m_i-1)^2(m_i-2)\cdots(m_i-N+1)\frac{t^{2m_i-N-2}}{2m_i-N-2}\Big]_0^T \\[1ex] \vdots & \vdots & & \vdots \\[1ex] \sum_{i=1}^{n} m_i^2(m_i-1)(m_i-2)\cdots(m_i-N+1)\frac{t^{2m_i-N}}{2m_i-N}\Big]_0^T & \sum_{i=1}^{n} m_i^2(m_i-1)^2(m_i-2)\cdots(m_i-N+1)\frac{t^{2m_i-N-1}}{2m_i-N-1}\Big]_0^T & \cdots & \sum_{i=1}^{n} m_i^2(m_i-1)^2(m_i-2)\cdots(m_i-N+1)\frac{t^{2m_i-2N+1}}{2m_i-2N+1}\Big]_0^T \end{vmatrix}$$

$$(j, k = 0, 1, 2, \cdots, N-1 \quad \text{and} \quad i = 1, 2, \cdots, n)$$

(7.35b)

above terms, i.e. t^{2m_i-1}, t^{2m_i-2}, $\cdots$, etc., to be negative. As a consequence, the singularity of these terms will prevent the use of the lower limit of the integration as zero. In the computational work, this problem has been solved by taking the lower limit of integration greater than zero, namely 0.09 h. A straightforward matrix inversion method (for instance, Jones, Smith and Welford, 1970) has been used to solve (7.14) for the coefficients a_j. The results of the computations are presented in Table 7.2 for $N = 3, 4, 5, \cdots, 8$.

Having determined the coefficients a_j, one continues, as indicated in section 7.2, to find the roots A_I ($I = 1, 2, \cdots, N$) of the characteristic equation (7.17). In this context, the secant method has been employed. The numerical values of the roots A_I are given in Table 7.3.

(b) Minimization of the objective function (7.22)

After we have determined the roots $A_1, A_2, \cdots, A_N$ of the characteristic equation (7.17), we continue to determine the values of the unknown constants appearing in the model equation (7.20) where the function $h(\cdot)$ appearing in the latter equation takes, for the present class of wood fibres, the form (Distéfano and Todeschini, 1973a, b; Haddad, 1987).

$$h(\cdot) = \exp(b\breve{\sigma}). \tag{7.36}$$

This involves the minimization of expression (7.22).

Thus, by substituting for $h(\cdot)$ and $\theta_i(\cdot)$ as given by (7.36) and (7.32), respectively, into (7.22), it follows that

$$\text{III}[b, D_I (I = 1, 2, \cdots, N)] = \sum_{i=1}^{n=5} \gamma_i \int_0^T \left\{ G_i t^{m_i} - \exp(b\breve{\sigma}) \sum_{I=1}^{N} D_I [1 - \exp(A_I t)] \right\}^2 dt$$

$$(i = 1, 2, \cdots, 5, \quad \text{and} \quad I = 1, 2, \cdots, N). \tag{7.37}$$

The computational analysis for minimizing the objective function (7.37) has been carried out using the steepest descent method (Haddad, 1975), whereby the lower limit of integration is taken as 0.09 h and γ_i assumed to be unity. The values of the parameters b and D_I which minimize (7.37) are shown in Table 7.4 for $N = 3, 4, 5, \cdots, 8$.

In order to test the predictive ability of the model, the predicted values of strain have been computed for each experiment i for $N = 3, 4, \cdots, 8$. This has been carried out by using the model equation (7.20) whereby the function $h(\cdot)$ appearing in this equation is given by equation (7.36). The predicted values of strain have been then compared with the corresponding experimental ones taken from Fig. 7.1 at corresponding stress level and time. The mean square error (MSE) was computed for each experiment i and different values of N through the relation

$$\text{MSE} = \frac{\sum_{l=1}^{\chi} [\varepsilon(t_l) - \breve{\varepsilon}(t_l)]^2}{\chi} \qquad (l = 1, 2, \cdots, \chi). \tag{7.38}$$

Table 7.2 Coefficients a_j ($j = 0, 1, \ldots, N-1$) for $N = 3, 4, \ldots, 8$, equation (7.14)

N	3	4	5	6	7	8
a_0	$0.420\ 050 \times 10^4$	$0.161\ 554 \times 10^6$	$0.791\ 297 \times 10^7$	$0.470\ 653 \times 10^9$	$0.328\ 268 \times 10^{11}$	$0.262\ 029 \times 10^{13}$
a_1	$0.269\ 029 \times 10^4$	$0.157\ 186 \times 10^6$	$0.104\ 922 \times 10^8$	$0.794\ 841 \times 10^9$	$0.676\ 065 \times 10^{11}$	$0.638\ 460 \times 10^{13}$
a_2	$0.129\ 167 \times 10^3$	$0.133\ 574 \times 10^5$	$0.135\ 210 \times 10^7$	$0.141\ 908 \times 10^9$	$0.157\ 559 \times 10^{11}$	$0.186\ 254 \times 10^{13}$
a_3		$0.240\ 356 \times 10^3$	$0.413\ 268 \times 10^5$	$0.642\ 202 \times 10^7$	$0.974\ 957 \times 10^9$	$0.149\ 517 \times 10^{12}$
a_4			$0.385\ 423 \times 10^3$	$0.991\ 915 \times 10^5$	$0.219\ 039 \times 10^8$	$0.453\ 662 \times 10^{10}$
a_5				$0.564\ 305 \times 10^3$	$0.202\ 941 \times 10^6$	$0.603\ 604 \times 10^8$
a_6					$0.776\ 941 \times 10^3$	$0.371\ 908 \times 10^6$
a_7						$0.102\ 323 \times 10^4$

Creep of dry summerwood fibres of longleaf pine holocellulose pulp after conditioning at 50% RH and 23 °C

Table 7.3 Roots A_l ($l = 1, 2, \ldots, N$, $N = 8$) (h^{-1}) of the characteristic equation (7.17)

N	3	4	5	6	7	8
A_1	$-0.760\ 54 \times 10^{-1}$	$-0.169\ 98 \times 10^{-1}$	$-0.174\ 52 \times 10^{-1}$	$-0.181\ 59 \times 10^{-1}$	$-0.192\ 38 \times 10^{-1}$	$-0.204\ 23 \times 10^{-1}$
A_2	$-0.119\ 45 \times 10^{-1}$	$-0.506\ 27 \times 10^{-2}$	$-0.531\ 17 \times 10^{-2}$	$-0.557\ 99 \times 10^{-2}$	$-0.589\ 84 \times 10^{-2}$	$-0.629\ 55 \times 10^{-2}$
A_3	$-0.362\ 55 \times 10^{-1}$	$-0.296\ 65 \times 10^{-2}$	$-0.316\ 80 \times 10^{-2}$	$-0.332\ 85 \times 10^{-2}$	$-0.343\ 48 \times 10^{-2}$	$-0.356\ 62 \times 10^{-2}$
A_4		$-0.481\ 98 \times 10^{-2}$	$-0.500\ 24 \times 10^{-2}$	$-0.512\ 06 \times 10^{-2}$	$-0.517\ 80 \times 10^{-2}$	$-0.524\ 76 \times 10^{-2}$
A_5			$-0.182\ 66 \times 10^{-3}$	$-0.301\ 17 \times 10^{-3}$	$-0.324\ 47 \times 10^{-3}$	$-0.349\ 38 \times 10^{-3}$
A_6				$-0.118\ 53 \times 10^{-3}$	$-0.141\ 94 \times 10^{-3}$	$-0.167\ 21 \times 10^{-3}$
A_7					$-0.234\ 17 \times 10^{-4}$	$-0.487\ 11 \times 10^{-4}$
A_8						$-0.252\ 94 \times 10^{-4}$

Creep of dry summerwood fibres of longleaf pine holocellulose pulp after conditioning at 50% RH and 23 °C

Table 7.4 Parameters D_l ($l = 1, 2, \ldots, N$; $N = 8$) ($\times 10^8$ dyn μm^{-2}) and b, equation (7.37)

N	3	4	5	6	7	8
D_1	$-1.697\ 959$	$-1.135\ 043$	$-0.843\ 524$	$-0.669\ 838$	$-0.554\ 878$	$-0.473\ 314$
D_2	$-23.881\ 838$	$-14.475\ 734$	$-9.887\ 900$	$-7.293\ 809$	$-5.676\ 604$	$-4.593\ 259$
D_3	$-103.587\ 203$	$-59.504\ 077$	$-39.375\ 231$	$-28.263\ 548$	$-21.434\ 256$	$-16.919\ 977$
D_4		$-165.241\ 146$	$-104.294\ 647$	$-73.314\ 790$	$-54.812\ 592$	$-42.745\ 777$
D_5			$-231.021\ 697$	$-155.184\ 215$	$-113.886\ 261$	$-87.865\ 797$
D_6				$-229.578\ 800$	$-210.430\ 911$	$-159.481\ 064$
D_7					$-370.145\ 498$	$-268.896\ 465$
D_8						$-422.254\ 346$
b	$0.267\ 11 \times 10^{-2}$	$0.119\ 87 \times 10^{-1}$	$0.123\ 08 \times 10^{-1}$	$0.124\ 98 \times 10^{-1}$	$0.125\ 05 \times 10^{-1}$	$0.124\ 97 \times 10^{-1}$

Creep of dry summerwood fibres of longleaf pine holocellulose pulp after conditioning at 50% RH and 23 °C.

Table 7.5 Material parameters characteristic of the suggested experimental relaxation equation (7.39)

i	$\tilde{\varepsilon}_i$ (cm cm^{-1})	$\tilde{\sigma}_i$ ($\times 10^8$ dyn cm^{-2})	g_i ($\tilde{h}^{zi}$) ($\times 10^8$ dyn cm^{-2})	z_i
1	0.01	4.5	1.919	0.064
2	0.02	9.7	4.285	0.064
3	0.03	13.6	5.682	0.066
4	0.04	17.9	6.908	0.058
5	0.05	22.05	6.906	0.065

Cotton fibres after conditioning at 65% RH and 25 °C.

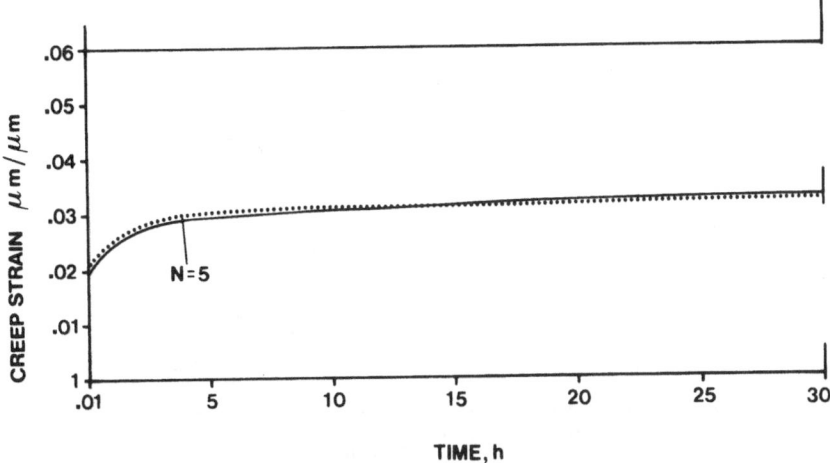

Fig. 7.2 Predictive capability of the proposed model in the case of first creep of pine pulp fibres. (applied stress, 38.4 dyn μm^2): ——, experiment; ······, prediction.

It was found that, for the considered fibres, the total MSE reaches a minimum value at $N = 5$. The predictive ability of the model is demonstrated in Fig. 7.2 for a stress level of 38.4 dyn μm^{-2}.

7.4.2 Relaxation response

The experimental curves due to Meredith (1954) showing the relaxation stress, i.e. the stress obtained under constant strain $\check{\varepsilon}_i$ ($i = 1, 2, \cdots, n; n = 5$), i indicating a relaxation experiment, against the logarithm of time are presented in Fig. 7.3 for the case of natural cellulose fibres (cotton).

With reference to Fig. 7.3 and expression (7.24), the stress relaxation for the above fibres is expressed in the form

$$\check{\sigma}_i(t) = \check{\sigma}_i - g_i t^{z_i} \qquad (i = 1, 2, \cdots, n) \tag{7.39}$$

where z_i are constants. The values of the parameters g_i and z_i ($i = 1, 2, \cdots, n; n = 5$) are given in Table 7.5 for the relaxation experiments presented in Fig. 7.3. In this case, the function $h_1(\cdot)$ of (7.25) is assumed to have the form

$$h_1(\cdot) = \exp(c\check{\varepsilon}_i) \tag{7.40}$$

where c is a constant. Thus, by combining equations (7.25) and (7.40) the former becomes

$$\sigma_i(t) = E\check{\varepsilon}_i + \exp(c\check{\varepsilon}_i) \sum_{l=1}^{N} \hat{D}_l[\exp(B_l t) - 1]$$

$$(i = 1, 2, \cdots, 5 \text{ and } I = 1, 2, \cdots, N). \tag{7.41}$$

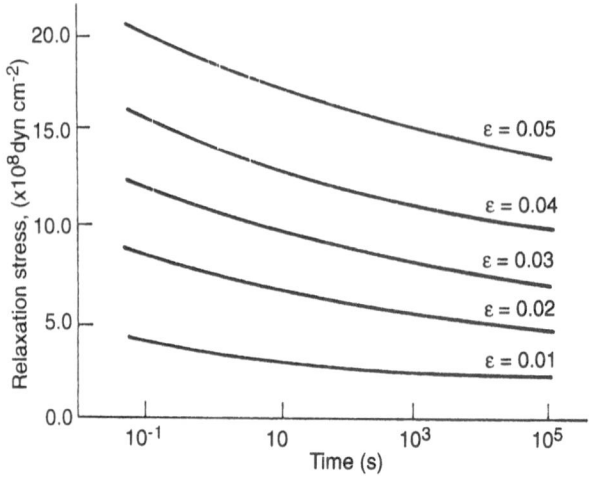

Fig. 7.3 Experimental relaxation curves of natural cellulose fibres (cotton) at 65% RH and 25 °C. (Source: Meredith, R. (1954) Relaxation of stress in stretched cellulose fibers, *J. Textile Inst.*, **45**, T438–T461. Reprinted by permission of British Textile Technology Group.)

With the information obtained from the experimental relaxation data of cotton fibres by means of equation (7.39) as illustrated above, one can proceed, following the analytical scheme introduced earlier, for modelling the creep response, to determine the parameters c, $\hat{D}_I$ and B_I $(I = 1, 2, \cdots, N)$ which characterize the relaxation response equation (7.41) (Table 7.6). In this regard, it was found that the total MSE for the number of experiments involved reaches a minimum value at $N = 4$. The predictive ability of the model is demonstrated in Fig. 7.4a for a relaxation strain input of 0.01 cm cm^{-1}.

The numerical values of the parameters characterizing the constitutive equation (7.41), as determined above for cotton fibres, may be used to examine further the predictive ability of the model for a strain level outside the strain range considered in the model. In this context, the experimental relaxation curve corresponding to an applied strain of 0.06 cm cm^{-1} (Meredith, 1954) was used. As shown in Fig. 7.4b, the stress relaxation values predicted by the model comply favourably well with the corresponding experimental ones even at this level of applied strain.

As demonstrated above, a common rheological model was used for the identification of both the creep and relaxation responses of a class of natural fibres by using experimental data from two different fibre sources. This was possible because of the generality introduced in the model in the form of mathematical functions that can be associated with the particular time-dependent performance of the class of fibres under consideration.

Table 7.6 Values of parameters pertaining to the relaxation constitutive equation (7.41)

N	3	4	5	6	7	8
$\hat{D}_l$ ($\times 10^8$ dyn cm^{-2})						
$\hat{D}_1$	64.848 773	−1.279 786	0.681 772	−0.191 855	−1.472 124	−2.125 503
$\hat{D}_2$	61.622 924	75.680 531	75.206 427	74.943 738	74.828 227	74.793 207
$\hat{D}_3$	85.071 193	99.129 241	98.657 634	98.424 634	98.368 221	98.355 531
$\hat{D}_4$		34.747 69	34.276 117	34.043 117	33.988 159	33.976 572
$\hat{D}_5$			17.350 358	17.350 358	17.295 414	17.283 834
$\hat{D}_6$				20.684 165	20.629 207	20.617 627
$\hat{D}_7$					24.108 952	24.097 379
$\hat{D}_8$						27.570 524
B_l (h^{-1})						
B_1	−1.099 525	−0.740 515	−0.554 445	−0.443 661	−0.370 971	−0.319 748
B_2	−10.089 640	−6.308 098	−4.413 529	−3.325 409	−2.640 529	−2.179 361
B_3	−38.468 035	−22.706 725	−15.347 084	−11.220 736	−8.658 291	−6.953 434
B_4		−59.359 363	−38.149 942	−27.226 782	−20.638 786	−16.316 638
B_5			−81.550 000	−55.507 862	−41.212 553	−32.142 377
B_6				−104.633 550	−74.250 533	−56.795 098
B_7					−128.372 337	−94.051 185
B_8						−152.616 160
c	−6.492 27	−7.011 54	−9.5185	−13.040 31	−14.585 07	−14.929 95

Cotton fibres.

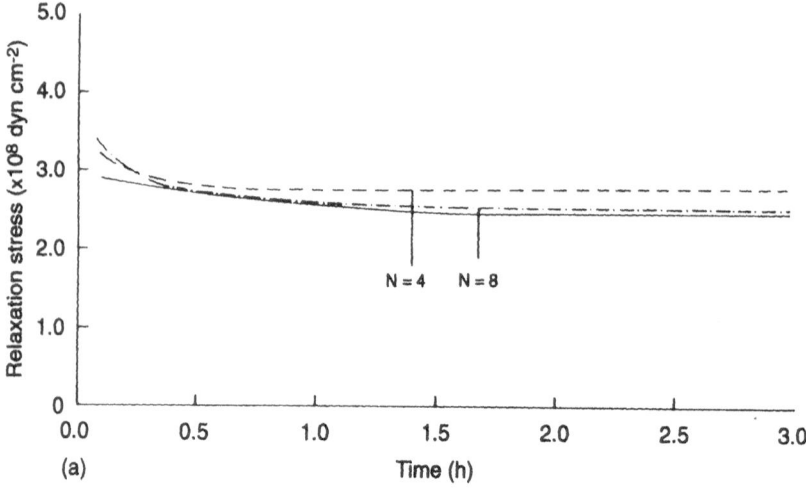

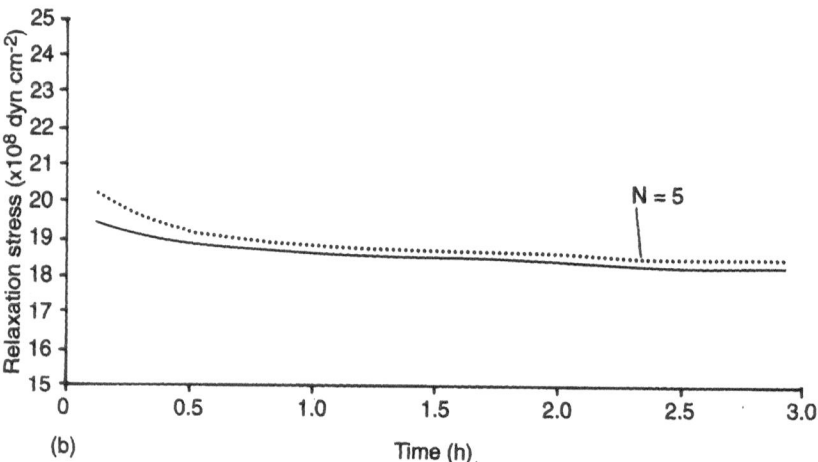

Fig. 7.4 Testing the predictive ability of the proposed model in the case of relaxation of cotton fibres: (a) applied strain 0.01 cm cm^{-1}; (b) applied strain 0.06 cm cm^{-1}; ———, experiment; – – –, – · –, · · · ·, prediction.

While the above formulation has been illustrated by a procedure applied to the case of single fibres of natural origin, it could be valid for other classes of viscoelastic materials provided that the forms of the experimental curves representing the rheological behaviour of such materials are observed. We demonstrate further below the possibility of application of the above model to the isothermal creep of a class of viscoelastic materials within a specified temperature range.

7.5 IDENTIFICATION OF THE ISOTHERMAL CREEP OF VISCOELASTIC MATERIALS WITHIN A SPECIFIED TEMPERATURE RANGE

In Fig. 7.5, the first creep strain is presented against the logarithm of time for a class of polyester resin (isophthalic polyester) at an applied stress of 4.55 MPa. Three creep experimental curves are shown at constant temperatures of 28.2 °C, 57.2 °C and 76.6 °C.

With reference to the shape of the experimental creep curves of Fig. 7.5, one may use equation (7.26) to express the creep response. The values of the parameters G_i and m_i ($i = 1, 2, \cdots, n; n = 3$) characterizing equation (7.26) for the present case of polyester resin are given in Table 7.7.

Referring to the creep constitutive equation (7.20), the function $h(\cdot)$ appearing in this equation may be assumed, for the case of creep at elevated (constant) temperatures, in a power-law form (e.g. Robotonov, 1969; Gittus, 1975; Haddad and Tanary, 1989) as

$$h(\cdot) = B\breve{\sigma}^k \tag{7.42}$$

in which k is a constant parameter which takes, usually, the value of 1 for a large number of polymeric materials, whereas B is a temperature-dependent parameter. The latter may be expressed in terms of the activation energy of the associated flow process (Haddad and Tanary, 1989) in the following form:

$$B = B_0 \exp\left(-\frac{U_0}{RT}\right). \tag{7.43}$$

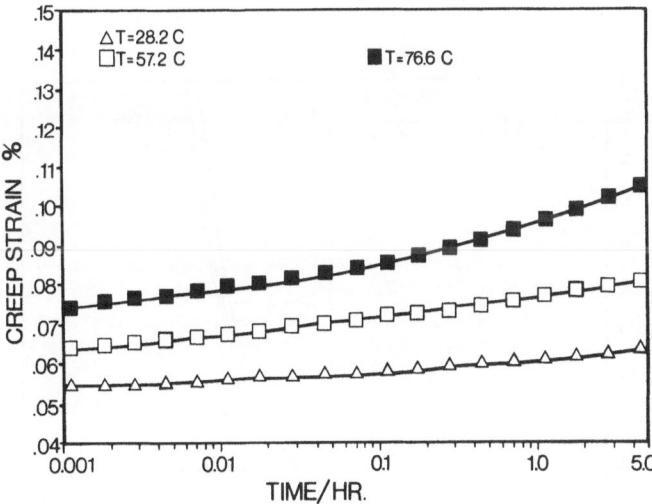

Fig. 7.5 Experimental creep curves of polyester resin (isophthalic polyester) at various base temperatures (applied stress of 4.55 MPa). Experimental data from Jerina *et al.* (1982).

Table 7.7 Material parameters pertaining to the suggested experimental creep expression (7.26)

i	T_i (°C)	$\breve{\sigma}$ (MPa)	$\breve{\varepsilon}_i$ (%)	$G_i(h^{m_i})$	m_i
1	28.2	4.522	0.0455	$0.149\,87 \times 10^{-1}$	0.087 06
2	57.2	4.552	0.0455	$0.310\,82 \times 10^{-1}$	0.079 34
3	76.6	4.552	0.0455	$0.504\,95 \times 10^{-1}$	0.090 23

Creep of isophthalic polyester resin at different base temperatures.

In the above equation, B_0 is a constant parameter to be determined from the experimental creep data, U_0 is the activation energy, R is the gas constant and T is the creep test temperature. Combining equations (7.20), (7.42) and (7.43), the creep constitutive equation (7.20) can be expressed for the present case as

$$\varepsilon_i(t) = \left\{ E^{-1} + B_0 \exp\left(-\frac{U_0}{RT_i} \right) \sum_{I=1}^{N} D_I[\exp(A_I t) - 1] \right\} \breve{\sigma}_i$$

$$(i = 1, 2, \cdots, n \text{ and } I = 1, 2, \cdots, N). \tag{7.44}$$

In equation (7.44), we first determine the value of the activation energy U_0 that is valid for the considered creep range, i.e. between 28.2 °C and 76.6 °C (Fig. 7.5). This may be done by using the expression for the function $h(\cdot)$ that is given by (7.42). Thus, with reference to (7.42) and (7.43), one can write

$$h_i(\cdot) = B_0 \exp\left(-\frac{U_0}{RT_i} \right) \breve{\sigma}_i = S_i$$

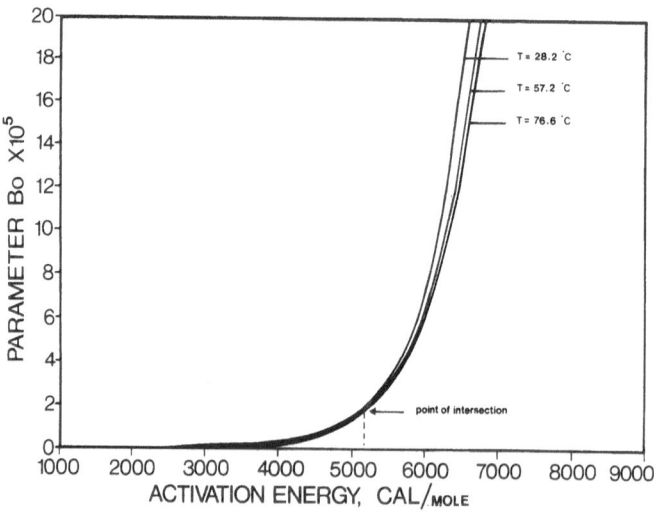

Fig. 7.6 Variation of material parameter B_0 with activation energy.

Table 7.8 Values of parameters pertaining to the creep constitutive equation (7.20) with the inclusion of equation (7.42)

N	3	4	5	6	7	8
D_1	-0.967×10^{-1}	-0.106	-0.104	-0.975×10^{-1}	-0.936×10^{-1}	-0.930×10^{-1}
D_2	-0.581×10^{-1}	-0.100	-0.902×10^{-1}	-0.418×10^{-1}	-0.381×10^{-1}	-0.356×10^{-1}
D_3	-0.581×10^{-1}	-0.101	-0.906×10^{-1}	-0.414×10^{-1}	-0.384×10^{-1}	-0.359×10^{-1}
D_4		-0.101	-0.907×10^{-1}	-0.414×10^{-1}	-0.384×10^{-1}	-0.350×10^{-1}
D_5			-0.907×10^{-1}	-0.415×10^{-1}	-0.385×10^{-1}	-0.360×10^{-1}
D_6				-0.415×10^{-1}	-0.385×10^{-1}	-0.360×10^{-1}
D_7					-0.385×10^{-1}	-0.361×10^{-1}
D_8						-0.360×10^{-1}
A_1	-6.831	-4.177	-2.843	-2.078	-1.592	-1.277
A_2	-75.357	-45.262	-10.282	-21.739	-16.407	-12.855
A_3	-301.811	-174.574	-115.313	-82.173	-61.648	-48.032
A_4		-468.735	-207.088	-208.752	-155.553	-120.675
A_5			-646.302	-435.321	-319.479	-246.079
A_6				-831.174	-584.847	-443.229
A_7					-1021.446	-743.112
A_8						-1215.767
B_0	0.973×10^{-4}	0.506×10^{-4}	0.446×10^{-4}	0.695×10^{-4}	0.658×10^{4}	0.627×10^{-4}

Isophthalic polyester resin.

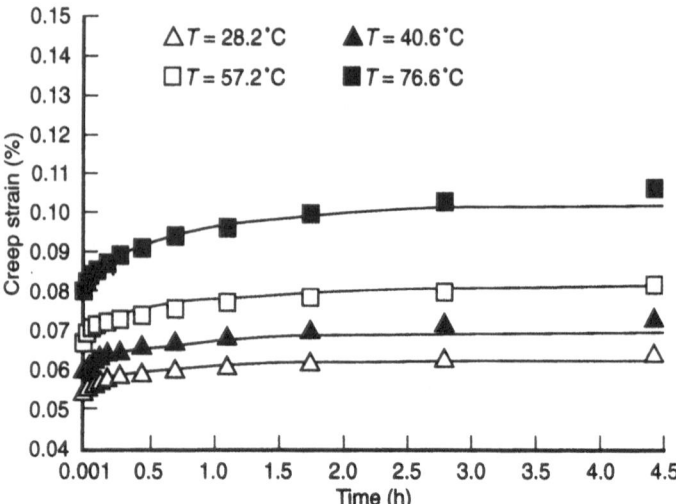

Fig. 7.7 Predictive capability of the proposed model for the creep strain of polyester resin (isophthalic polyester) for a temperature range of 28 °C–76.6 °C (applied stress of 4.55 MPa).

or

$$B_0 = \frac{S_i}{\breve{\sigma}_i} \exp\left(\frac{U_0}{RT_i}\right) \tag{7.45}$$

where the exponent k appearing in (7.42) has been taken as unity and S_i is the numerical value of $h_i(\cdot)$ for each experiment i ($i = 1, 2, \cdots, n$). Thus, the value of U_0 valid for the creep range considered may be defined by the intercept value of the corresponding curves on the B–U_0 graph (Fig. 7.6) that represent the number of experiments ($i = 1, 2, \cdots, n; n = 3$) through the use of equation (7.45).

From Fig. 7.6, it is determined, for the present case, that U_0 has a value of 5.2 kcal mol^{-1}. Having determined the value of U_0, one may proceed to determine the values of the remaining parameters in the creep constitutive equation (7.44), i.e. B_0, D_I and A_I ($I = 1, 2, \cdots, N$). This can be carried out using the differential approximation and minimization procedure outlined earlier in section 7.2. In this context, the values of these parameters for the case of polyester resin are given in Table 7.8 for $N = 3, 4, \cdots, 8$. In this case, the MSE for the total number of creep experiments considered reaches a minimum at $N = 8$. The predictive ability of the model is demonstrated in Fig. 7.7.

REFERENCES

Bellman, R. (1970) *Methods of Nonlinear Analysis*, Vol. 1, Academic Press, New York.
Distéfano, N. (1970) On the identification problem in linear viscoelasticity. *Z. Angew. Math. Mech.*, **50**, 683–90.

Distéfano, N. (1971) System identification problems in hereditary biomechanical processes, in Proc. 5th Asilomar Conf. on Circuits and Systems, Pacific Grove, CA, pp. 248–51.

Distéfano, N. and Todeschini, R. (1973a) Modeling, identification and prediction of a class of nonlinear viscoelastic materials (I). *Int. J. Solids Struct.*, **9**, 805–18.

Distéfano, N. and Todeschini, R. (1973b) Modeling, identification and prediction of a class of nonlinear viscoelastic materials (II). *Int. J. Solids Struct.*, **12**, 1431–8.

Gittus, J. (1975) *Viscoelasticity and Creep Fracture in Solids*, Wiley, New York, pp. 8–12.

Haddad, Y. M. (1975) Response behaviour of a two-dimensional fibrous network. PhD Thesis, McGill University, Montreal, Canada.

Haddad, Y. M. (1987) Un modèle de réponse en viscoelasticité non-linéaire de matériaux. *Res Mech.*, **20**, 235–53.

Haddad, Y. M. and Tanary, S. (1987) Characterization of the rheological response of a class of single fibers. *J. Rheol.*, **31**(7), 515–26.

Haddad, Y. M. and Tanary, S. (1989) On the micromechanical characterization of the creep response of a class of composite systems. *J. Pressure Vessel Technol.*, **3**, 177–82.

Hill, R. L. (1967) The creep behaviour of individual pulp fibres under tensile stress. *Tappi*, **50**(8), 432–40.

Jerina, K. L., Schapery, R. A., Jung, R. W. and Sanders, B. A. (1982) Viscoelastic characterization of a random fibre composite material employing micromechanics, in *ASTM Spec. Tech. Publ. 772* (ed. B. A. Sanders), ASTM, Philadelphia, PA, pp. 225–50.

Jones, M. L., Smith, G. M. and Welford, J. C. (1970) *Applied Numerical Methods for Digital Computation with Fortran*, International Textbook Company, PA.

Meredith, R. (1954) Relaxation of stress in stretched cellulose fibres. *J. Textile Inst.*, **45**, T438–T461.

Mikhlin, S. G. (1965) *The Problems of the Minimum of a Quadratic Functional*, Holden Day, San Francisco.

Robotonov, Y. N. (1969) *Creep Problems in Structural Members*, North-Holland, London, pp. 178–215.

FURTHER READING

Abadie, J. (ed.) (1967) *Nonlinear Programming*, North-Holland, Amsterdam.

Bellman, R. and Kalaba, R. (1965) *Quasilinearization and Nonlinear Boundary-value Problems*, Elsevier, London.

Christensen, R. M. (1968) On obtaining solutions in nonlinear viscoelasticity. *J. Appl. Mech.*, **35**, 129–33.

Distéfano, N. and Pister, K. (1970) On modeling and identification in biophysics with application to the rheology of the red cell membrane, in Proc. 23rd Annu. Conf. on Engineering in Medicine and Biology, p. 110.

Hadley, D. W. and Ward, I. M. (1965) Non-linear creep and recovery behaviour of polypropylene fibres. *J. Mech. Phys. Solids*, **13**, 397.

Hopkins, I. L. and Hamming, R. W. (1957) On creep and relaxation. *J. Appl. Phys.*, **28**(8), 906–9.

Lai, J. S. Y. and Findley, W. H. (1968a) Prediction of uniaxial stress relaxation from creep of nonlinear viscoelastic material. *Trans. Soc. Rheol.*, **12**, 243–57.

Lai, J. S. Y. and Findley, W. N. (1968b) Stress relaxation of nonlinear viscoelastic material under uniaxial strain. *Trans. Soc. Rheol.*, **12**, 259–80.

Lee, E. H. (1955) Stress analysis in viscoelastic bodies. *Q. Appl. Math.*, **13**, 183–90.

Lee, E. H. (1958) Viscoelastic stress analysis, in Proc. 1st Symp. on Naval Structural Mechanics, Pergamon, London, pp. 456–82.

Mitsoulis, E. (1988) A heuristic approach to modeling viscoelasticity in polymer processing, Society of Plastics Engineers Technical Papers, Vol. XXXIV, ANTEC 1988, Atlanta, GA, pp. 140–4.

Morland, L. W. and Lee, E. H. (1960) Stress analysis for linear viscoelastic materials with temperature variation. *Trans. Soc. Rheol.*, **4**, 233–63.

Pister, K. and Distéfano, N. (1970) On some modelling and identification problems in biomechanics. *J. Biomed. Syst.*, **1**(2), 32–47.

Roesler, F. C. and Twyman, W. A. (1955) An iteration method for the determination of relaxation spectra. *Proc. Phys. Soc. London B*, **68**(2), 97–105.

Schapery, R. A. (1969) On the characterization of nonlinear viscoelastic materials. *Polym. Eng. Sci.*, **9**, 295–310.

Taylor, R. L., Pister, K. and Goureau, G. (1970) Thermomechanical analysis of viscoelastic solids. *Int. J. Numer. Methods Eng.*, **2**, 45–60.

Zakhariev, G., Khadzhikov, L. and Marinov, P. (1971) A rheological model for polymers and glass reinforced plastics. *Polym. Mech.*, **7**, 761–6.

8

Wave propagation

8.1 INTRODUCTION

When a localized disturbance is applied suddenly in a medium, it will soon propagate to other parts of this medium. This simple fact constitutes a general basis for the interesting subject of 'wave propagation'. Well-cited examples of wave propagation in different media include, for instance, the transmission of sound in air, the propagation of seismic disturbances in the earth and the transmission of radio waves. In the particular case when the suddenly applied disturbance is mechanical, e.g. a suddenly applied force, the resulting waves in the medium are due to stress effects and, thus, these waves are referred to as 'stress waves'. Our attention in this chapter is focused on the propagation of stress waves in viscoelastic solid media. In our representation, we consider the solid medium to be a continuum. Hence, the mechanics of wave motion in the medium will be dealt with from a continuous mechanics point of view. The basic concepts of continuum mechanics have been presented in Chapter 1. In such a continuum, the solid medium, the disturbance is generally considered to spread outward in a three-dimensional sense (Graff, 1975). A wavefront is considered to be associated with the outwardly propagating disturbance. Consequently, particles of the medium that are located ahead of the wavefront are assumed to have experienced no motion; meantime, particles that are located behind the front are visualized to have experienced motion and may continue to vibrate for some time.

For a continuum solid, two distinct effects due to a disturbance input may be generally encountered:

1. the solid may transit tensile and compressive stresses and the motion of particles would be generally in the direction of the wave motion;
2. in addition, the solid may transit shear stress, and thus the motion of the particles would be in a direction transverse to the direction of wave motion.

During their motion, waves propagating in a solid may encounter or interact with boundaries of the medium. On striking a boundary, part or whole of an incident wave may be reflected and the mode of propagation of the wave may change.

Although, in this chapter, our main interest is to study the propagation of stress waves in viscoelastic materials, we devote the next section to the consideration of the

motion of elastic waves. This assists as an introductory step to the subject of viscoelastic waves with which we deal in the remainder of the chapter. The comparison
between the strain responses to a pulse of constant stress input, of a specific time
duration, applied to initially undisturbed elastic and viscoelastic material specimens
has been discussed briefly in the introduction to linear viscoelasticity in Chapter 2.

In recent years, there has been considerable interest in the subject of wave
propagation from both theoretical and experimental points of view. Such interest
was motivated primarily by the advancement in testing and measurement techniques.
With the recent progress in fields such as electronics and laser optics (e.g. holographic
interferometry), stress waves of high frequency can be now produced and detected
easily. This has been particularly pronounced in important domains of nondestructive
testing such as ultrasonics and acoustic emission. The combination of these two
techniques has led further to the newly developed acousto-ultrasonic technique as a
modern nondestructive tool for the evaluation and prediction of the mechanical
properties of engineering materials (Vary, 1988; Tanary and Haddad, 1988). Another
equally important cause for the ensuing interest in the subject of wave propagation
is the continuous emerging of newly developed industrial materials such as plastics
and polymeric composite material systems. In this, the study of the phenomenon of
wave motion has been able to identify microstructural problems and to assist in the
development of homogeneous and inhomogeneous material systems. Further, any
new development within the realm of smart materials and structures (Rogers, Barker
and Jaeger, 1988; Tanagi, 1990; Srinivasan and Haddad, 1992) is expected to depend
on the ability to employ wave propagation as a successful detecting mechanism for
a feedback concerning the status of such materials.

For a historical background of the subject of wave propagation, the reader is
referred to Kolsky (1963), Graff (1975) and others. For a review of the common
methods for producing and detecting stress waves in solids, reference is made, for
instance, to the books by Hetenyi (1950), Dove and Adams (1964), Dally and Riley
(1965), Keast (1967), and Magrab and Blomquist (1971). Comprehensive review
articles in this area are due to Hillier (1960), Kolsky (1958, 1960), Worely (1962)
and others. Experimental studies of the dynamic stress–strain relations of materials
under the conditions of shock loading have become significantly important in recent
years. The increased interest in the subject matter has been motivated by the
increasing number of applications and, as well, by the contributions provided by such
studies to a better understanding of the mechanisms of deformation of engineering
materials (for example, Barker and Hollenbach, 1964, 1970 and Frederick, 1965).

8.2 WAVE PROPAGATION IN ELASTIC SOLIDS

8.2.1 Wave propagation in unbounded elastic solids

An unbounded solid is considered to extend indefinitely in the three dimensions of
space so that the complications which might arise from reflections of waves at the
boundaries of the medium might be disregarded.

The equations of motion of a continuum have been derived in Chapter 1 (section 1.3). These equations (1.22) were presented in terms of the stress components acting on a small parallelepiped of the continuum without the inclusion of the response behaviour of the medium. However, in order to employ these equations in the study of wave propagation, one may substitute the stress components by the corresponding components of strain through the use of the constitutive relationships of the particular medium under consideration.

For an isotropic elastic solid, the stress–strain relations can be expressed (e.g. Sokolnikoff, 1956) in component form as

$$\sigma_{11} = \lambda\Delta + 2\mu\varepsilon_{11}, \quad \sigma_{22} = \lambda\Delta + 2\mu\varepsilon_{22}, \quad \varepsilon_{33} = \lambda\Delta + 2\mu\varepsilon_{33},$$

$$\sigma_{23} = \mu\varepsilon_{23}, \quad \sigma_{31} = \mu\varepsilon_{31}, \quad \sigma_{12} = \mu\varepsilon_{12}. \tag{8.1}$$

In the above relations, $\Delta = \varepsilon_{kk} = \varepsilon_{11} + \varepsilon_{22} + \varepsilon_{33}$ is the dilatation which represents the change in volume of unit cube of the solid and λ and μ are Lamé's elastic constants. In the theory of elasticity, four elastic (material) constants are usually used. These are Young's modulus E, Poisson's ratio v, the bulk modulus K and the rigidity (shear) modulus which is Lamé's constant μ. From the definitions of these constants and using equations (8.1) the following relations between the constants, in the case of an isotropic elastic solid, can be determined:

$$E = \frac{\mu(3\lambda + 2\mu)}{\lambda + \mu}, \quad v = \frac{\lambda}{2(\lambda + \mu)}, \quad K = \lambda + \frac{2\mu}{3}. \tag{8.2}$$

Substituting from the constitutive relations (8.1) for the stress components in the equations of motion (1.22), the equation of motion for an isotropic elastic solid, in the absence of body forces, can be written in the x_1 direction in terms of the strain as

$$\rho\frac{\partial^2 u_1}{\partial t^2} = \frac{\partial}{\partial x_1}(\lambda\Delta + 2\mu\varepsilon_{11}) + \frac{\partial}{\partial x_2}(\mu\varepsilon_{12}) + \frac{\partial}{\partial x_3}(\mu\varepsilon_{13}) \tag{8.3}$$

where u_1 is the displacement component in the x_1 direction. Replacing the strain components in (8.3) by the corresponding displacement components from (1.73), it follows that

$$\rho\frac{\partial^2 u_1}{\partial t^2} = (\lambda + \mu)\frac{\partial\Delta}{\partial x_1} + \mu\nabla^2 u_1. \tag{8.4a}$$

∇^2 is Laplace's operator defined by

$$\nabla^2 = \frac{\partial^2}{\partial x_1^2} + \frac{\partial^2}{\partial x_2^2} + \frac{\partial^2}{\partial x_3^2}.$$

Similar relations to (8.4a) can be established for the other two components of the displacement vector, namely

$$\rho\frac{\partial^2 u_2}{\partial t^2} = (\lambda + \mu)\frac{\partial\Delta}{\partial x_2} + \mu\nabla^2 u_2 \tag{8.4b}$$

and

$$\rho \frac{\partial^2 u_3}{\partial t^2} = (\lambda + \mu) \frac{\partial \Delta}{\partial x_3} + \mu \nabla^2 u_3. \qquad (8.4c)$$

Equations (8.4) are the equations of motion, in terms of the displacement, for an isotropic elastic solid in the absence of body forces. These equations may be expressed conveniently in a vector form as

$$\rho \frac{\partial^2 u_2}{\partial t^2} = (\lambda + \mu) \nabla \nabla \cdot \mathbf{u} + \mu \nabla^2 \mathbf{u} \qquad (8.5)$$

which is the form of the well-known Navier equation of motion. The latter is conveniently adopted as the governing equation for the motion of an isotropic, elastic solid. Equation (8.5) corresponds to the propagation of two types of waves through an unbounded isotropic, elastic solid, namely 'dilatational' and 'rotational' waves.

Differentiating (8.4a) with respect to x_1, (8.4b) with respect to x_2 and (8.4c) with respect to x_3 and adding the resulting derivations, one obtains the following 'wave equation' for an unbounded isotropic, elastic medium:

$$\rho \frac{\partial^2 \Delta}{\partial t^2} = (\lambda + 2\mu) \nabla^2 \Delta \qquad (8.6)$$

The above wave equation indicates that the dilatation Δ propagates through the medium with a velocity of magnitude $[(\lambda + 2\mu)/\rho]^{1/2}$. Denoting the latter by c_1, then $c_1 = [(\lambda + 2\mu)/\rho]^{1/2}$.

In view of equations (8.2), the magnitude of the dilatational wave velocity c_1 may be expressed further by

$$c_1 = \left[\frac{(\lambda + 2\mu)}{\rho} \right]^{1/2} = \left[\frac{E(1 - v)}{\rho(1 + v)(1 - 2v)} \right]^{1/2} = \left[\frac{K + 4\mu/3}{\rho} \right]^{1/2}. \qquad (8.7)$$

It is noticed from (8.7) that the velocity c_1 is dependent only on the elastic constants as well as the density of the elastic material. In an operational form, the wave equation (8.6) can be written as

$$\Gamma_1^2 \Delta = 0 \qquad (8.8)$$

where Γ_1^2 is a dilatational wave operator expressed by (Chou, 1968)

$$\Gamma_1^2 = \nabla^2 - \frac{1}{c_1^2} \frac{\partial^2}{\partial t^2} \qquad (8.9)$$

and $\Delta = \nabla \cdot \mathbf{u}$ is the dilatation.

A 'dilatational' wave, corresponding to the wave equation (8.8), is also referred to as 'irrotational' since the propagation of such a wave involves no rotation of an elemental volume of the solid. A dilatational wave is also known as a 'bulk wave' or 'primary (P) wave'.

On the other hand, if we eliminate the dilatation Δ between (8.4b) and (8.4c), that is by differentiating (8.4b) with respect to x_3 and (8.4c) with respect to x_2, and subtracting, then

$$\rho \frac{\partial^2}{\partial t^2}\left(\frac{\partial u_3}{\partial x_2} - \frac{\partial u_2}{\partial x_3}\right) = \mu \nabla^2\left(\frac{\partial u_3}{\partial x_2} - \frac{\partial u_2}{\partial x_3}\right).$$

This equation can be written, in view of (1.69), as

$$\rho \frac{\partial^2 \omega_1}{\partial t^2} = \mu \nabla^2 \omega_1$$

where ω_1 is the rotation about the x_1 axis. Similar relations can be obtained for ω_2 and ω_3 (the rotations about the x_2 and x_3 axes, respectively). Thus, in generalized notation, one can write

$$\rho \frac{\partial^2 \boldsymbol{\omega}}{\partial t^2} = \mu \nabla^2 \boldsymbol{\omega} \tag{8.10}$$

where $\boldsymbol{\omega} = \nabla \times \mathbf{u}/2$ is the rotation vector.

It follows from (8.10) that the rotational wave propagates in an isotropic, elastic solid with a velocity of magnitude $(\mu/\rho)^{1/2}$. We denote the magnitude of the rotational wave velocity by c_2; then

$$c_2 = (\mu/\rho)^{1/2}. \tag{8.11}$$

It is noticed, from the above expression, that the velocity c_2 is dependent only on the elastic constants as well as the density of the elastic material.

Applying the vector operator curl (Appendix A) to (8.5), it can be shown that the vector form of the wave equation (8.10) can be written as

$$\Gamma_2^2 \boldsymbol{\omega} = 0 \tag{8.12}$$

where Γ_2^2 is a rotational wave operator of the form (Chou, 1968)

$$\Gamma_2^2 = \nabla^2 - \frac{1}{c_2^2}\frac{\partial^2}{\partial t^2} \tag{8.13}$$

and $\boldsymbol{\omega} = \nabla \times \mathbf{u}$ is the rotation.

A 'rotational' wave is also called an 'equivoluminal' wave since there is no volume change during the propagation of the wave. A rotational wave is also known as a 'distortional' wave or 'secondary (S) wave'.

Equation (8.8), or (8.12), is a necessary, but not a sufficient, condition for the satisfaction of the Navier governing equation of motion (8.5). Thus, for every displacement field that satisfies (8.5), the corresponding Δ and $\boldsymbol{\omega}$ will satisfy (8.8) and (8.12), respectively. On the other hand, a displacement field with a dilatation satisfying (8.8), or a rotation satisfying (8.12), would not be necessarily a solution of the Naviers' governing equation (8.5).

As mentioned previously, the particle motion in a dilatational wave is longitudinal, i.e. along the direction of wave propagation. In case of a rotational wave, the particle motion is transverse, that is perpendicular to the direction of propagation of the wave. Experimentally, one would generally attempt to generate one type of wave with the exclusion of the other. However, it should be emphasized that, in the propagation of dilatational waves in an unbounded solid, the medium would not be simply subjected to pure compression, but to a combination of compression and shear. This is supported by the physical situation and mathematically by the appearance of both the bulk modulus and the shear modulus in the expression of the dilatational velocity (equation (8.7)), as shown for instance in Kolsky (1963) and Tschoegl (1989).

(a) Irrotational and rotational displacement fields

Consider the displacement vector field **u**. In dynamic elasticity, u may be decomposed into an irrotational field, say $\mathbf{u}_{IR}$, associated with a scalar potential ϕ and a rotational field, $\mathbf{u}_R$, associated with a vector potential $\boldsymbol{\psi}$. Thus, according to the Helmholtz theorem (Morse and Feshbach, 1953), for any displacement field, subject to mild continuity and boundary conditions, one may find at least one set of functions ϕ and $\boldsymbol{\psi}$ such that

$$\mathbf{u} = \nabla\phi + \nabla \times \omega, \quad \nabla\cdot = 0. \tag{8.14}$$

The condition $\nabla \cdot \boldsymbol{\psi} = 0$ is necessary to determine uniquely the three components of the displacement vector u from the four components of ϕ, $\boldsymbol{\psi}$.

Substituting (8.14) into Navier's equation (8.5) yields

$$c_1^2\nabla\nabla^2\phi + c_2^2\nabla \times (\nabla^2\boldsymbol{\psi}) = (\nabla\phi + \nabla \times \boldsymbol{\psi})\mathbf{u}. \tag{8.15}$$

Every solution of (8.14), or (8.15), is always a solution of (8.5). Accordingly, equations (8.14) and (8.15) are also governing equations of the induced motion in an isotropic, elastic solid and each constitutes an exact equivalence to (8.5) (Chou, 1968).

A particular class of solutions of (8.15) is

$$\nabla\Gamma_1^2\phi = \mathbf{0}, \quad \nabla \times \Gamma_2^2\boldsymbol{\psi} = \mathbf{0}, \tag{8.16}$$

with a particular solution

$$\Gamma_1^2\phi = 0; \quad \Gamma_2^2\boldsymbol{\psi} = \mathbf{0}. \tag{8.17}$$

This is with the understanding that the class of solutions presented by (8.16) and (8.17) is sufficient, but not necessary, for the satisfaction of (8.5). In equation (8.17), Γ_1^2 and Γ_2^2 are the dilatational and rotational wave operators introduced earlier by equations (8.9) and (8.13), respectively.

(b) An irrotational field

A displacement field, **u**, is referred to as 'irrotational' if

$$\nabla \times \boldsymbol{\psi} = \mathbf{0}, \quad \mathbf{u} = \mathbf{u}_{IR}. \tag{8.18}$$

For an irrotational wave, one has, following equation (8.5)

$$\Gamma_1^2 u_{IR} = 0 \qquad (8.19)$$

or, alternatively, according to potential theory,

$$\mathbf{u}_{IR} = \nabla\phi \qquad (8.20)$$

where ϕ is a scalar potential function. Equation (8.20) implies that, for an irrotational wave, the rotational vector $\boldsymbol{\omega}$ is equal to zero in magnitude. Following (8.17), then, for an irrotational field

$$\Delta = \nabla^2\phi \quad \text{and} \quad \frac{\partial\Delta}{\partial x_i} = \nabla^2 u_i \qquad (i = 1, 2, 3). \qquad (8.21)$$

Accordingly, the scalar potential ϕ is seen to be associated with the dilatational (irrotational) part of the disturbance. Substituting (8.21) into (8.4), one has, for an irrotational field,

$$\rho \frac{\partial^2 u_i}{\partial t^2} = (\lambda + 2\mu)\, \nabla^2 u_i \qquad (i = 1, 2, 3). \qquad (8.22)$$

(c) Rotational field

A displacement field, $\mathbf{u}$, is called rotational if

$$\nabla \cdot \mathbf{u} = 0, \quad \mathbf{u} = \mathbf{u}_R. \qquad (8.23)$$

For a rotational field, the Navier governing equation (8.5) results in

$$\Gamma_2^2 \mathbf{u} = 0 \qquad (8.24)$$

and

$$\mathbf{u} = \nabla \times \boldsymbol{\psi}, \qquad (8.25)$$

i.e. the vector potential $\boldsymbol{\psi}$ is associated with the rotational part of the disturbance.

The above conditions for a rotational wave translate into, in this case, the dilatation $\Delta = 0$. Hence, the set of equations (8.4) reduces, for a rotational wave, to

$$\rho \frac{\partial^2 u_i}{\partial t^2} = \mu\, \nabla^2 u_i \qquad (i = 1, 2, 3). \qquad (8.26)$$

Combining equations (8.14), (8.20) and (8.25), it follows that, in an isotropic, elastic solid, a displacement field $\mathbf{u}$ is decomposed vectorially into an irrotational field $\mathbf{u}_{IR}$ and a rotational one $\mathbf{u}_R$. That is,

$$\mathbf{u} = \mathbf{u}_{IR} + \mathbf{u}_R. \qquad (8.27)$$

Further, in view of (8.17), (8.20) and (8.25), it may be concluded that, for every displacement field that satisfies (8.5), there exists a set of functions $\mathbf{u}_{IR}$ and $\mathbf{u}_R$ such that (equations (8.19) and (8.24))

$$\Gamma_1^2 \mathbf{u}_{IR} = 0 \quad \text{and} \quad \Gamma_2^2 \mathbf{u}_R = 0. \tag{8.28}$$

This translates, physically, into the following.

A disturbance in an isotropic, elastic solid would generate two waves, one dilatational, involving no rotation, with velocity c_1 and the other is rotational, involving no volume changes, that propagates at velocity c_2. The ratio of the two speeds may be expressed, with reference to (8.7) and (8.11), as

$$\frac{c_1}{c_2} = \frac{2 - 2v}{1 - 2v}$$

Since $0 \le v \le 1$, then $c_1 > c_2$.

In view of (8.28), these two waves are not coupled within the continuous solid (except perhaps on the boundary where the prescribed boundary conditions must be satisfied).

Table 8.1 summarizes the relationships given in the foregoing, in terms of displacements, while Table 8.2 gives such relationships in terms of potentials. Chou (1968) should also be referred to.

8.2.2 Plane waves in unbounded elastic media

Plane waves are propagating disturbances in two or three dimensions where the motion of every particle in planes perpendicular to the direction of propagation is the same. An example of a propagating (three-dimensional) plane disturbance is given in Fig. 8.1. As shown in the figure, the magnitude of the propagation velocity of the plane is denoted by c while the normal to the plane is designated by $\mathbf{n}$. The position of an arbitrary point P on the plane is indicated by $\mathbf{r}$.

For the plane wave illustrated in Fig. 8.1, the motion of every particle along the plane is defined by

$$\mathbf{u} \cdot \mathbf{r} - ct = \text{constant}. \tag{8.29}$$

Consider now the plane wave

$$\mathbf{u} = \mathbf{A}f(\mathbf{n} \cdot \mathbf{r} - ct) \tag{8.30}$$

where $\mathbf{A}$ is the displacement vector of the particle along the plane of the wave and $f(\cdot)$ indicates an appropriate function of the shown argument. Substituting (8.30) in Navier's governing equation of motion (8.5), it can be shown that

$$(\lambda + \mu)A_j n_j n_i + \mu A_i = \rho c^2 A_i. \tag{8.31}$$

Relation (8.31) represents three homogeneous equations in the amplitude components A_1, A_2, A_3. This leads, on expanding the determinant of coefficients, to

$$(\lambda + 2\mu - \rho c^2)(\mu - \rho c^2)^2 = 0. \tag{8.32}$$

Table 8.1 Wave propagation in an isotropic, elastic (unbounded) solid: Pertaining relations in terms of displacement

---

Displacement field **u**
General governing equation: Navier's

$$\rho\frac{\partial^2 \mathbf{u}}{\partial t^2} = \mu\nabla^2\mathbf{u} + (\lambda + \mu)\,\nabla\nabla\cdot\mathbf{u}$$

Two propagating waves
  Dilatational (irrotational), $(\mathbf{u}_{IR},\, c_1)$
  Rotational (distortional), $(\mathbf{u}_R,\, c_2)$

Necessary and sufficient relations for the satisfaction of Navier's governing equation (above)

$$\mathbf{u} = \mathbf{u}_{IR} + \mathbf{u}_R$$

$$\left(\nabla^2 - \frac{1}{c_1^2}\frac{1}{\partial t^2}\right)\mathbf{u}_{IR} = 0$$

$$\left(\nabla^2 - \frac{1}{c_2^2}\frac{1}{\partial t^2}\right)\mathbf{u}_{R} = 0$$

Necessary but not sufficient relations for the satisfaction of Navier's governing equation

  Dilatation Δ:

$$\Delta = \nabla\cdot\mathbf{u};\ \left(\nabla^2 - \frac{1}{c_1^2}\frac{\partial^2}{\partial t^2}\right)\Delta = 0$$

  Rotation $\boldsymbol{\omega}$:

$$\boldsymbol{\omega} = \nabla\times\mathbf{u};\ \left(\nabla^2 - \frac{1}{c_2^2}\frac{\partial^2}{\partial t^2}\right)\boldsymbol{\omega} = 0$$

Sufficient but not necessary relations for the satisfaction of Navier's governing equation
  Dilatational (irrotational)

$$\nabla\times\mathbf{u} = 0;\ \left(\nabla^2 - \frac{1}{c_1^2}\frac{\partial^2}{\partial t^2}\right)\mathbf{u} = 0$$

  Rotational (distortional)

$$\nabla\cdot\mathbf{u} = 0;\ \left(\nabla^2 - \frac{1}{c_2^2}\frac{\partial^2}{\partial t^2}\right)\mathbf{u} = 0$$

where

$$c_1^2 = \frac{\lambda + 2\mu}{\rho} = \frac{E(1-v)}{\rho(1+v)(1-2v)} = \frac{K + \frac{4}{3}\mu}{\rho}$$

$$c_2^2 = \mu/\rho$$

Table 8.2 Wave propagation in an isotropic, elastic (unbounded) solid: Relationships between Navier's equation and other related governing equations in terms of potential

General governing equation: Navier's

$$\rho \frac{\partial^2 \mathbf{u}}{\partial t^2} = \mu \, \nabla^2 \mathbf{u} + (\lambda + \mu) \, \nabla \nabla \cdot \mathbf{u}$$

Displacement field $\quad\quad\quad\quad\quad\quad \mathbf{u}(\phi, \boldsymbol{\psi})$

Necessary and sufficient relations for the satisfaction of Navier's governing equation (above)

$$\mathbf{u} = \nabla \phi + \nabla \times \boldsymbol{\psi}; \; \nabla \cdot \boldsymbol{\psi} = 0$$

$$\nabla \left(\nabla^2 - \frac{1}{c_1^2} \frac{\partial^2}{\partial t^2} \right) \phi + \nabla \times \left(\nabla^2 - \frac{1}{c_2^2} \frac{\partial^2}{\partial t^2} \right) \boldsymbol{\psi} = 0$$

$$\left(\nabla^2 - \frac{1}{c_1^2} \frac{\partial^2}{\partial t^2} \right) \phi = 0; \left(\nabla^2 - \frac{1}{c_2^2} \frac{\partial^2}{\partial t^2} \right) \boldsymbol{\psi} = 0$$

Necessary but not sufficient relations for the satisfaction of Navier's governing equation

$$\left(\nabla^2 - \frac{1}{c_1^2} \frac{\partial^2}{\partial t^2} \right) \nabla^2 \phi = 0$$

$$\left(\nabla^2 - \frac{1}{c_2^2} \frac{\partial^2}{\partial t^2} \right) \nabla^2 \boldsymbol{\psi} = 0$$

where

$$c_1^2 = \frac{\lambda + 2\mu}{\rho} = \frac{E(1 - v)}{\rho(1 + v)(1 - 2v)} = \frac{K + \frac{4}{3}\mu}{\rho}$$

$$c_2^2 = \mu/\rho$$

This equation gives the two roots

$$c_1 = \left(\frac{\lambda + 2\mu}{\rho} \right)^{1/2}$$

and $\hspace{20em}$ (8.33)

$$c_2 = (\mu/\rho)^{1/2}$$

which again are, respectively, the magnitudes of the velocities of dilatational and rotational waves. Accordingly, plane waves may propagate at one or the other velocity (i.e. c_1 or c_2) in the isotropic, elastic medium.

8.2.3 Wave propagation in semi-infinite elastic media

When a stress wave encounters a boundary between two media, energy is reflected and transmitted from and across the boundary. On the other hand, if the boundary

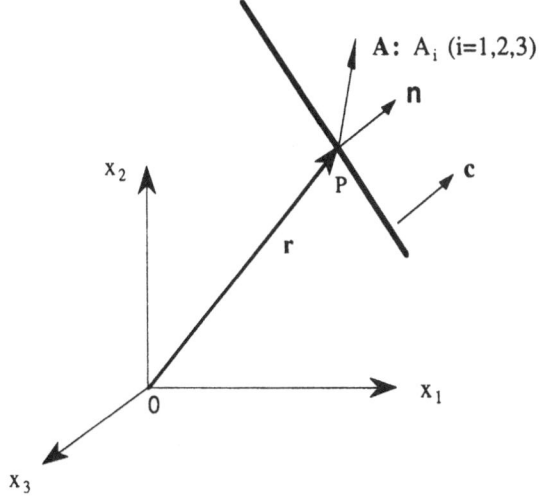

Fig. 8.1 Plane wave motion in an unbounded, elastic medium.

is a free surface, reflection of the waves will be much more pronounced. It is well recognized that a characteristic phenomenon of the elastic wave–boundary interaction in solids is that of mode coonversion. In this, an incident wave, either pressure or shear, on the boundary will be converted into two waves on reflection. Such a mode-conversion phenomenon along with the fact that two types of waves may exist in an elastic solid, as discussed earlier, accounts for the relative complexity of wave propagation in solids in general as compared with equivalent problems in acoustics and electromagnetics (e.g. Graff, 1975).

(a) Governing equations

With reference to Fig. 8.2, we consider, following Graff (1975), plane harmonic waves propagating in the half-space $x_2 \geq 0$. It is assumed that the wave normal n lies in the x_1x_2 plane. This plane will be referred to as the vertical plane while the x_1x_3 plane, the surface of the half-space, will be referred to as the horizontal plane. Recalling the previous discussion concerning the propagation of plane waves in infinite media (section 8.2.2), it is recognized that the particle motion due to dilatation will be in the direction of the wave normal and will, thus, be in the vertical plane only. The transverse particle motion, however, is due to shear and will have components both in the vertical plane and parallel to the horizontal plane. In Fig. 8.2, the normal displacement component is designated by u_n and the transverse components are denoted by u_v and u_3 which are, respectively, in the vertical and horizontal planes. As every particle along the plane of the wave is acquiring the same motion, the motion will be invariant with respect to x_3 if the wave normal is in the vertical plane. In terms of the potentials ϕ and ψ, the governing equations

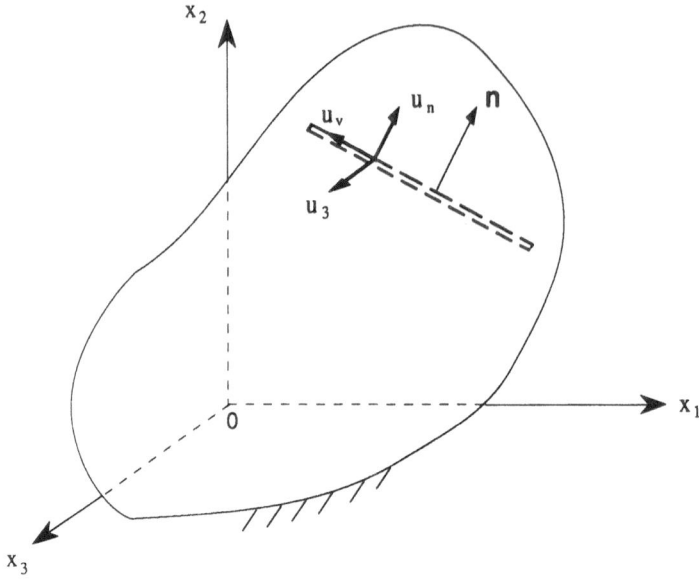

Fig. 8.2 Wave motion in a semi-infinite elastic medium.

can be expressed as

$$u_1 = \frac{\partial \phi}{\partial x_1} + \frac{\partial \psi_3}{\partial x_2},$$

$$u_2 = \frac{\partial \phi}{\partial x_2} - \frac{\partial \psi_3}{\partial x_1},$$

$$u_3 = -\frac{\partial \psi_1}{\partial x_2} + \frac{\partial \psi_2}{\partial x_1},$$

$$\frac{\partial \psi_1}{\partial x_1} + \frac{\partial \psi_2}{\partial x_2} = 0$$

and   (8.34)

$$\nabla^2 \phi = \frac{1}{c_1^2} \frac{\partial^2 \phi}{\partial t^2}, \quad \nabla^2 \psi_i = \frac{1}{c_2^2} \frac{\partial^2 \psi_i}{\partial t^2}$$

where ψ_i $(i = 1, 2, 3)$ are the components of the vector function $\boldsymbol{\psi}$. In deriving the above governing equations both the postulate $\nabla \cdot \boldsymbol{\psi} = 0$ and the x_3 independence of all quantities have been used.

Combining the displacement expressions in (8.34) with the stress–displacement constitutive relations for the isotropic elastic solid, the stress components can be established in terms of the potentials ϕ and ψ, i.e.

$$\sigma_{11} = (\lambda + 2\mu)\left(\frac{\partial^2 \phi}{\partial x_1^2} + \frac{\partial^2 \phi}{\partial x_2^2}\right) - 2\mu\left(\frac{\partial^2 \phi}{\partial x_2^2} - \frac{\partial^2 \psi_3}{\partial x_1 \partial x_2}\right),$$

$$\sigma_{22} = (\lambda + 2\mu)\left(\frac{\partial^2 \phi}{\partial x_1^2} + \frac{\partial^2 \phi}{\partial x_2^2}\right) - 2\mu\left(\frac{\partial^2 \phi}{\partial x_1^2} + \frac{\partial^2 \psi_3}{\partial x_1 \partial x_2}\right)$$

$$\sigma_{12} = \mu\left(2\frac{\partial^2 \phi}{\partial x_1 \partial x_2} + \frac{\partial^2 \psi_3}{\partial x_2^2} - \frac{\partial^2 \psi_3}{\partial x_1^2}\right),$$

$$\sigma_{23} = \mu\left(-\frac{\partial^2 \psi_1}{\partial x_2^2} + \frac{\partial^2 \psi_2}{\partial x_1 \partial x_2}\right) \quad \text{and} \quad \sigma_{13} = 0$$

(8.35)

with boundary conditions

$$\sigma_{22} = \sigma_{21} = \sigma_{23} = 0, \quad x_2 = 0. \tag{8.36}$$

Experimental studies on wave propagation in semi-infinite media may vary considerably in scope. Ultrasonic excitation is often used as an impulsive surface force; meantime, photoelasticity has been conventionally adopted as a recording technique for the patterns of wave motion in elastic materials. Dally, Durelli and Riley (1960), for instance, used small explosive charges of lead azide (PbN_6) to load dynamically a low-modulus urethane rubber plate and the dynamic fringe propagation patterns were recorded by a high-speed camera (Dally, 1968 and Graff, 1975). Dally and Riley (1967) used an embedded polariscope technique to study experimentally the three-dimensional problem of a point load on a half-space using a photo-elastic method (e.g. Pindera, 1986). Riley and Dally (1966) considered the application of the photoelastic recording technique to study the wave motion in layered media.

(b) Surface waves

When the solid has a free surface, 'Rayleigh' surface waves can also exist. These waves were first introduced by Rayleigh (1887) (Lamb (1904) should also be referred to) who showed that their effect decays rapidly with depth and that their velocity is less than that of body waves c_1 and c_2. It is shown by Kolsky (1963) that Rayleigh waves do, in fact, travel with a fraction ξ of the velocity c_2 of distortional waves where ξ is obtained from the equation

$$\xi^6 - 8\xi^4 + (24 - 16b^2)\xi^2 + 16b^2 - 16 = 0. \tag{8.37}$$

In the above equation, b is an elastic constant of the material expressed by

$$b = [(1 - 2v)/(2 - 2v)]^{1/2} \tag{8.38}$$

where v is Poisson's ratio.

In Rayleigh waves, the particle motion is parallel to the direction of wave propagation and it is in a plane perpendicular to the surface containing the waves during travel.

In case of an elastic solid, the velocity of a surface wave is independent of the frequency and depends, similarly to the body waves, on the elastic constants of the material. In other words, there is no dispersion (change of form) of these waves, i.e. a plane surface wave will travel without change in form.

When a dilatational wave is incident on a free surface with an angle α (Fig. 8.3(a)), two waves are generated on reflection: one is a dilatational wave reflected at an angle equal to the angle of incidence α, while the other is a distortional wave reflected at a smaller angle β where $\sin \beta / \sin \alpha = c_2/c_1$.

Similarly, if a distortional wave is incident on a free surface at an angle γ (Fig. 8.3(b)), both distortional and dilatational waves are generally reflected. The distortional wave is reflected at the same angle γ while the dilatational wave is reflected at a generally smaller angle δ where $\sin \gamma / \sin \delta = c_2/c_1$.

Tatel (1954) used a 'model seismogram' approach to study surface wave motion on a half-space. In this, Tatel induced a point impulse on the surface of a large block of steel using a piezo-electric transducer. The vertical components of displacement were detected by a receiving transducer a few centimetres away. Dally and Thau (1967) considered the application of the photoelastic technique to study surface wave propagation (Thau and Dally (1969) should also be referred to). Viktorov (1967) reported a number of studies, using ultrasonics, on the effects of surface defects on Rayleigh waves. Other work in this area is, for instance, due to Goodier, Jahsman

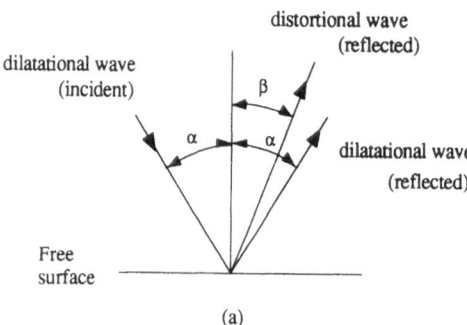

(a)

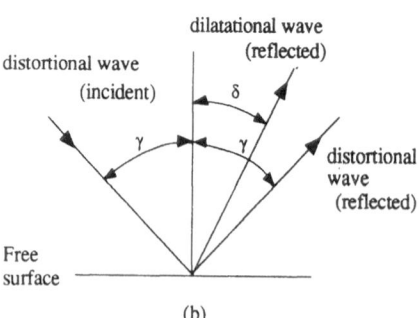

(b)

Fig. 8.3 Reflection of (a) dilatational and (b) distortional waves.

and Ripperger (1959), Dally and Lewis (1968) and Reinhardt and Dally (1970), amongst others.

8.2.4 Wave propagation in bounded elastic solids

A typical representation of a bounded medium is a material specimen of finite dimensions. In the study of the wave propagation in bounded solids, the geometry of the material specimen would be kept simple such that the complex end effects that could be caused by the boundaries are minimized. In rods, a thin rod of uniform circular cross-section is often taken as a simple representation of a bounded medium. In this case, the disturbance would be applied at one end of the rod and propagate along its length whereby the ratio between the dimension of the transverse direction (the radius) and the wave length becomes an important factor in the mode of travel of the propagating wave. The wave equations for rods and plates infinite in length have been known for a long period of time. Chree (1889), among others, developed the results for elastic rods and Rayleigh (1887) and Lamb (1917) established the wave equations for elastic plates. Davies (1948) studied the propagation characteristics of a transient disturbance in rods. Considerable research efforts have been undertaken subsequently whereby particular attention has been given to the determination of the frequency spectra, analysis of transient disturbances and applications concerning wave propagation in plates, rods and shells.

Stress wave propagation in thin rods and bars are conventionally classified in three categories, namely longitudinal, torsional and lateral.

In a **longitudinal** wave, the disturbance may be in the form of a uniaxial force or displacement applied in the direction of the longitudinal axis of the rod. In this case, segments of the rod undergo extension and compression without rotation and with no lateral displacement of the axis of the rod.

If the wavelength is long compared with the lateral dimensions of the rod, longitudinal waves travel with constant velocity given by

$$c_1 = (E/\rho)^{1/2} \tag{8.39}$$

where E is Young's modulus.

When the wavelength is comparable with the lateral dimensions of the rod, the velocity of the longitudinal waves would depend on the wavelength and may approach the velocity of Rayleigh surface waves at very short wavelengths.

A large number of experimental studies on the propagation of longitudinal waves in rods are directed towards the determination of material properties of the rod. In case of elastic solids, the method of mechanical impact is considered to be the most conventional testing method. It is carried out by impacting one solid against the other (Barton, Volterra and Citron, 1958; Ripperger, 1953). Goldsmith, Polivka and Yang (1966), for instance, considered pulse propagation in concrete rods with the aim of evaluating the effects of an inhomogeneous microstructure on wave propagation. Other work of interest on longitudinal waves in elastic rods includes that of Fisher (1954), Becker and Carl (1962) and Lindholm (1964) among others.

The early developments in the subject of wave propagation in plates are due to Rayleigh (1887) and Lamb (1917). The Rayleigh–lamb work is concerned with the propagation of continuous, straight crested waves in plates infinite in length and have traction-free surfaces.

In the case of stress wave propagation in plates and when the wavelength is long compared with the thickness of the plate, the magnitude of the velocity of propagation of a plane longitudinal wave is constant. It is given by

$$c_L = \left[\frac{4\mu(\lambda + \mu)}{(\lambda + 2\mu)\rho}\right]^{1/2} = \left[\frac{E}{\rho(1 - v^2)}\right]^{1/2} \tag{8.40}$$

On the other hand, if the wavelength is short compared with the thickness of the plate, plane longitudinal waves would travel with the velocity of Rayleigh surface waves. Experimental studies of the propagation of elastic waves in plates were carried out, for instance, by Dohrenwend, Drucker and Moore (1944), Press and Oliver (1955) and Medick (1961), among others.

In the case of **torsional** waves, the disturbance may be in the form of a torque or an angular displacement applied at the end of the rod. Here, each transverse section of the rod would rotate in its plane while the axis of the rod encounters no motion.

If the wavelength is long compared with the lateral dimensions of the rod, torsional waves would travel with constant velocity. Its magnitude is given by the following relation for a rod of circular cross-section:

$$c_t = (\mu/\rho)^{1/2} \tag{8.41}$$

where μ is the shear modulus. In view of the definition of the torsional wave propagation mentioned above, it is frequently assumed that such motion is independent of the wavelength.

In **lateral** 'flexural' wave propagation, flexure of portions of the rod occurs whereby segments of the central axis move laterally during wave propagation. In this motion, the velocity of wave propagation depends on the wavelength. For very long wavelengths, the velocity magnitude is given by

$$c_f = 2\pi c_1 k/\Lambda \tag{8.42}$$

where $c_1 = (E/\rho)^{1/2}$ (equation (8.39)), k is the radius of gyration of a cross-section of the rod about its neutral axis and Λ is the wavelength.

The velocity c_f of a flexural wave could also approach that of a Rayleigh surface wave when the wavelength becomes small in comparison with the lateral dimensions of the rod.

In the treatment of propagation of elastic disturbances along a cylindrical bar, it is often assumed in the literature that dispersion will occur only during the travel of flexural pulses, whilst longitudinal and torsional pulses will propagate along the bar without dispersion.

For a review of the phenomenon of wave motion in elastic solids and discussion of problems involved, the reader is further referred to Timoshenko (1921), Prescott

(1942), Hudson (1943), Davies (1948), Kolsky (1954a, b, 1963), Evans *et al.* (1954), Oliver (1957), Hsieh and Kolsky (1958), Graff (1975) and others.

8.3 TRANSITION TO WAVE MOTION IN VISCOELASTIC MATERIALS

8.3.1 Internal friction and dissipation

Real materials are never perfectly elastic. Thus, when a material specimen is subjected to dynamic loading, part of its mechanical energy is converted into heat. The various microstructural mechanisms by which the mechanical energy is converted into heat are conventionally referred to as 'internal friction' (Kolsky, 1963). Because of the complexity of the microstructure, several microscopic and macroscopic dissipative mechanisms exist in the material. The extent of energy loss would generally depend on the input load characteristics and the environmental conditions, as well as the inherent and macroscopic properties of the material specimen.

An internal dissipative mechanism in the case of polycrystalline solids, for instance, is due to the variation in crystallographic orientation of neighbouring grains. This results in nonuniformity of the distribution of local strains when the material specimen is loaded. This is in addition to the nonuniformity of local strains that may be caused by imperfections in the material (e.g. microcracks, fissures, flaws, foreign inclusions and grain boundaries). Consequently, a nonuniform temperature field may exist and thermal currents of varying magnitudes would flow within the crystal lattice. Other microscopic mechanisms could be also responsible for the transfer of energy into heat. One mechanism is due to dislocations, that is the movement of regions of disarray in the crystals (e.g. Orowan, 1934; Polanyi, 1934; Bradfield, 1951). An additional microscopic mechanism is due to the motion of solute atoms in the crystal lattice on the application of external loading (e.g. Gorsky, 1936; Snoek, 1941). A possible microscopic mechanism which attenuates stress waves in polycrystalline solids is 'scattering' (e.g. Kolsky, 1963). This mechanism may occur in a polycrystalline solid when the incident wavelength becomes comparable with the grain size. In this, Mason and McSkimin (1947), for instance, found that when the wavelength is long compared with the grain size, the attenuation is inversely proportional to fourth power of the wavelength (see Rayleigh, 1894).

On the macroscopic level, the following effects of internal friction are particularly important.

(a) Static hysteresis

This is primarily due to the anelastic characteristics of the material. In this case, when a material specimen is taken through a stress cycle, it would show a 'hysteresis loop', that is the stress–strain curve for an increasing stress input does not retrace its earlier downward path, if the material specimen is reloaded in an exact manner reflecting the unloading. The area enclosed by this loop represents mechanical energy which has been dissipated into heat. Although this effect may seem to be insignificant for some materials under static loading, it could be a pronounced factor in the attenuation of stress waves travelling in such materials. In the latter case, each layer of the material is taken through a loading cycle. For sinusoidal oscillations, for example, the number of hysteresis cycles is dependent on the frequency and the latter may be of the order of millions per second.

(b) Viscous loss

Such a loss is particularly noticeable in case of polymers with organic long-chain molecules. The internal forces here are of a viscous nature and imply that the mechanical behaviour of such materials is a function of the rate of strain (e.g. Tobolosky, Powell and Eyring, 1943; Alfrey, 1948; Kolsky, 1963). In the case of viscoelastic materials, it is recognized that stress waves whose periods are close to the relaxation times of the material are severely attenuated when passing through it (Kolsky, 1963). In metallic materials, however, the dissipative mechanism tends to be more related to their macroscopic thermal properties (Zener, 1948).

(c) Stress wave motion effect

In this, the compression and dilatation due to the stress wave motion in the material produces temperature gradients. Thus, the finite thermal conductivity of the solid would be an influential mechanism by which the mechanical energy of waves may dissipate as thermal energy.

8.3.2 Evaluation of internal friction

Internal friction in solids is often defined by the so-called 'specific loss' or, alternatively, 'specific damping' of the specimen. It is denoted by the symbol D and is conventionally expressed by

$$D = \frac{\Delta W}{W}. \tag{8.43}$$

In the above relation, ΔW is the energy dissipated on subjecting the specimen to a stress cycle and W is the elastic energy stored in the specimen during this cycle (Chapter 3). The magnitude of D depends on the amplitude and the speed of the cycle and other boundary conditions, as well as the past history of the specimen. The reader is referred to Kolsky (1963) for other definitions of internal friction and its measurement.

Mechanical dissipation is particularly pronounced in the case of viscoelastic materials, particularly those of high polymeric origin. In most of these materials, the presence of mechanical dissipation can effectively change the nature of wave motion in them. In addition to the significant mechanical dissipation that can occur in viscoelastic materials, it is well recognized that these materials are 'dispersive'. In view of the latter property, phase velocity of a wave propagating in a viscoelastic material will depend on wave frequency. More specifically, waves of high frequency will propagate in viscoelastic materials with a greater phase velocity than if these waves have a low frequency. Consequently, a mechanical disturbance would continually change in shape during its motion in a viscoelastic medium. Further, the attenuation of high-frequency waves in viscoelastic materials is greater than that of waves of low frequency. In the case of sinusoidal waves, for instance, the above two characteristics of wave motion in a viscoelastic medium would translate into a differential absorption as well as a differential dispersion of the Fourier components of the pulse (Kolsky, 1963).

8.4 VISCOELASTIC WAVE MOTION

As realized in section 8.2, the constitutive equation for a particular material must be combined with the equations of motion in order to solve a specific problem concerning the wave propagation in such material. In contrast to the situation in linear elasticity, the viscoelastic constitutive equation, even in the linear case, is complex by virtue of the existence of integrodifferential terms in this equation and the time dependency of the viscoelastic material functions involved. This added complexity has limited quite significantly the progress in dynamic viscoelasticity in general. Consequently, the majority of problems that have been successfully treated concerning viscoelastic wave phenomena have been limited to simple material representation. A large number of viscoelastic wave propagation problems, within the linear response behaviour of the material, have been attempted by different researchers using a correspondence with an available or deductible solution of an analogous linear elastic problem.

Kolsky (1956, 1960) presented a comprehensive review of the subject of viscoelastic waves in solids from both theoretical and experimental points of view. In his treatment of the subject matter, Kolsky employed the superposition property of solutions in linear viscoelasticity through the application of Fourier analysis. Kolsky (1960) considered, for instance, the motion of a longitudinal disturbance along a thin filament. In this context, the equation of motion along the filament is expressed by

$$\frac{\partial \sigma}{dx} = \rho\left(\frac{\partial^2 u}{dt^2}\right) \tag{8.44}$$

where σ is the longitudinal stress, x is the distance along the filament, u is the longitudinal displacement and ρ is the density. For a sinusoidal wave propagating in a linear viscoelastic solid, the stress is related to the strain through a complex modulus representation (Chapter 3):

$$\sigma = (E_1 + iE_2)\varepsilon = (E_1 + iE_2)\frac{\partial u}{\partial x}. \tag{8.45}$$

Combining (8.44) and (8.45), then

$$(E_1 + iE_2) \frac{\partial^2 u}{\partial x^2} = \rho \frac{\partial^2 u}{\partial t^2}. \tag{8.46}$$

The solution of (8.46) for a propagating sinusoidal wave of frequency $\omega/2\pi$, whose displacement at the origin is $u_0 \cos \omega t$, is expressed as

$$u(x) = u_0 \exp(-\alpha t) \cos[\omega(t - x/c)] \tag{8.47}$$

where

$$c = (E^*/\rho)^{1/2} \sec \delta/2, \tag{8.48a}$$

$$\alpha = (\omega/c) \tan \delta/2, \tag{8.48b}$$

$$E^* = E_1^2 + E_2^2 \tag{8.48c}$$

and

$$\tan \delta = E_2/E_1. \tag{8.48d}$$

On the assumption that, for most polymers, $\tan \delta \ll 1$, then $\sec \delta/2 \approx 1$ and $\tan \delta/2 \approx \frac{1}{2} \tan \delta$. Thus, (8.48a) and (8.48b) are, respectively, reduced to

$$c = \left(\frac{E^*}{\rho}\right)^{1/2} \quad \text{and} \quad \alpha = \frac{\omega}{2c} \tan \delta \tag{8.49}$$

where c and α are referred to as 'propagation constants'. Accordingly, if the values of the moduli E_1 and E_2 (or E^* and $\tan \delta$) are known from experiment over a sufficient frequency range, the displacement of the disturbance along the filament may be calculated by (8.47) with the use of (8.49).

From an experimental point of view, two types of disturbance inputs are often considered for the study of wave propagation in materials, i.e. sinusoidal waves and pulse inputs (for example, Hillier, 1949, 1960; Hillier and Kolsky, 1949; Kolsky, 1960).

8.4.1 Sinusoidal inputs

For this type of disturbance input, continuous trains, of small amplitude of vibration, are propagated along filaments of the material. As introduced in the foregoing, if the displacement input on one end of the specimen is $u_0 \cos \omega t$, then the displacement at a distance x along the filament is given by (8.47). Hence, by measuring the amplitude and phase of the vibration at different points along the filament, the propagation constants c and α can be determined from (8.47). Consequently E^* and $\tan \delta$ (or E_1 and E_2) as functions of frequency $\omega/2\pi$ are found from (8.48) or (8.49). Hillier and Kolsky (1949) and Ballon and Smith (1949), for instance, have used this method for the determination of the dynamic properties of viscoelastic materials such as rubber and plastics in the range of 10^2–10^3 cycles s^{-1} (Kolsky (1960) should also be referred to).

For a linear viscoelastic solid, provided that E_1 is not changing too rapidly with frequency, one may write (Ferry and Williams, 1952)

$$\frac{dE_1}{d\omega} \approx \frac{2E_2}{\pi\omega} \tag{8.50}$$

which can be written in view of (8.48d) as

$$\frac{d(\log E_1)}{d(\log \omega)} \approx \frac{2}{\pi} \tan \delta. \tag{8.51}$$

For most polymers, at temperatures near their transition from the rubber-like to the glassy-like temperature, $\tan \delta$ varies comparatively little with frequency (e.g. Nashif, Jones and Henderson, 1965). For this case, one may assume that $\tan \delta$ is constant (i.e. independent of frequency). Under the latter assumption, equation (8.51) may be integrated to give

$$E_1 \approx E_1(\omega_0) \exp\left[\frac{2}{\pi} \tan \delta \log\left(\frac{\omega}{\omega_0}\right)\right] \tag{8.52}$$

where $E_1(\omega_0)$ is the value of E_1 at a fixed reference frequency $\omega_0/2\pi$. Further, if one assumes, as mentioned before, that $\tan \delta \ll 1$, one can express the propagation constants α and c, with reference to (8.48) and (8.49), as

$$c \approx \left(\frac{E_1}{\rho}\right)^{1/2} \quad \text{and} \quad \alpha = \frac{\omega}{2c} \tan \delta. \tag{8.53}$$

Meantime, equation (8.52) is approximated further as

$$E_1 \approx E_1(\omega_0)\left[1 + \frac{2}{\pi} \tan \delta \log\left(\frac{\omega}{\omega_0}\right)\right] \tag{8.54}$$

whereby the exponential term in (8.52) has been expanded asymptotically and the first two terms in the expansion are retained. Accordingly, one writes with reference to (8.53) that

$$c \approx c_0\left[1 + \frac{2 \tan \delta}{\pi} \log\left(\frac{\omega}{\omega_0}\right)\right]^{1/2} \tag{8.55a}$$

where

$$c_0 = [E_1(\omega_0)/\rho]^{1/2}. \tag{8.55b}$$

Figure 8.4 (Kolsky, 1960; experimental results after Hillier, 1949) supports a linear relation between c and $\log \omega$ for polyethylene in the frequency range shown.

With reference to (8.49), if $\tan \delta$ is constant, the attenuation constant α would be proportional to ω/c. Further, since the phase velocity c varies comparatively slowly with frequency, over a limited frequency range, equations (8.55), one may expect the attenuation constant α, equation (8.53), to be proportional to the frequency.

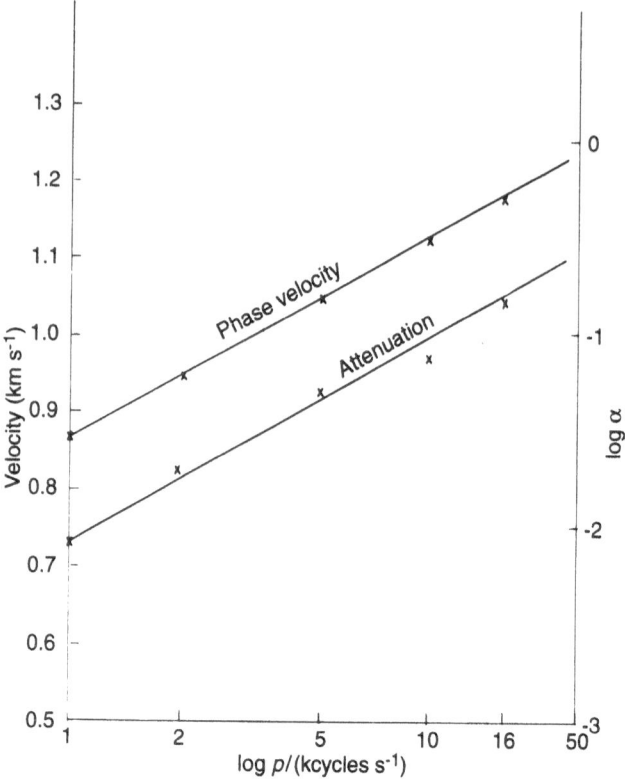

Fig. 8.4 Experimental values of phase velocity c and log α (Hillier, 1949) plotted against log frequency, for polyethylene. (Source: Kolsky, H. (1960) Viscoelastic waves, in Int. Symp. on Stress Wave Propagation in Materials (Ed. N. Davids), Interscience Publishers, New York, pp. 59–90. Reprinted with permission.)

Accordingly, log α should vary linearly with log ω as shown by the graph by Hillier (1949) in Fig. 8.4.

For the study of viscoelastic wave propagation at higher frequencies, pulsed ultrasonic methods are used (for example, Ivey, Mrowea and Guth, 1949; Cunningham and Ivey, 1956). The experimental practice of the technique could vary quite significantly; however, in principle, a finite number of sinusoidal cycles of frequency $\omega/2\pi$ are introduced in one end of the specimen and the resultant wave motion is recorded at a number of points along the length of the specimen. From a measurement of transient time and amplitude ratio, estimates of the phase velocity $c(\omega)$ and the attenuation constant $\alpha(\omega)$ can be made. The ultrasonic technique has the advantage of being relatively simple. It is particularly powerful for investigating the wave propagation properties in elastic materials. In case of viscoelastic materials, however, the technique unfortunately suffers from certain theoretical difficulties of interpretation as pointed out by Kolsky (1960): the time of transit of the pulse depends on the group velocity of the wave packet and, for a dispersive medium, this is, in general, different from the phase velocity c. In the absence of attenuation, these two velocities can be related (Kolsky, 1963); however, in a medium which is dissipative as well as dispersive, the relation between group velocity and phase velocity is not clear yet.

8.4.2 Pulse inputs

Few experimental research efforts have been focused on the study of pulse propagation in viscoelastic materials. In the early work of Hillier (1949) and Hillier and Kolsky (1949) steady-state longitudinal vibrations were induced in prestretched filaments (0.06 cm in diameter) of polythene, neoprene and nylon by means of a transducer element attached to one end of the material specimen. The experimental studies were carried out within low frequency range $< 16 \times 10^3$ cycles s^{-1}. The response of the filament was determined at various points along its length by means of a crystal pick-up. In this, measurements were taken of the variations in the vibration amplitude and the phase. After allowing for the effect of pick-up (Hunter, 1960), the experimental results included both phase velocity and attenuation at a number of frequencies. Kolsky (1956) presented experimental results after Hillier (1949) which show the phase velocity and attenuation in polythene (ICI Alkathene grade 20) against frequency for experiments carried out at 10 °C.

Kolsky (1954a, b, 1956) carried out a number of experiments on the change of the shape of longitudinal stress pulses as they travel along rods of various plastics. These pulses were produced by the detonation of small explosive charges with initial durations of about 2–3 μs. Figure 8.5 shows oscillograph records which were obtained by Kolsky (1960) with rods of polymethyl methacrylate and polyethylene. As noted by Kolsky (1960), with the polyethylene specimen, after two or three reflections, the length of the pulse had become more than twice the length of the specimen, with the result that the movement of the ends of the specimen become continuous. Figure 8.6 (due to Kolsky, 1960) shows the curves of particle velocity with the passage of time for pulses which had propagated in polyethylene rods 30, 60 and 90 cm in length. It can be seen in the figure that the pulses become progressively flatter, but retain an asymmetrical shape.

8.5 VISCOELASTIC WAVE MOTION IN SEMI-INFINITE LINEAR MEDIA

In this section, we deal with the problem of determining the stress distribution in a semi-infinite viscoelastic rod subject to dynamic loading. The problem was examined by Lee and Morrison (1956). In Lee and Morrison's work, the stress and velocity distributions associated with the propagation of an impulsively applied velocity and stress along viscoelastic rods, as presented by different mechanical models, were determined. Morrison (1956) also considered analytically the wave propagation in a viscoelastic rod of the Voigt model type and also studied viscoelastic materials with three-parameter models. In an earlier work, Hillier (1949) (Hillier and Kolsky (1949) should also be referred to) studied the motion of longitudinal sinusoidal waves along a viscoelastic filament assuming a Maxwell solid, a Voigt solid and a three-element

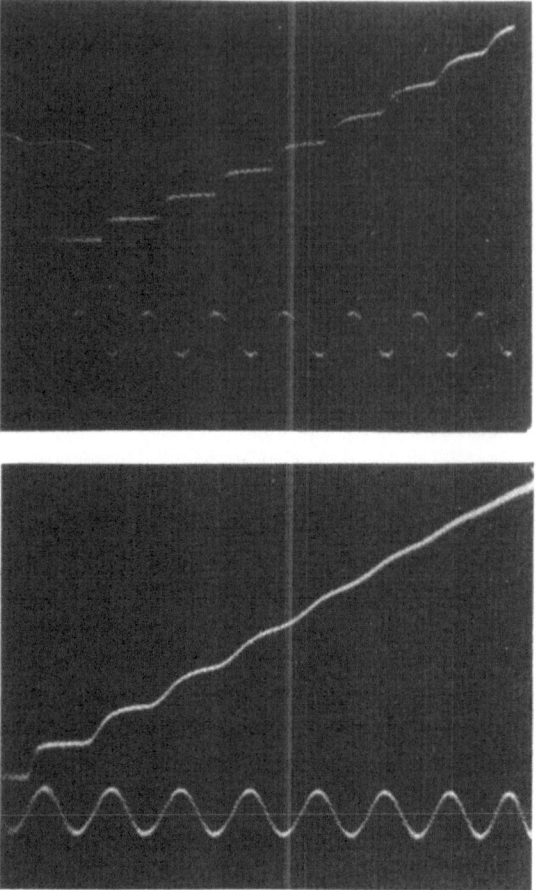

Fig. 8.5 (a) Oscillograph record of displacement at end of polymethyl methacrylate rod 46 cm long and 1.25 cm in diameter when 5 mg charge of lead has been detonated at opposite end. Period of timing wave is 500 μs. (b) Oscillograph record, similar to (a), for polyethylene rod 20 cm long and 1.25 cm in diameter. (Source: Kolsky, H. (1960) Viscoelastic waves, in Int. Symp. on Stress Wave Propagation in Materials (Ed. N. Davids), Interscience Publishers, London, pp. 59–90. Reprinted with permission.)

model representations. Lee and Kanter (1953) considered the stress distribution in a rod of Maxwell material subjected to a mechanical impact. Glauz and Lee (1954), on the other hand, used the method of characteristics to determine the stress in a viscoelastic material made of four-parameter model.

Consider a semi-infinite rod as shown in Fig. 8.7 where $x \geq 0$, with the x coordinate measured along the length of the rod. In this figure, $x(t)$ denotes the position of a section of the rod at time t and $u(x, t)$ is the displacement of this section in the direction of increasing x.

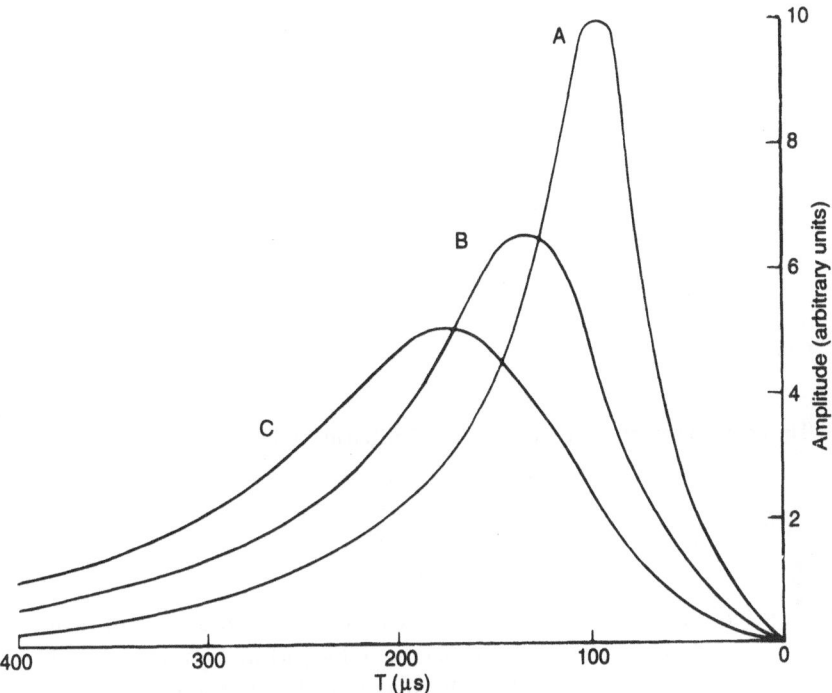

Fig. 8.6 Curves of particle velocity distributions for pulses that have travelled through different lengths of polyethylene rods: curve A, 30 cm; curve B, 60 cm; curve C, 90 cm. (Source: Kolsky, H. (1960) Viscoelastic waves, in Int. Symp. on Stress Wave Propagation in Materials (Ed. N. Davids), Interscience Publishers, London, pp. 59–90. Reprinted with permission.)

Let

- $\sigma(x, t)$ denote the nominal comprehensive stress transmitted across the section x of the rod at time t,
- $\varepsilon(x, t)$ designate the nominal compressive strain corresponding to $\sigma(x, t)$ and
- ρ be the mass density of the material.

The governing equation of motion, in the absence of body forces, in the x direction is

$$\rho \frac{\partial^2 u}{\partial t^2} = - \frac{\partial \sigma}{\partial x}$$

or, in a more compact form,

$$\rho u_{tt} = -\sigma_x \tag{8.56}$$

where a subscript denotes partial differentiation with respect to the corresponding variable.

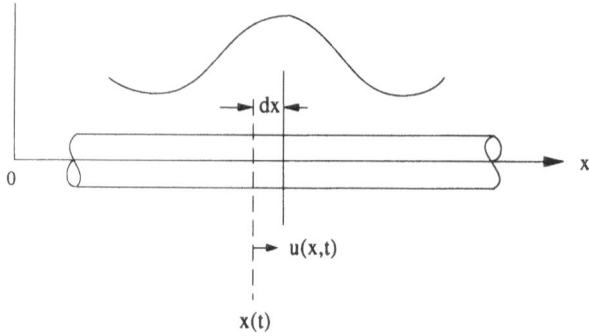

Fig. 8.7 Stress wave propagation in a semi-infinite rod.

The nominal compressive strain $\varepsilon(x, t)$ is written in terms of the displacement u as

$$\varepsilon = -u_x. \tag{8.57}$$

The particle velocity $v(x, t)$ is expressed in terms of the displacement u as

$$v = u_t. \tag{8.58}$$

The semi-infinite rod is considered to be initially unstrained and at rest when, at $t = 0$, the end $x = 0$ is subjected to a mechanical impact (disturbance) with either a constant stress or a constant velocity. In either case, the stress or the velocity at the end $x = 0$ is specified. The governing boundary conditions are

$$\sigma = \sigma_0 H(t) \quad \text{or} \quad v = v_0 H(t), \quad \text{at} \quad x = 0, \tag{8.59}$$

where σ_0 is the applied constant stress, v_0 is the applied constant velocity and $H(t)$ is the Heaviside step function, that is

$$H(t) = \begin{cases} 1 & \text{for } t > 0 \\ 0 & \text{for } t < 0 \end{cases}.$$

In addition to equations (8.56) to (8.59), the constitutive equation for the particular viscoelastic material must be included in the process of determining the stress distribution in the semi-infinite rod. In this section, we consider the representation of the viscoelastic material by different mechanical models; hence, the corresponding constitutive relations are provided accordingly. In this, we follow closely the work of Lee and Morrison (1956). The reader is referred also to Morrison (1956) and Lee and Kanter (1953).

8.5.1 General representation of viscoelastic models

(a) Elastic model: stress–strain law

The stress–strain law is given by

$$\sigma = E\varepsilon \tag{8.60}$$

where E is Young's modulus. Combining equation (8.60) with equations (8.56) to (8.58), then

$$\sigma_{xx} = (\rho/E)\sigma_{tt},$$
$$u_{xx} = (\rho/E)u_{tt},$$
$$\varepsilon_{xx} = (\rho/E)\varepsilon_{tt},$$
$$v_{xx} = (\rho/E)v_{tt}. \tag{8.61}$$

Each of equations (8.61) may be written in the form of the partial differential equation

$$f_{xx} = (\rho/E)f_{tt} \tag{8.62}$$

in which f is an arbitrary variable and whereby equation (8.62) is satisfied by σ, u, ε and v.

(b) Viscous model: stress–strain law

The stress–strain law is given by

$$\sigma = \eta\varepsilon_t \tag{8.63}$$

where η is the viscosity coefficient. Combining (8.63) with equations (8.56) to (8.58), it follows that

$$\sigma_{xx} = (\rho/\eta)\sigma_t,$$
$$u_{xx} = (\rho/\eta)u_t,$$
$$\varepsilon_{xx} = (\rho/\eta)\varepsilon_t,$$
$$v_{xx} = (\rho/\eta)v_t. \tag{8.64}$$

Equations (8.64) lead to the partial differential equation

$$f_{xx} = (\rho/\eta)f_t \tag{8.65}$$

which is satisfied by σ, u, ε and v.

From the above relations for the elastic and viscous models, it is seen that, if we determine the stress solution $\sigma(x, t)$ for the problem in which the end $x = 0$ of the rod is given a constant stress σ_0 at $t = 0$ and this stress is maintained constant afterwards, then we can determine the velocity solution $v(x, t)$ for the alternate problem in which the end of the rod $x = 0$ is given an impulsive constant velocity v_0 which is subsequently maintained, and vice versa. In effect, we have

$$\frac{\sigma(x, t)}{\sigma_0} \equiv \frac{v(x, t)}{v_0}. \tag{8.66a}$$

For convenience, Lee and Morrison (1956) introduced the dimensionless variables

$$\tau = (E/\eta)t \quad \text{and} \quad \xi = \frac{(\rho E)^{1/2}}{\eta}x \tag{8.66b}$$

so that the partial differential equation (8.62) for the elastic model becomes

$$f_{\xi\xi} = f_{\tau\tau} \tag{8.67}$$

and the partial differential equation (8.65) for the viscous element can be written as

$$f_{\xi\xi} = f_{\tau}. \tag{8.68}$$

The partial differential equations satisfied by f for other mechanical models are given in the appendix of the paper by Lee and Morrison (1956).

The stress solutions for the rod when subjected to a constant applied stress and to a constant applied velocity are to be determined. For convenience, the stress solutions are represented (Lee and Morrison, 1956) by the following dimensionless variables:

$$\Sigma(\xi, \tau) = \frac{\sigma(x, t)}{\sigma_0}$$

for constant applied stress and

$$\Sigma'(\xi, \tau) = \frac{\sigma(x, t)}{(\rho E)^{1/2} v_0}$$

for constant applied velocity.

(c) Elastic models: stress distribution

Combining equations (8.59) and (8.67) and using Laplace transform, the stress distribution for the elastic model is

$$\Sigma(\xi, \tau) = H(t - \xi) \tag{8.69}$$

for constant applied stress and

$$\Sigma'(\xi, \tau) = H(\tau - \xi) \tag{8.70}$$

for constant applied velocity.

Figures 8.8(a) and 8.8(b) show, respectively, schematics of the stress distributions, in the case of an elastic rod subjected to a suddenly applied constant end stress σ_0, against τ for a fixed ξ and against ξ for a fixed τ.

The stress discontinuity in the rod subjected to a sudden dynamic loading travels in the form of a wavefront. Hence, for a fixed ξ, that part of the rod represents the time before the wave front passes. Similarly, for a fixed τ, it contains that part of the bar where the wavefront has not yet reached. In case of the elastic rod, as shown in Fig. 8.8(a), for $\xi = 2\sqrt{2}$, the stress at this section of the elastic rod, when subjected to constant applied stress, is zero as it represents the time before the wavefront passes and, as soon as the wavefront reaches $\tau = \tau_1$, the stress jumps to the value of unity and remains constant afterwards. Similarly for $\xi = 4\sqrt{2}$, as soon as the wavefront reaches $\tau = \tau_2$, the stress jumps to the value of unity and remains

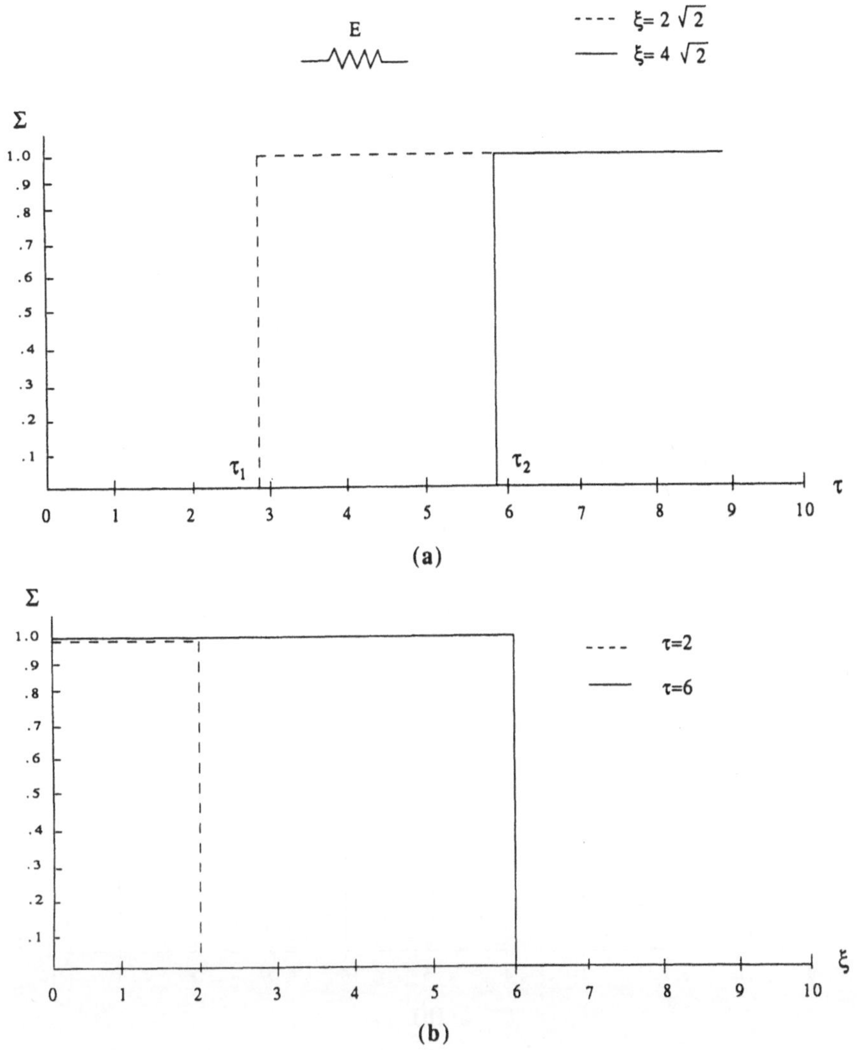

Fig. 8.8 Schematics of stress distributions for an elastic rod under suddenly applied constant end stress σ_0.

subsequently constant. As shown in Fig. 8.8(b), for $\tau = 2$, the stress jumps to the value of unity and remains constant until $\xi = 2$ when the stress instantaneously becomes zero. Similarly for $\tau = 6$, the stress jumps to the value of unity and remains constant until $\xi = 6$ when the stress instantaneously becomes zero.

Similar stress distributions are exhibited by the elastic model in the case of constant end velocity loading as shown in Fig. 8.9.

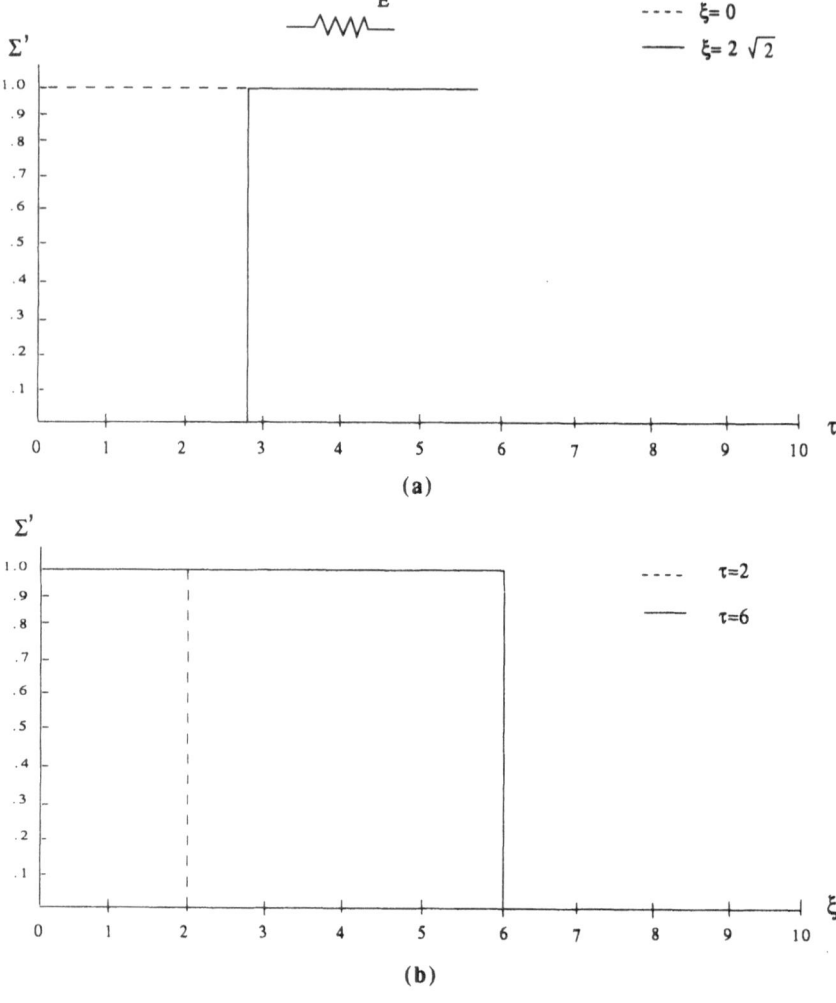

Fig. 8.9 Schematics of stress distributions for an elastic rod under suddenly applied constant end velocity v_0.

(d) Viscous model: stress distribution

Similarly, for the viscous model, the stress distributions are determined from (8.59) and (8.68) as

$$\Sigma(\xi, \tau) = Z\left(\frac{\xi}{2\tau^{1/2}}\right) \tag{8.71}$$

for constant applied stress and

$$\Sigma'(\xi, \tau) = \frac{\exp(-\xi^2/4\tau)}{(\pi\tau)^{1/2}} \tag{8.72}$$

for constant applied velocity where, in (8.71),

$$Z(z) = (2/\pi^{1/2}) \int_z^\infty \exp(-\mu^2) \, d\mu. \tag{8.73}$$

For a viscous rod under suddenly applied constant end stress, as shown in Fig. 8.10(a), for $\xi = 1$, the stress increases asymptotically as soon as the constant stress loading is applied and continues to increase, ultimately reaching unity. For $\xi = 4$,

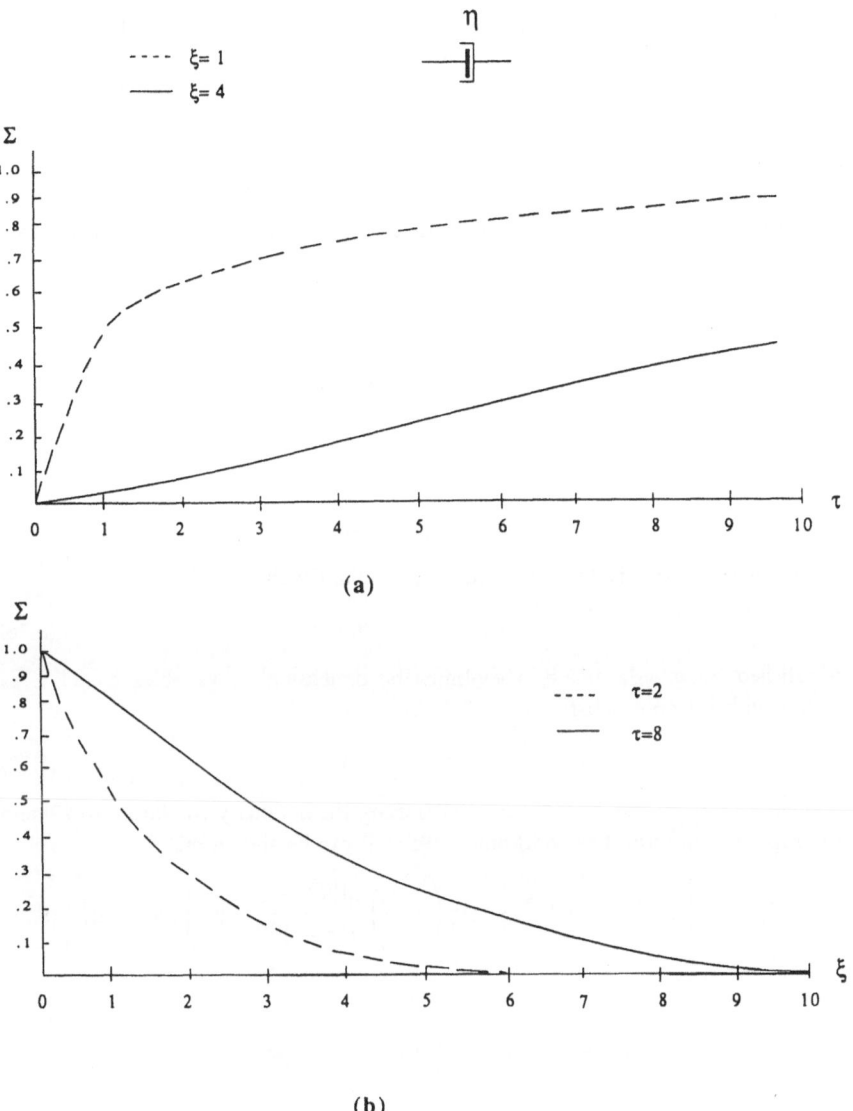

Fig. 8.10 Schematics of stress distributions for a viscous element under suddenly applied constant end stress σ_0.

the stress starts increasing more gradually, but continues to increase asymptotically, ultimately approaching a value less than unity. As shown in Fig. 8.10(b), for $\tau = 2$, the stress continues to decrease rapidly and vanishes. For $\tau = 8$, the stress starts to decrease less rapidly, but ultimately vanishes.

In case of constant applied velocity as shown in Fig. 8.11(a), the stress at the end of the viscous rod, $\xi = 0$, becomes instantaneously infinite as the viscous element resists the deformation due to sudden impact, but it rapidly decreases and may finally vanish. For $\xi = 4$, the stress tends to increase gradually after impact and after some time decreases and may finally vanish.

For a fixed τ, as shown in Figs. 8.11(b) and 8.11(c), the stress jumps to a value of less than unity and then ultimately vanishes. As τ increases, Fig. 8.11(c), the jump value reduces and the stress continues to decrease asymptotically and finally vanishes.

(e) Maxwell model

The stress–strain law for the Maxwell model, in view of (2.15), can be written as

$$\varepsilon_t = (1/E)\sigma_t + (1/\eta)\sigma. \tag{8.74}$$

Combining (8.74) with equations (8.56) to (8.58), it follows that

$$
\begin{aligned}
\sigma_{xx} - (\rho/E)\sigma_{tt} - (\rho/\eta)\sigma_t &= 0, \\
u_{xx} - (\rho/E)u_{tt} - (\rho/\eta)u_t &= 0, \\
\varepsilon_{xx} - (\rho/E)\varepsilon_{tt} - (\rho/\eta)\varepsilon_t &= 0, \\
v_{xx} - (\rho/E)v_{tt} - (\rho/\eta)v_t &= 0.
\end{aligned}
\tag{8.75}
$$

Equations (8.75) signify that the partial differential equation

$$f_{xx} - (\rho/E)f_{tt} - (\rho/\eta)f_t = 0 \tag{8.76}$$

is satisfied by σ, u, ε and v. Meantime, the dimensionless variables ξ and τ, as expressed by (8.66b), satisfy

$$f_{\xi\xi} - f_{\tau\tau} - f_\tau = 0. \tag{8.77}$$

Solving for $\Sigma(\xi, \tau)$ and $\Sigma'(\xi, \tau)$ from (8.77) using the boundary conditions (8.59) and the Laplace transform (Lee and Kanter, 1953), it can be shown that

$$\Sigma(\xi, \tau) = \left[\exp\left(-\frac{\xi}{2} \right) + \frac{\xi}{2} \int_\xi^\tau \exp\left(-\frac{\zeta}{2} \right) \frac{I_1[\frac{1}{2}(\zeta^2 - \xi^2)]}{(\zeta^2 - \xi^2)^{1/2}} \, d\zeta \right] H(\tau - \xi) \tag{8.78}$$

and

$$\Sigma'(\xi, \tau) = \exp\left(-\frac{\tau}{2} \right) I_0[\frac{1}{2}(\tau^2 - \xi^2)^{1/2}] H(\tau - \xi) \tag{8.79}$$

where, in the above two equations, I_1 and I_0 are Bessel functions of the associated arguments. For the evaluation of these functions the reader is referred, for instance,

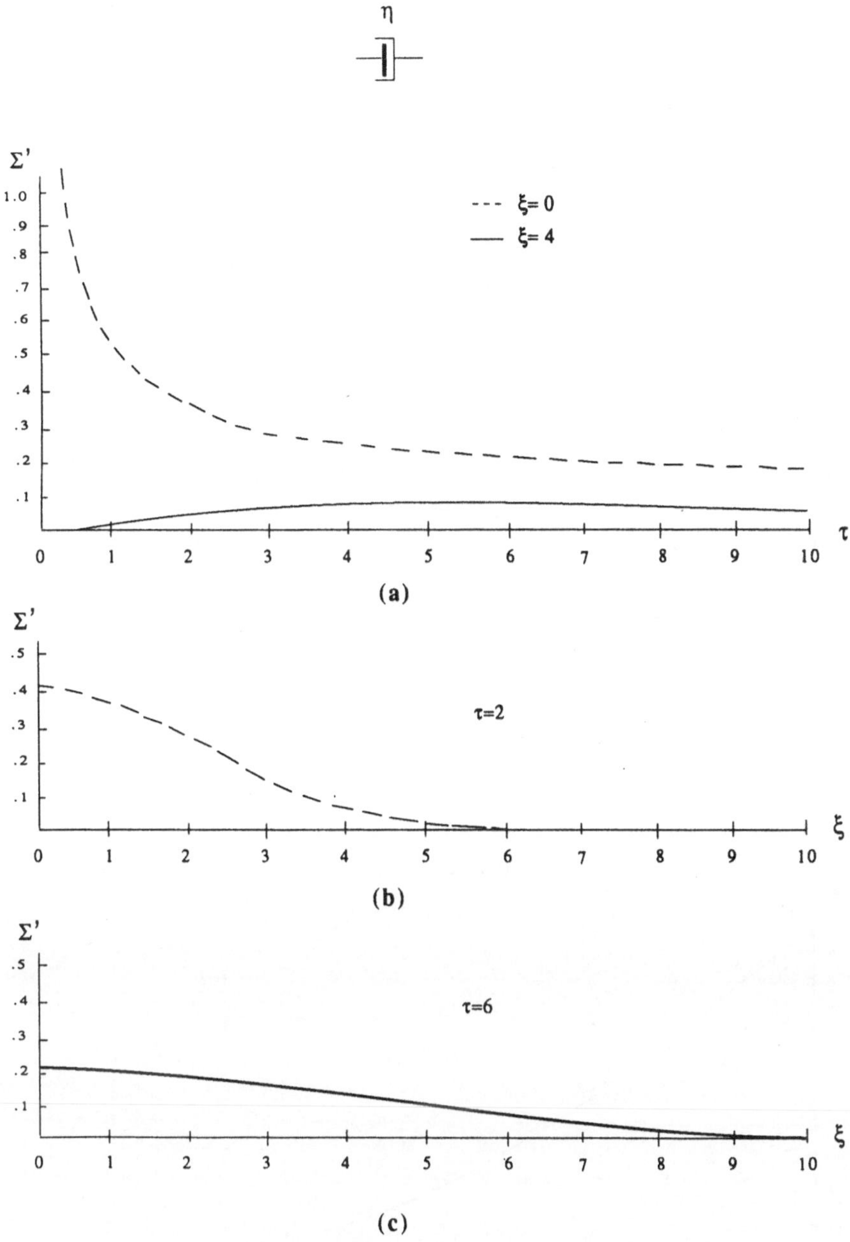

Fig. 8.11 Schematics of stress distributions for a viscous element under suddenly applied constant end velocity v_0.

to Janke and Emde (1945). The stress distributions for the Maxwell model are schematically shown in Fig. 8.12 in the case of constant stress and in Fig. 8.13 in the case of suddenly applied constant velocity with reference, respectively, to expressions (8.78) and (8.79).

As shown in Fig. 8.12(a), for $\xi = 1$, the stress at this section of the Maxwell rod when subjected to a constant applied stress is zero as it represents the time before

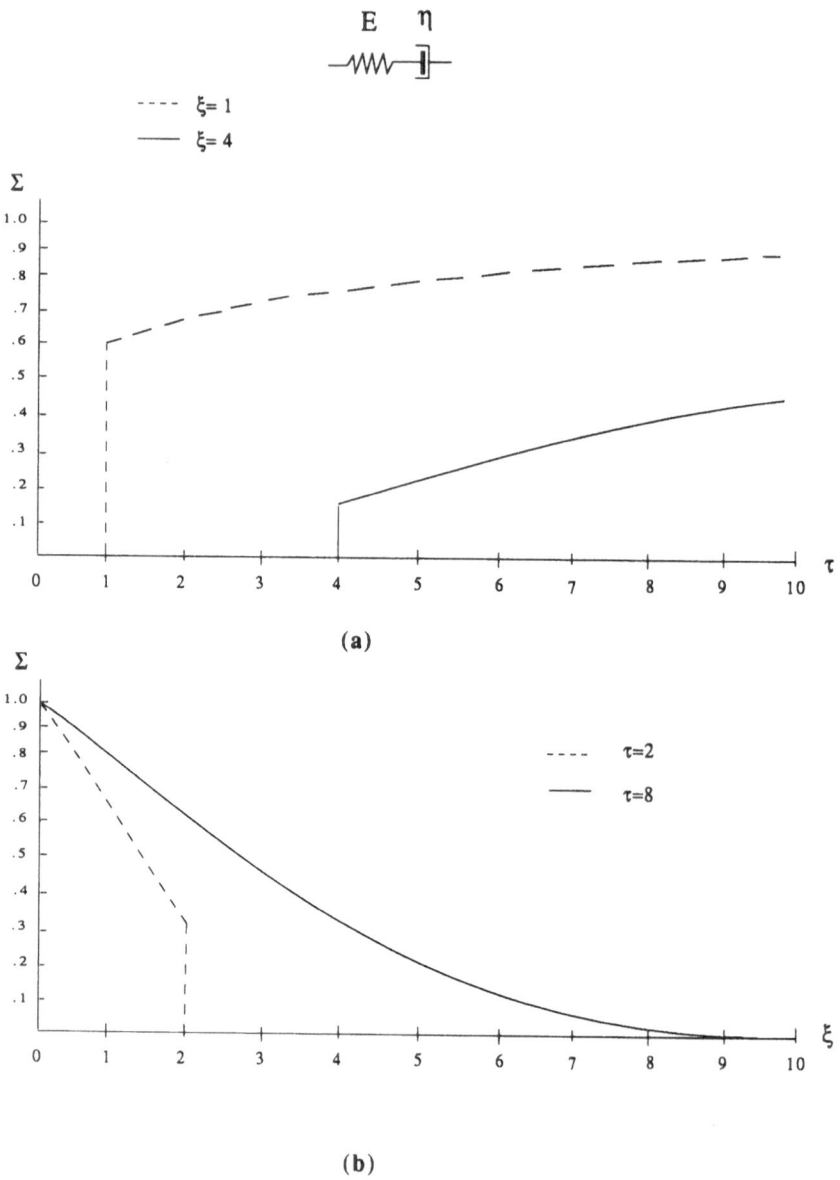

(a)

(b)

Fig. 8.12 Schematics of stress distributions for a Maxwell model under suddenly applied constant end stress σ_0.

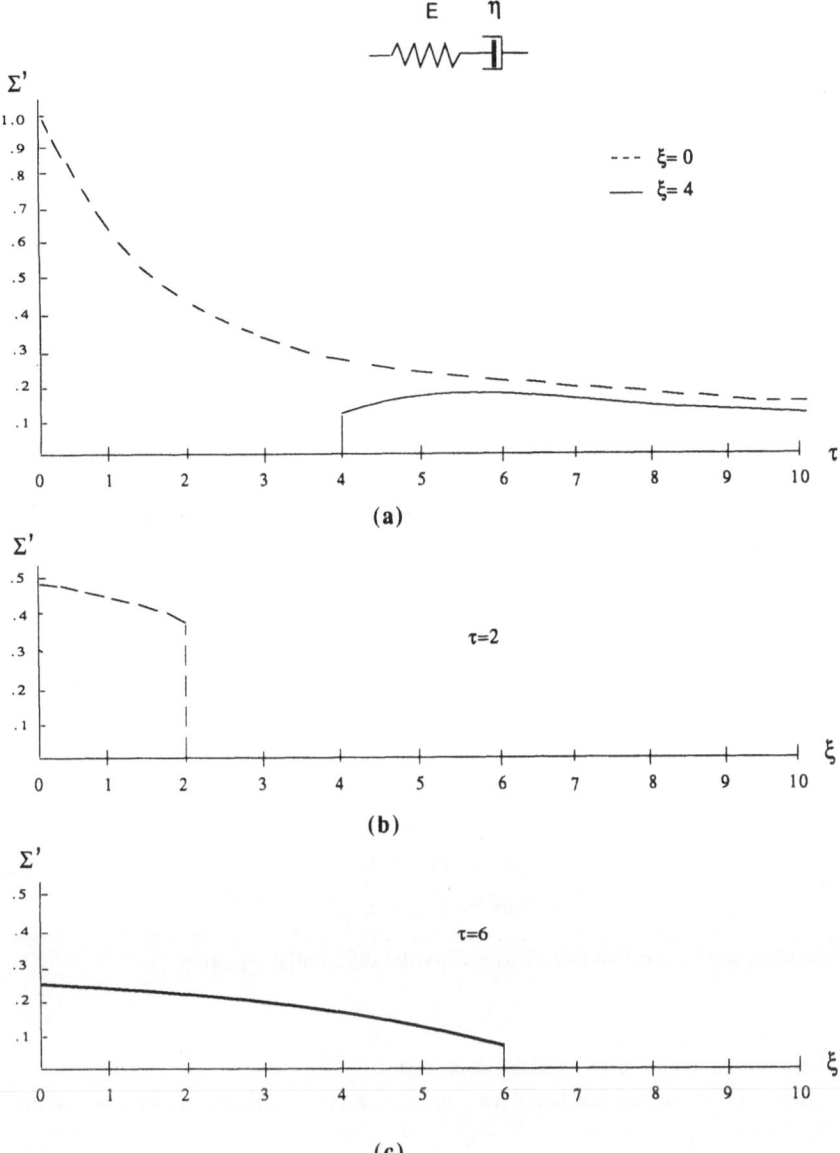

Fig. 8.13 Schematics of stress distributions for a Maxwell model under suddenly applied constant end velocity v_0.

the wavefront passes, but, as soon as the wavefront reaches $\tau = 1$, the stress jumps instantaneously to a value less than unity and continues to increase, ultimately approaching unity. With the increase in ξ, the stress jump reduces drastically, but the stress continues to increase afterwards. For a fixed τ, Fig. 8.12(b), the stress reduces rapidly for lower values of τ, but, as τ increases, the stress decreases more gradually.

In a case of a constant applied end velocity, Fig. 8.13(a), the stress at the end of the rod, $\xi = 0$, jumps instantaneously to the value of unity, but decreases monotonically and becomes eventually zero. At $\xi = 4$, the stress is zero until $\tau = 4$, where the stress jumps to a value much less than unity and continues to increase first and then decreases until ultimately it vanishes. It can be seen that the behaviour of this model is similar to that of a viscous model.

For a fixed τ, Fig. 8.13(b), the stress jumps to a value less than unity immediately following the impact and starts decreasing and vanishes with an instantaneous drop in stress. As time increases, Fig. 8.13(c), the stress jump reduces and the stress decreases more gradually and finally vanishes. This behaviour indicates further that the Maxwell model ultimately behaves like a viscous model for times long compared with the relaxation times.

(f) Voigt (Kelvin) model

The stress–strain law for the Voigt (Kelvin) model is expressed, with reference to (2.28) as

$$\sigma = E\varepsilon + (1/\eta)\varepsilon_t. \tag{8.80}$$

Combining (8.56), (8.57) and (8.80),

$$\rho\sigma_{tt} = E\sigma_{xx} + \eta\sigma_{xxt}. \tag{8.81}$$

Similarly, it can be shown that

$$\begin{aligned}
\rho u_{tt} &= Eu_{xx} + \eta u_{xxt}, \\
\rho\varepsilon_{tt} &= E\varepsilon_{xx} + \eta\varepsilon_{xxt}, \\
\rho v_{tt} &= Ev_{xx} + \eta v_{xxt}
\end{aligned} \tag{8.82}$$

Equations (8.81) and (8.82) lead to the partial differential equation

$$\rho f_{tt} = Ef_{xx} + \eta f_{xxt} \tag{8.83}$$

where the variable f is satisfied by σ, u, ε and v.

Hence, the dimensionless variables τ and ξ, as given by (8.66b), satisfy the relation

$$f_{\tau\tau} = f_{\xi\xi} + f_{\xi\xi\tau}. \tag{8.84}$$

One may solve for the stress distributions using equations (8.59) and (8.84) and the Laplace transform.

The stress distributions for the Voigt (Kelvin) model, as determined by Lee and Morrison (1956), are shown in Figs. 8.14 and 8.15.

Figures 8.14(a) and 8.14(b) show, respectively, the stress distributions, for a constant applied end stress, against τ for a fixed ξ and against ξ for a fixed τ.

As shown in Fig. 8.14(a), for $\xi = 2\sqrt{2}$, the stress commences to increase immediately after loading owing to the stress application and continues to increase

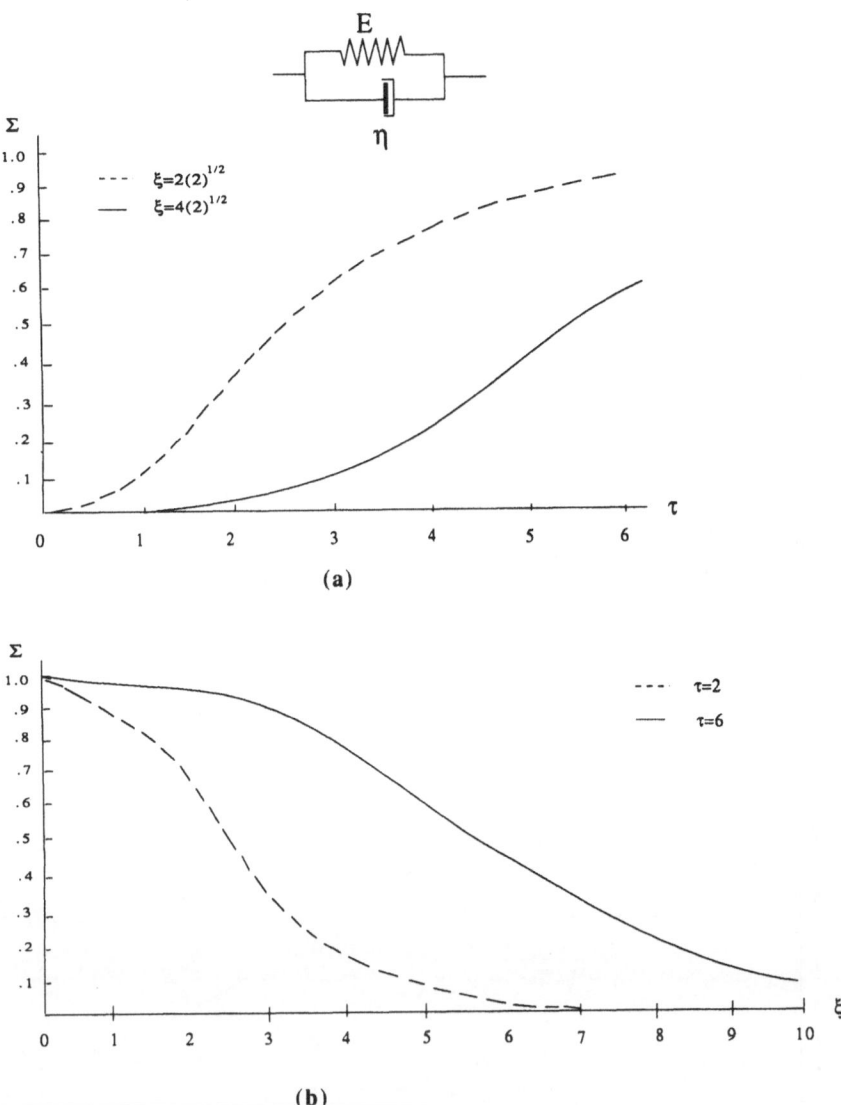

Fig. 8.14 Schematics of stress distributions for a Voigt model under suddenly applied constant end stress σ_0.

asymptotically approaching ultimately the value unity. As ξ increases, the stress commences to increase gradually approaching ultimately the value of unity.

For a fixed τ, at $\tau = 2$, the stress decreases quite rapidly and, as τ increases, the stress decreases more gradually, approaching ultimately the value of zero as shown in Fig. 8.14(b).

In the case of a constant applied end velocity, as shown in Fig. 8.15(a), the stress at the end of the rod, $\xi = 0$, becomes infinite at the moment of the impact owing

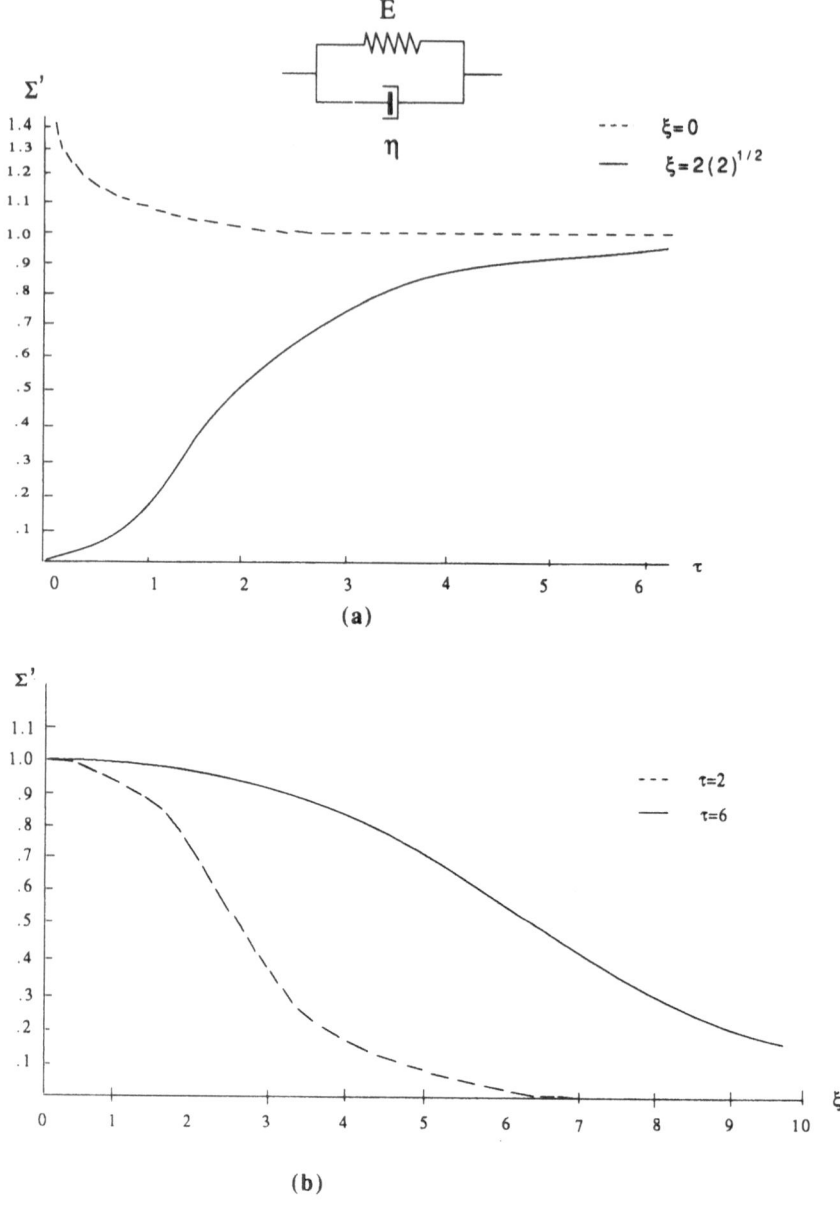

Fig. 8.15 Schematics of stress distributions for a Voigt model under suddenly applied constant end velocity v_0.

to the resistance of the viscous element to deformation but rapidly decreases asymptotically and ultimately reaches the value of unity. At $\xi = 2\sqrt{2}$, the stress commences to increase as soon as the impact takes place and asymptotically approaches the value of unity.

For a fixed τ, as the value of ξ increases, the stress decreases more gradually, approaching ultimately the value of zero as shown in Fig. 8.15(b).

It can be seen from Fig. 8.15 that the Voigt (Kelvin) model behaves ultimately like an elastic model. The spread of the stress distribution in the Voigt (Kelvin) model as compared with the sharp wave front occurring in the case of the elastic model is due to the presence of the viscous element which is in parallel with the elastic element in the Voigt model. However, as the time increases, viscous behaviour tends to die out and the rod behaves as an elastic model. This demonstrates further the fact that Voigt material exhibits a delayed elastic response.

(g) Three-element model (one viscous and two elastic elements)

The model consists of two springs and a dashpot in the configuration shown in Fig. 8.16.

For the model shown in Fig. 8.16, the stress–strain relation is

$$\sigma_t/E' + (1/\eta)\sigma = \varepsilon_t(1 + E/E') + (E/\eta)\varepsilon. \tag{8.85}$$

Combining equations (8.57) and (8.85),

$$\sigma_t/E' + (1/\eta)\sigma = -u_{xt}(1 + E/E') - (E/\eta)u_x. \tag{8.86}$$

Combining further equations (8.56) and (8.86), it can be shown that, after carrying out the appropriate differentiation,

$$\sigma_{tt}/E' + (1/\eta)\sigma_t = -(\sigma_{xx}/\rho)(1 + E/E') - (E/\eta)u_{xt}. \tag{8.87}$$

Differentiating (8.87) with respect to t and combining it with the derivative of (8.56) with respect to x, it follows that

$$\rho\sigma_{ttt}/E' + (\rho/\eta)\sigma_{tt} = \sigma_{xxt}(1 + E/E') + (E/\eta)\sigma_{xx}. \tag{8.88}$$

Equation (8.88) can be written in the following form of the partial differential equation:

$$\rho f_{ttt}/E' + (\rho/\eta)f_{tt} = f_{xxt}(1 + E/E') + (E/\eta)f_{xx} \tag{8.89}$$

which is satisfied by σ. It can also be shown that (8.89) is satisfied by the variables u, ε and v. Hence, the dimensionless variables ξ and τ, as given earlier by (8.66b), will satisfy the partial differential equation

$$kf_{\tau\tau\tau} + f_{\tau\tau} = (1 + k)f_{\xi\xi\tau} + f_{\xi\xi} \tag{8.90}$$

where $k = E/E'$. One may solve for the stress distributions $\Sigma(\xi, \tau)$ and $\Sigma'(\xi, \tau)$ for the three-element model using (8.90) and the boundary conditions (8.59) (Lee and Morrison, 1956). Schematics of stress distributions as determined by Lee and Morrison (1956) are shown, respectively, in Figs. 8.16 and 8.17 for the case when $E = E'$, i.e. when $k = 1$.

It is seen from the schematics of stress distributions (Figs. 8.16 and 8.17) that as the value of τ increases the one-viscous and two-elastic elements model has a stress distribution similar to that of the Voigt (Kelvin) model except for the sharp discontinuity wavefront in case of the three-element model. Such a sharp discontinuity

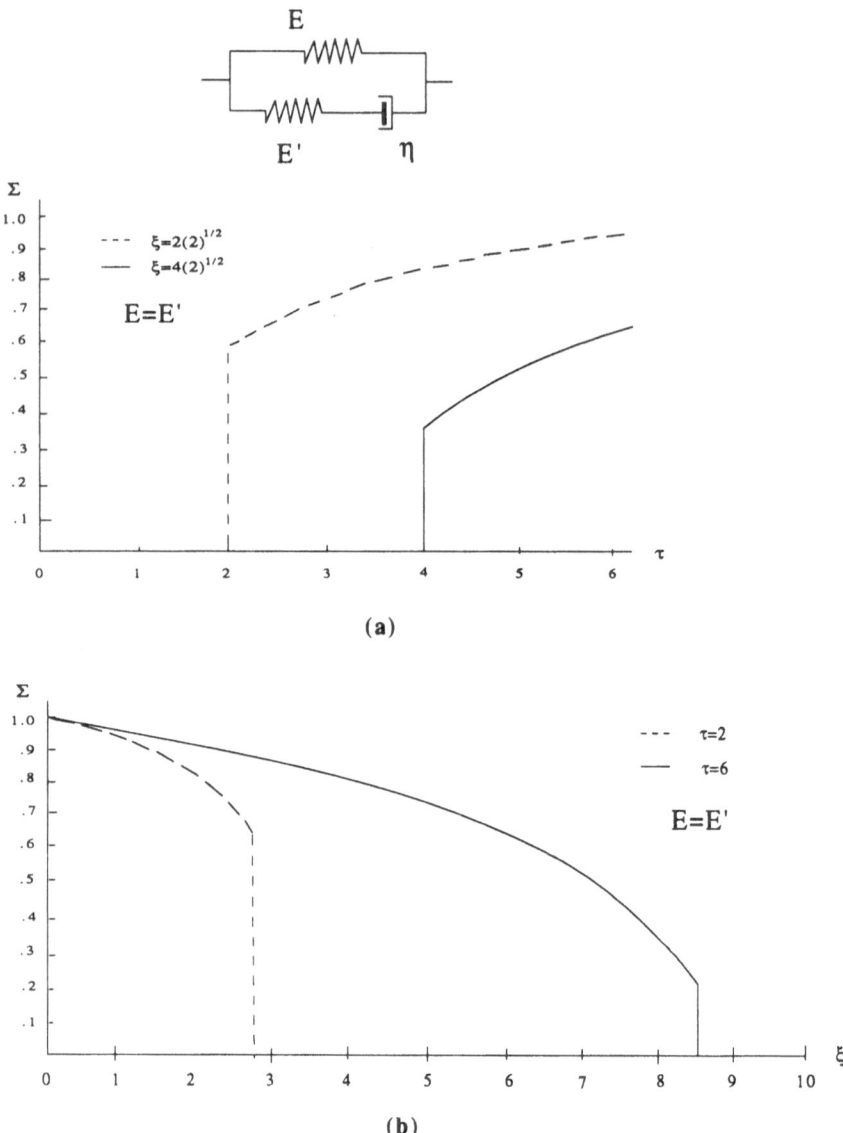

Fig. 8.16 Schematics of stress distributions for a three-element model (one viscous and two elastic elements) under suddenly applied constant end stress σ_0. $E = E'$.

results from the predominance of the two elastic elements. As the value of τ increases, the sharp discontinuity reduces and the stress behaviour of the one-viscous and two-elastic elements model has little or no difference from the stress behaviour of the Voigt model. It is noticed that, for very short times, the dealt-with three-element model behaves more like a Maxwell model than a Voigt (Kelvin) model. This type of

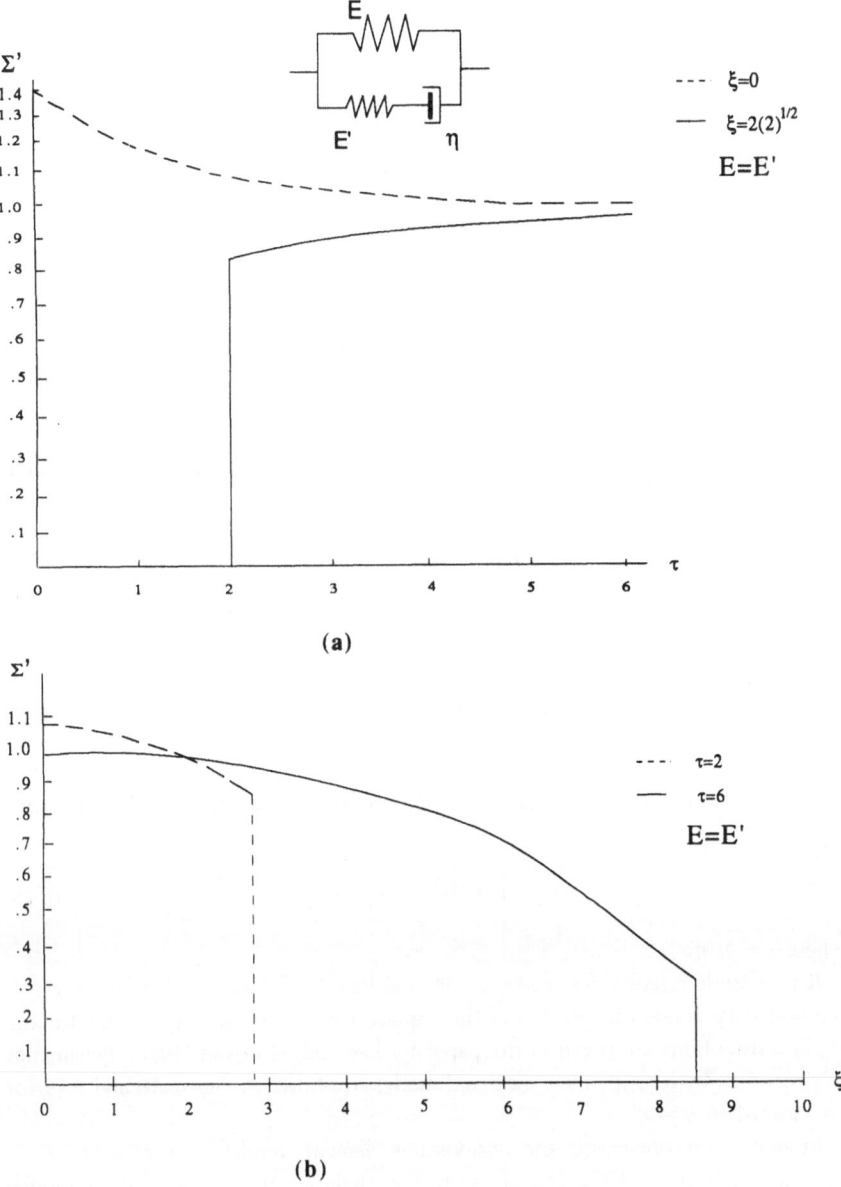

Fig. 8.17 Schematics of stress distributions for a three-element model (one viscous and two elastic elements) under suddenly applied constant end velocity v_0. $E = E'$.

behaviour is seen for both the constant applied end stress (Fig. 8.16) as well as for the constant applied end velocity (Fig. 8.17).

(h) Three-element model (one elastic and two viscous elements)

This model consists of a spring and two dashpots in the configuration shown in Fig. 8.18.

For the model shown in Fig. 8.18, the stress–strain equation is

$$E\varepsilon_t + \eta'\varepsilon_{tt} = (E/\eta)\sigma + (1 + \eta'/\eta)\sigma_t. \tag{8.91}$$

Recalling equations (8.57) and (8.91), it can be shown that

$$-Eu_{xt} - \eta'u_{xtt} = (E/\eta)\sigma_t + (1 + \eta'/\eta)\sigma_t. \tag{8.92}$$

Differentiating (8.56) with respect to x and substituting the resulting differentiation into (8.92),

$$(\rho E/\eta)\sigma_t + \rho(1 + \eta'/\eta)\sigma_{tt} = E\sigma_{xx} + \eta'\sigma_{xxt}. \tag{8.93}$$

It can be also shown that u, ε and v satisfy the same form of the partial differential equation, i.e.

$$(\rho E/\eta)f_t + \rho(1 + \eta'/\eta)f_{tt} = Ef_{xx} + \eta'f_{xxt} \tag{8.94}$$

which is already satisfied by σ in view of (8.93). Accordingly, the dimensionless variables ξ and τ, expressed earlier by (8.66b), will satisfy the partial differential equation

$$f_\tau + (1 + \lambda)f_{\tau\tau} = f_{\xi\xi} + \lambda f_{\xi\xi\tau} \tag{8.95}$$

where $\lambda = \eta/\eta'$.

It is possible to solve for the stress distributions $\Sigma(\xi, \tau)$ and $\Sigma'(\xi, \tau)$ using (8.95), the boundary conditions (8.59) and the Laplace transform. The expressions for the stress distributions are given in the paper by Lee and Morrison (1956). Schematics of stress distributions for the considered model are shown in Figs. 8.18 and 8.19 for the case when $\eta = \eta'$.

In case of the one-elastic and two-viscous elements model, Figs. 8.18 and 8.19, the stress behaviour at the end of the rod is similar to that of the viscous model. Both models exhibit very large stresses at the end of the rods owing to the role played by the viscous element in resisting the sudden deformation caused by the impact. As this deformation is instantaneous, it corresponds to an infinite strain rate which results in an infinite stress to produce it. After a considerable length of time, so that the effects of the sudden impact have delayed, it is seen that the one-elastic and two-viscous elements model behaves like the Maxwell or a viscous model. Thus, the one-elastic and two-viscous elements model can be analysed on the basis of a

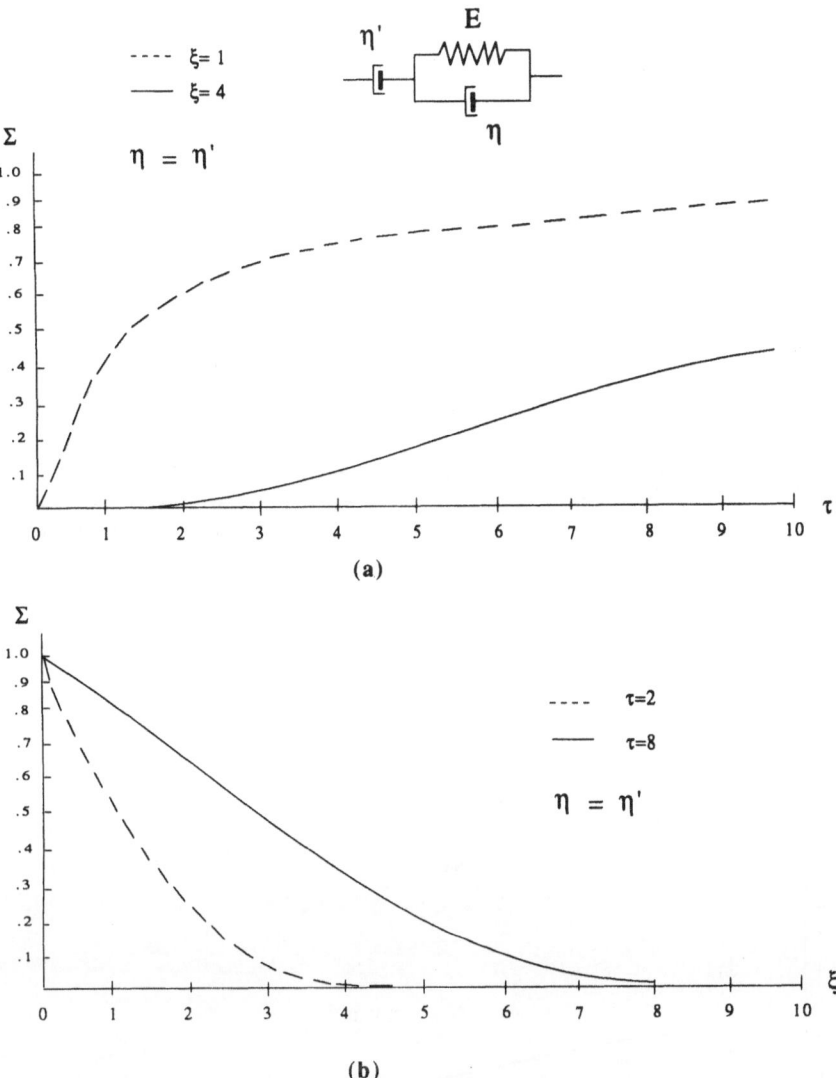

Fig. 8.18 Schematics of stress distributions for a three-element model (one elastic and two viscous elements) under suddenly applied constant end stress σ_0. $\eta = \eta'$.

Maxwell model or a viscous model at large values of τ when the elastic behaviour completely vanishes and the unconstrained viscous element governs the stress distribution. The similarity in stress distributions mentioned above is more pronounced in the case of the constant stress impact than in the case of the constant velocity impact.

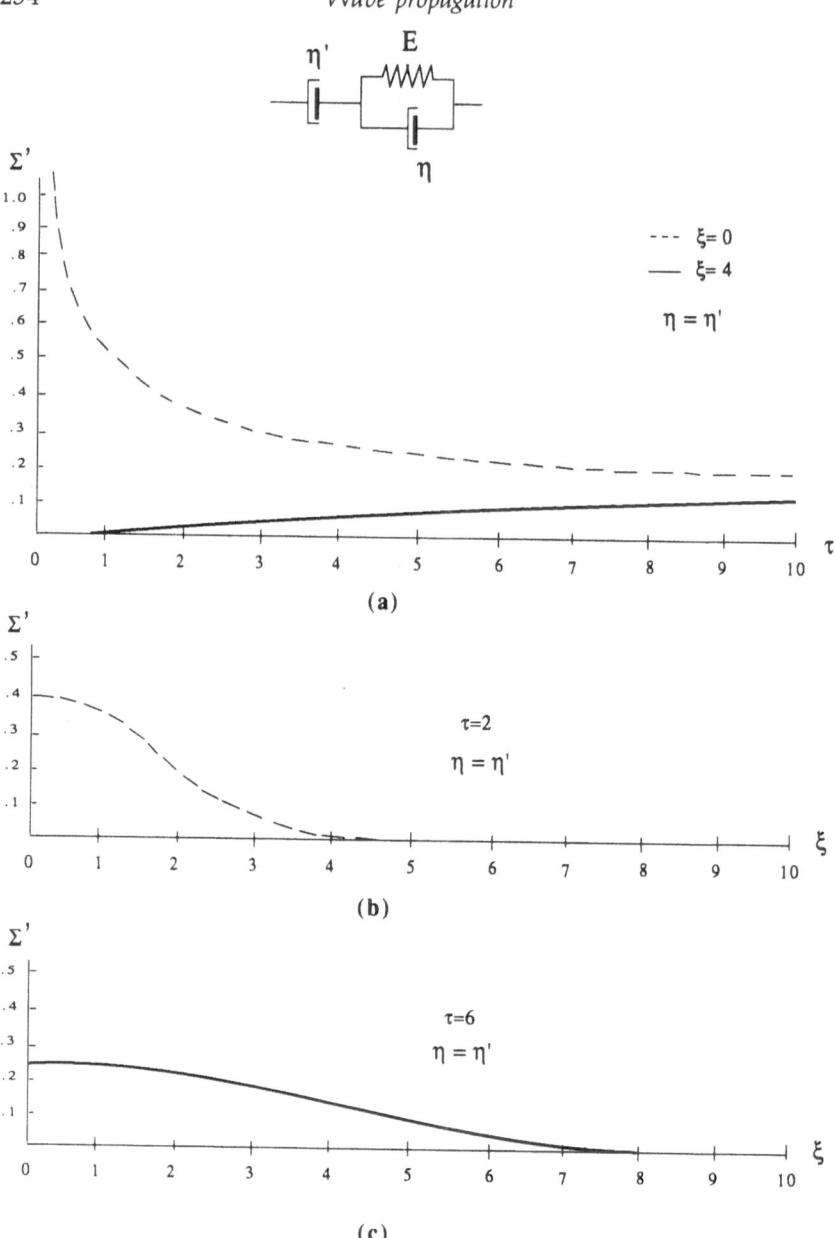

Fig. 8.19 Schematics of stress distributions for a three-element model (one elastic and two viscous elements) under suddenly applied constant end velocity v_0. $\eta = \eta'$.

8.5.2 Comparison of stress distributions for viscoelastic models with ultimate elastic behaviour

This group includes, for example, the elastic element, the Voigt model and the three-element model (one viscous and two elastic elements). We present, in this context, the results obtained by Lee and Morrison (1956).

Figure 8.20 shows comparison of stress distributions for viscoelastic models with ultimate elastic behaviour under constant end stress σ_0. Figure 8.20(a) illustrates the variations in the stress distributions $\Sigma(\xi, \tau)$, where

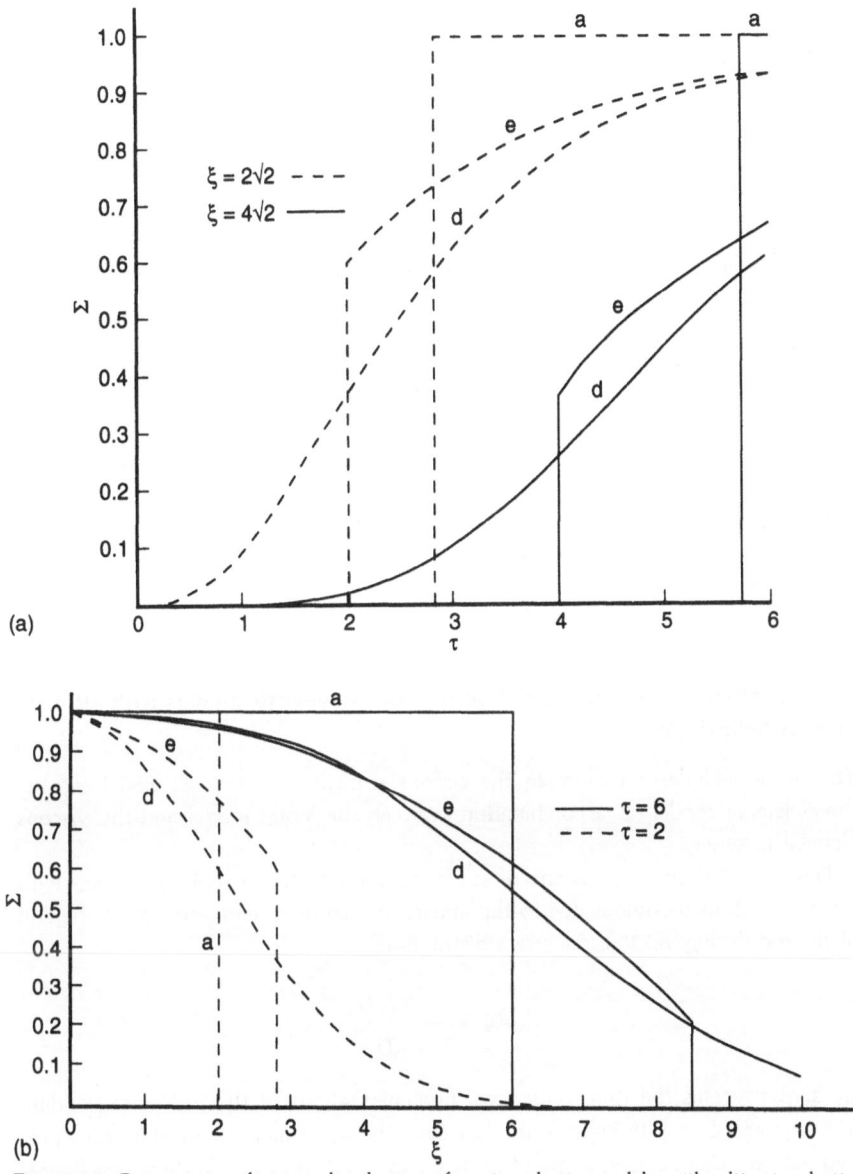

Fig. 8.20 Comparison of stress distributions for viscoelastic models with ultimate elastic behaviour (constant end stress σ_0): curves a, elastic element; curves d, Voigt solid; curves e, three-element model (one viscous and two elastic elements, $E = E'$). (Source: Lee, E. H. and Morrison, J. A. (1956) A comparison of the propagation of longitudinal waves in rods of viscoelastic materials. *J. Polym. Sci.*, **19**, 93–110 (Interscience Publishers, Inc., Copyright © 1956). Reprinted by permission of John Wiley & Sons, Inc.)

$$\Sigma(\xi, \tau) = \frac{\sigma(x, t)}{\sigma_0},$$

with τ for two different sections of the rod identified by $\xi = 2\sqrt{2}$ and $\xi = 4\sqrt{2}$. On the other hand, Fig. 8.20(b) demonstrates the stress distributions along the rod at two times corresponding to $\tau = 2$ and $\tau = 6$. In both figures, the impact is due to an impulsively applied constant end stress σ_0.

Most apparent from Fig. 8.20 is that both the Voigt and the three-parameter models exhibit very similar distributions of stress for large times. However, such behaviour would generally differ from that of the elastic rod owing to the viscous flow occurring in the former two models.

Figure 8.21 shows comparison of stress distributions for viscoelastic models with ultimate elastic behaviour under constant end velocity v_0. Figure 8.21(a) shows the variations of the stress distribution

$$\Sigma'(\xi, \tau) = \frac{\sigma(x, t)}{(\rho E)^{1/2} v_0}$$

with τ for different values of ξ, namely $\xi = 0$ and $\xi = 2\sqrt{2}$. In Fig. 8.21(b), however, the stress solutions $\Sigma'(\xi, \tau)$ are plotted against ξ for two different values of τ, namely $\tau = 2$ and $\tau = 6$, measured from the instant $\tau = 0$ when the impact was applied at the end of the rod. In both figures, the impact, as indicated earlier, is due to a constant end velocity v_0.

8.5.3 Comparison of stress distributions for viscoelastic models with ultimate viscous behaviour

This group includes, for example, the viscous element, the Maxwell model and the three-element model (as a combination between the Voigt model and the viscous element in series).

Figure 8.22 shows comparison of stress distributions for viscoelastic models with ultimate viscous behaviour due to the impact of a constant end stress σ_0 at the end of the rod. In Fig. 8.22(a), the stress distributions

$$\Sigma(\xi, \tau) = \frac{\sigma(x, t)}{\sigma_0}$$

are shown versus the time τ for two different sections of the rod corresponding to $\xi = 1$ and $\xi = 4$. In Fig. 8.22(b), however, the stress along the rod is shown for two different times corresponding to $\tau = 2$ and $\tau = 8$. It is seen in these two figures that the stress distributions in the mechanical models dealt with are close to each other at several delay times after the wave front arrival time.

Figure 8.23 shows a comparison of stress distributions for viscoelastic models with ultimate viscous behaviour when the impact is due to a constant end velocity v_0.

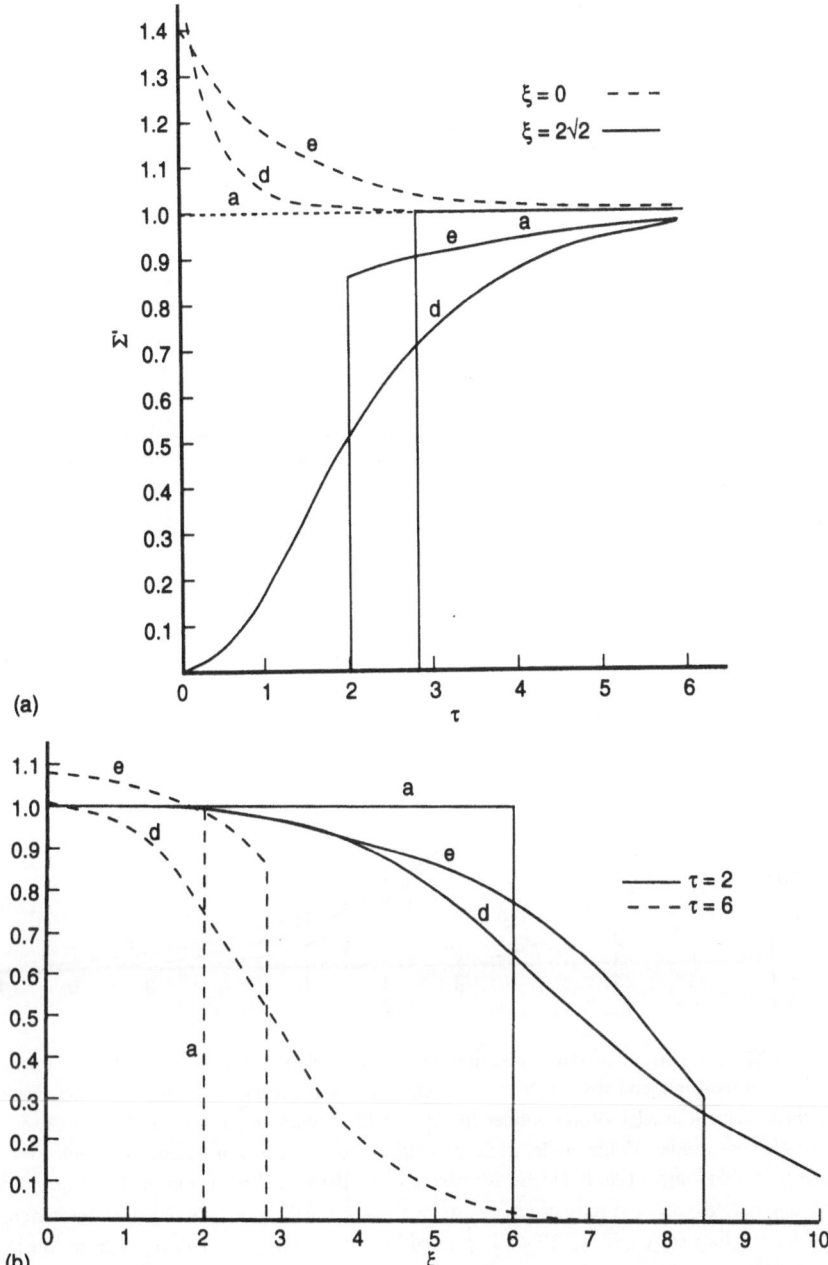

Fig. 8.21 Comparison of stress distributions for viscoelastic models with ultimate elastic behaviour (constant end velocity v_0): curves a, elastic element; curves d, Voigt solid; curves e, three-element model (one viscous and two elastic elements $E = E'$). (Source: Lee, E. H. and Morrison, J. A. (1956) A comparison of the propagation of longitudinal waves in rods of viscoelastic materials. *J. Polym. Sci.*, **19**, 93–110 (Interscience Publishers, Inc., copyright © 1956). Reprinted by permission of John Wiley & Sons, Inc.)

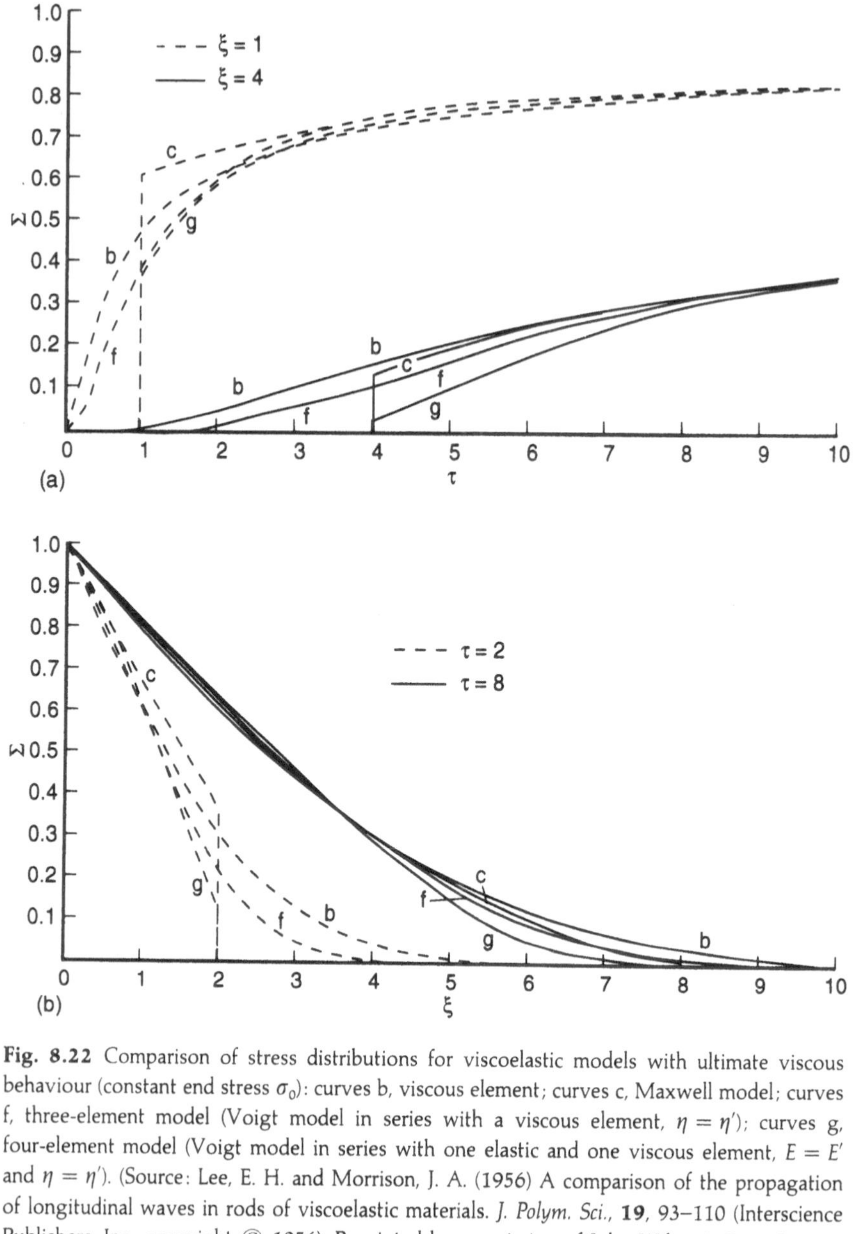

Fig. 8.22 Comparison of stress distributions for viscoelastic models with ultimate viscous behaviour (constant end stress σ_0): curves b, viscous element; curves c, Maxwell model; curves f, three-element model (Voigt model in series with a viscous element, $\eta = \eta'$); curves g, four-element model (Voigt model in series with one elastic and one viscous element, $E = E'$ and $\eta = \eta'$). (Source: Lee, E. H. and Morrison, J. A. (1956) A comparison of the propagation of longitudinal waves in rods of viscoelastic materials. *J. Polym. Sci.*, **19**, 93–110 (Interscience Publishers, Inc., copyright © 1956). Reprinted by permission of John Wiley & Sons, Inc.)

Figure 8.23(a) shows the stress solutions

$$\Sigma'(\xi, \tau) = \frac{\sigma(x, t)}{(\rho E)^{1/2} v_0}$$

against τ for two different values of ξ, namely $\xi = 0$ (i.e. at the end of the rod

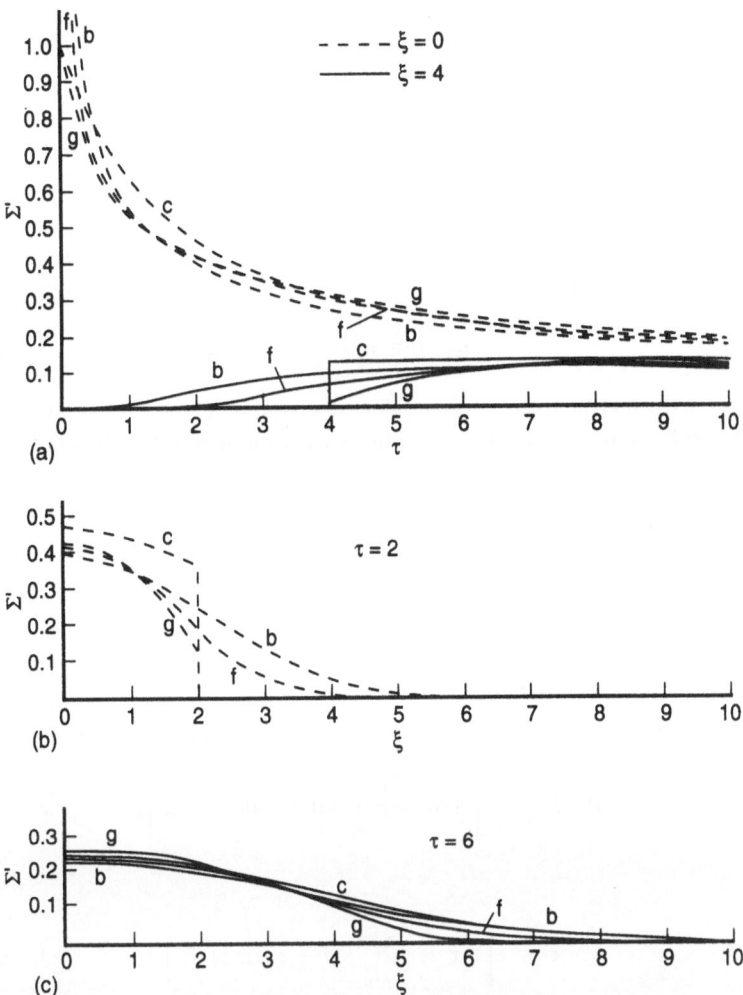

Fig. 8.23 Comparison of stress distributions for viscoelastic models with ultimate viscous behaviour (constant end velocity v_0): curves b, viscous element; curves c, Maxwell model; curves f, three-element model (Voigt model in series with a viscous element, $\eta = \eta'$); curves g, four-element model (Voigt model in series with one elastic and one viscous element, $E = E'$ and $\eta = \eta'$). (Source: Lee, E. H. and Morrison, J. A. (1956) A comparison of the propagation of longitudinal waves in rods of viscoelastic materials. *J. Polym. Sci.*, **19**, 93–110 (Interscience Publishers, Inc., copyright (1956). Reprinted by permission of John Wiley & Sons, Inc.)

$x = 0$) and $\xi = 4$. Meantime, Figs. 8.23(b) and 8.23(c) illustrate the stress distributions along the rod, that is $\Sigma'(\xi, \tau)$ against ξ, for $\tau = 2$ and $\tau = 6$, respectively. The results shown in Fig. 8.23 are, as mentioned, due to an impulsively constant velocity v_0 applied at the end of the rod at $\tau = 0$.

Figures 8.22 and 8.23, introduced above, are due to Lee and Morrison (1956).

8.6 SOLUTION OF THE WAVE EQUATION IN LINEAR VISCOELASTICITY AS BASED ON BOLTZMANN'S SUPERPOSITION PRINCIPLE

Consider a homogeneous, isotropic rod and let x_i ($i = 1, 2, 3$) denote the Cartesian coordinates of any material particle p in the deformed (current) state. For the longitudinal motion of the rod in the x_1 direction, the displacement is expressed as

$$\mathbf{u}(\mathbf{x}, t) = u_1(x_1, t)\mathbf{e}_1 \tag{8.96}$$

where $\mathbf{e}_1$ is the unit vector component along the x_1 axis. It is assumed here that the displacement $\mathbf{u}(\mathbf{x}, t)$ is a continuous function of $\mathbf{x}, t$ for all $\mathbf{x}, t$. In this case, $\varepsilon_{11}(x_1, t)$ and $\sigma_{11}(x_1, t)$ will be the only corresponding nonvanishing components of the strain and stress, respectively, where

$$\varepsilon_{11}(x_1, t) = \frac{\partial u_1(x_1, t)}{\partial x_1} \tag{8.97}$$

and the stress is connected to the strain via Boltzmann's hereditary creep and relaxation constitutive equations introduced earlier by (2.76) and (2.83), respectively. Recalling the latter two equations, one may write in the same order that

$$\varepsilon(x, t) = E^{-1}\left[\sigma(x, t) + \int_0^t F(t - \tau)\sigma(x, \tau)\,d\tau\right] \tag{8.98}$$

and

$$\sigma(x, t) = E\left[\varepsilon(x, t) + \int_0^t R(t - \tau)\varepsilon(x, \tau)\,d\tau\right]. \tag{8.99}$$

Meantime, the equation of motion can be written, in the absence of body forces, in view of (1.22) as

$$\rho\frac{\partial^2 u(x, t)}{\partial t^2} = \frac{\partial \sigma(x, t)}{\partial x}. \tag{8.100}$$

Combining now (8.98) and (8.100), the creep wave equation, in the absence of body forces, can be written as

$$\frac{\partial^2 u(x, t)}{\partial x^2} - \frac{1}{c^2}\frac{\partial^2 u(x, t)}{\partial t^2} = \frac{1}{c^2}\int_0^t F(t - \tau)\frac{\partial^2 u(x, t)}{\partial \tau^2} \tag{8.101}$$

where

$$c^2 = E/\rho. \tag{8.102}$$

Further, with reference to (8.99) and (8.100), the relaxation wave equation, in the absence of body forces, is

$$\frac{1}{c^2} \frac{\partial^2 u(x, t)}{\partial t^2} - \frac{\partial^2 u(x, t)}{\partial x^2} = \int_0^t R(t - \tau) \frac{\partial^2 u(x, t)}{\partial x^2} d\tau \tag{8.103}$$

with $c^2 = E/\rho$.

The Laplace transforms of the wave equations (8.101) and (8.103) read, respectively, as follows:

$$\frac{d^2}{dx^2} \bar{u}(x, s) = \left(\frac{s}{c}\right)^2 [1 + \bar{F}(s)] \bar{u}(x, s) \tag{8.104}$$

and

$$\frac{d^2}{dx^2} \bar{u}(x, s) = \left(\frac{s}{c}\right)^2 [1 + \bar{R}(s)]^{-1} \bar{u}(x, s). \tag{8.105}$$

Considering, for instance, equation (8.104) corresponding to the creep case, the general transform solution can be written (Graffi, 1982) as

$$\bar{u}(x, s) = A(s) \exp\left\{\frac{xs}{c} [1 + \bar{F}(s)]^{1/2}\right\}$$

$$+ B(s) \exp\left\{-\frac{xs}{c} [1 + \bar{F}(s)]^{1/2}\right\} \tag{8.106}$$

where $A(s)$ and $B(s)$ are functions of the Laplace parameter s. Both $A(s)$ and $B(s)$ are to be determined.

At this point, we shall assume that the rod is semi-infinite in extent and initially undisturbed in the sense that

$$u(x, 0) = \frac{\partial u(x, 0)}{\partial t} = 0. \tag{8.107}$$

The following boundary conditions are further assumed:

$$u(0, t) = u_0(t), \qquad t \geq 0;$$

$$\lim_{x \to \infty} u(x, t) = 0, \qquad t \geq 0. \tag{8.108}$$

In this case, one must impose that $A(s) = 0$ in (8.106) in order to avoid the exponential increase with x of the first term in this equation. Thus, $B(s)$ in (8.106) will assume the value of the Laplace transform of the input $u_0(t)$ and (8.106) becomes

$$\bar{u}(x, s) = \bar{u}_0(s) \exp\left\{-\frac{xs}{c} [1 + \bar{F}(s)]^{1/2}\right\}. \tag{8.109}$$

The inversion of the Laplace transforms in (8.109) leads to (Graffi, 1982)

$$u(x, t) = H(t - x/c) \exp(\alpha x/c)\left[u_0(t - x/c) + \int_0^{t-x/c} F(x, t - \tau - x/c)u_0(\tau) \, d\tau \right]$$

(8.110)

in which $H(\cdot)$ is the Heaviside step function and where the translation and convolution formulae for the Laplace transform (Appendix C) were used. Another representation of the solution (8.110) is due to Mainardi and Turchetti (1975, 1979). Mainardi and Nervosi (1980) have also considered the inclusion of such a presentation in their treatment of transient waves in a viscoelastic rod. A similar treatment may be considered for the relaxation case based on equation (8.105).

8.7 SOLUTION OF THE WAVE PROPAGATION PROBLEM IN A LINEAR VISCOELASTIC SOLID USING THE CORRESPONDENCE PRINCIPLE

In this section, a presentation is given, following Chao and Achenbach (1964), on the utilization of the correspondence principle to solve wave propagation problems in linear viscoelasticity when the solutions of the corresponding elastic problems are known.

The constitutive equations for an isotropic, elastic solid are given in section 8.2 by the set of equations (8.1). With reference to these equations, it is recognized that for an isotropic, elastic solid, two independent constants completely define the stress–strain relations. If the shear modulus μ and the bulk modulus K, for instance, are chosen, the constitutive equations (8.1) can be written in the following tensorial form:

$$\sigma_{ij} = [K - (2/3)\mu]\varepsilon_{kk}\delta_{ij} + 2\mu\varepsilon_{ij}.$$

(8.111)

On the other hand, the constitutive relations for an isotropic, linear viscoelastic material are time dependent. These relations were given for the creep case by (4.27) and for the relaxation case by (4.29). Recalling, for instance, equations (4.27), the following constitutive relations can be written for the creep of an isotropic, viscoelastic solid:

$$2\mu\varepsilon'_{ij}(t) = \sigma'_{ij}(t) + \int_{-\infty}^{t} F_1(t - \tau) \frac{d\sigma'_{ij}(t)}{d\tau} \, d\tau$$

(8.112a)

and

$$2K\varepsilon_{ij}(t) = \sigma_{ij}(t) + \int_{-\infty}^{t} F_2(t - \tau) \frac{d\sigma_{ij}(t)}{d\tau} \, d\tau$$

(8.112b)

in which $F_1(\cdot)$ and $F_2(\cdot)$ are the creep functions governing, respectively, the shear and dilatational behaviours of the medium. The treatment of stress-wave propagation in a viscoelastic solid which obeys the constitutive relations (8.112) leads to complicated mathematical analysis in that the solution of partial integrodifferential equations is involved. Volterra (1931) considered the problem by adopting a functional analysis approach, but it seems that the results of the theory have found, so far, little application in the study of the dynamic behaviour of viscoelastic materials (e.g. Kolsky, 1963).

Chao and Achenbach (1964) discussed the application of the Laplace transform to viscoelastic wave propagation problems using the well-known correspondence principle (Bland, 1960; Schapery, 1974). It was shown by these authors that, under the restricted condition of constant Poisson's ratio, a class of viscoelastic problems may be solved provided that the solution of the corresponding elastic problem is known. Applying the Laplace transform to (8.112a) and (8.112b), with some additional manipulation, yields

$$\bar{\sigma}_{ij} = [\bar{K}(s) - \tfrac{2}{3}\bar{\mu}(s)]\bar{\varepsilon}_{kk}\delta_{ij} + 2\bar{\mu}(s)\bar{\varepsilon}_{ij} \tag{8.113a}$$

where

$$\bar{\mu}(s) = \frac{\mu}{1 + s\bar{F}_1(s)} = \mu\beta(s) \tag{8.113b}$$

and

$$\bar{K}(s) = \frac{K}{1 + s\bar{F}_2(s)} = K\gamma(s) \tag{8.113c}$$

where s is the Laplace transform parameter.

Meantime, the Laplace transform of the stress equation of motion (1.22), in the absence of body forces, can be written as

$$\bar{\sigma}_{ij,j} = \rho s^2 \bar{u}_i \tag{8.114}$$

where ρ is the mass density of the material.

Combining (8.113) and (8.114) yields the governing differential equations for the transformed displacements of a viscoelastic medium, that is

$$(\bar{K} + \tfrac{1}{3}\bar{\mu})\bar{u}_{j,ji} + \bar{\mu}u_{i,jj} = \rho s^2 \bar{u}_i. \tag{8.115}$$

We decompose the displacement vector **u** into dilatational and rotational parts, i.e.

$$\bar{\mathbf{u}} = \bar{\mathbf{v}} + \bar{\boldsymbol{\omega}} \tag{8.116}$$

where

$$\bar{v}_{i,i} = 0 \quad \text{and} \quad \bar{\omega}_{i,j} = \omega_{j,i}. \tag{8.117}$$

Accordingly, the transformed equations (8.115) will be satisfied if

$$\bar{v}_{i,jj} = \frac{s^2}{\bar{c}_1^2}\,\bar{v}_i \tag{8.118}$$

and

$$\bar{\omega}_{i,jj} = \frac{s^2}{\bar{c}_2^2} \bar{\omega}_i \qquad (8.119)$$

where $\bar{c}_1$ and $\bar{c}_2$ are the transformed velocities for the dilatational and rotational waves respectively, i.e.

$$\bar{c}_1^2 = \bar{K}(s) + \tfrac{4}{3}\bar{\mu}(s) \qquad (8.120)$$

and

$$\bar{c}_2^2 = \frac{\bar{\mu}(s)}{\rho}. \qquad (8.121)$$

The same treatment may be applied for the isotropic, elastic medium if the constitutive equation (8.111) is used instead of (8.113a). On the other hand, the analogous equations to (8.118) and (8.119) for the isotropic, elastic body are obtained if $\bar{\mu}(s)$ and $\bar{K}(s)$ in (8.113b) and (8.113c) are replaced by the elastic moduli μ and K respectively. The above treatment was presented by Chao and Achenbach (1964) with the following conclusion: the Laplace transforms of the solutions for a viscoelastic wave propagation problem can be obtained from the Laplace transforms of the solutions for the elastic problem with the same boundary and initial conditions by replacing the shear modulus μ by its Laplace transform $\bar{\mu}(s)$ and the bulk modulus K by its Laplace transform $\bar{K}(s)$.

The above conclusion is, in essence, a form of the well-known **correspondence principle**; that is the problem of obtaining solutions concerning the response behaviour of a linear viscoelastic solid is reduced to a problem of inverting the Laplace transforms of the corresponding elastic solution.

Chao and Achenbach (1964) (Achenbach and Chao (1962) should also be referred to) considered the application of the above approach to the study of the displacement and stress fields inside an infinite, viscoelastic body of a constant Poisson's ratio. In their treatment, the authors assumed the input force to be time dependent and concentrated at one point. Two illustrative examples were subsequently given. In the first example, the displacement components in the radial and the vertical directions on the surface of a viscoelastic half-space loaded suddenly by a vertical force of constant magnitude were evaluated. In the second example, the radial stress for the problem of the expanding spherical cavity in an infinite viscoelastic medium was dealt with.

8.8 NONLINEAR VISCOELASTIC WAVE PROPAGATION

The subject of characterization of the nonlinear viscoelastic response of materials has been dealt with in some detail in Chapter 6. As mentioned in Chapter 6, although considerable research efforts have been made over recent decades towards characterization of the nonlinear viscoelastic nature of materials, interest in the study of wave propagation in such materials did not develop until recently. Most of the studies on wave propagation in nonlinear viscoelastic materials dealt essentially with the one-dimensional motion within the context of the general constitutive theory of materials with fading memory. These studies have considered the propagation of both acceleration and shock waves in viscoelastic media with the objective of establishing the governing conditions for their growth or decay. Such governing conditions implied the existence of steady waves in the dissipative viscoelastic media.

An initial study in the area of nonlinear wave propagation is due to Malvern (1951). Malvern's approach is concerned with the motion of a plastic wave in a ductile material (e.g. a metal with a strain memory effect). As a special case, however, Malvern considered the motion of such a type of wave in a model of a viscoelastic solid. The modes of propagation of acceleration waves in different media have been studied by, among others, Truesdell and Toupin (1960), Thomas (1961), Hill (1962), Varley and Cumberbatch (1965), Coleman, Gurtin and Herrera (1965), Coleman and Gurtin (1965a–c) and Bailey and Chen (1971). Varley (1965) discussed the mode of propagation of an arbitrary acceleration wave as it advances into a finitely strained viscoelastic material which, until the arrival of the front, is undergoing any admissible deformation. The viscoelastic material is seen in Varley's work to be generally inhomogeneous and anisotropic. Coleman, Gurtin and Herrera (1965) and Coleman and Gurtin (1965a–c) dealt comprehensively with the theory of nonlinear viscoelastic wave propagation in a series of research papers. In the first two papers of the series, the authors dealt with the propagation of shock and acceleration fronts in materials with memory resting on the assumption that the stress is a functional of the history of the deformation gradient with the exclusion of any thermal influences. In the subsequent two papers (Parts III and IV of the series), Coleman and Gurtin (1965b, c) have allowed the stress to be affected not only by the history of strain, but also by the history of a thermodynamic variable such as the temperature (Coleman (1964) and Coleman and Gurtin (1966) should also be referred to). An extension of this work to include mild discontinuities was considered by Coleman, Greenberg and Gurtin (1966). The problem of propagation of steady shock waves in nonlinear thermoviscoelastic solids has been also considered, for instance, by Ahrens and Duvall (1966), Greenberg (1967), Chen, Gurtin and Walsh (1970), Schuler and Walsh (1971), Dunwoody (1972), Huilgol (1973) and Nunziato and Walsh (1973). In this, Nunziato and Walsh (1973) expressed the governing equations in terms of material response functions which can be determined from shock wave, thermophysical and bulk response data. The results of the analysis were compared with experimental steady-wave studies concerning the solid polymer polymethyl methacrylate (PMMA). The existence and propagation of steady waves in a class of dissipative materials were considered also by, among others, Greenberg (1968) and Schuler (1970).

On the experimental side, research in the field of shock wave physics has made it possible to produce high amplitude strain waves. Barker and Hollenbach (1970) and Schuler (1970) used a gas gun (Barker and Hollenbach, 1964, 1965) to produce a planar impact between two plates. This has been parallel with the development of advanced recording and measurement techniques such as laser interferometry (Barker and Hollenbach, 1964, 1965, 1970; Barker, 1968). Such experimental efforts were particularly effective in the production of one-dimensional strains of very large amplitude; meantime, they allowed wave motion to be observed with high resolution and accuracy. Chen and Gurtin (1972a, b) discussed the use of experimental results concerning steady shock waves to predict the acceleration wave response of nonlinear viscoelastic materials. Meantime, Nunziato and Sutherland (1973) used acoustic waves for the determination of stress relaxation functions of a class of polymeric materials.

Schuler, Nunziato and Walsh (1973) presented a comprehensive review of some theoretical and experimental developments in the domain of nonlinear viscoelastic wave propagation. Confining their attention to the case of one-dimensional strain, they reviewed theories of shock and acceleration wave propagation in materials with memory and discussed the theoretical predictions with some experimental results for the polymeric solid PMMA. In this, these authors were particularly influenced by the work of Coleman, Gurtin and Herrera (1965), Coleman and Gurtin (1965a–c) and Chen and Gurtin (1970). We follow closely the work of these authors in the following presentation.

8.8.1 Kinematics and balance laws in one-dimensional motion

(a) Kinematics

In the case of one-dimensional motion, we identify the spatial position of a material point (particle) at time t by the coordinate $x(X, t)$. The counterpart of this position coordinate in the reference configuration, R, is $X(x, t)$. It is assumed that the coordinate function $x(X, t)$ is continuous for all X and t. The corresponding displacement function $u(X, t)$ is, thus, a continuous function of X, t for all X and t. Assuming suitable smoothness of the motion (Schuler, Nunziato and Walsh, 1973) the particle velocity is expressed by

$$v(t) = \frac{\partial}{\partial t} u(X, t) = \partial_t u(X, t) \tag{8.122}$$

and the compressive strain is expressed by

$$\varepsilon(t) = -\frac{\partial}{\partial X} u(X, t) = -\partial_x u(X, t). \tag{8.123}$$

A wave propagating in such continuous medium may be seen (Coleman and Gurtin, 1965a–c) as a family of points $\hat{X}(t)$, $-\infty < t < \infty$, where $\hat{X}(t)$ is the material point in the reference configuration R at which the wave front is to be found at time t. Thus, the spatial position of the wave may be expressed by

$$\hat{x}(t) = x[\hat{X}(t), t] \tag{8.124}$$

with $\hat{x}(t)$ designating the spatial position of the wave at time t.

The wave velocity $V(t)$ at time t is defined by

$$V(t) = \frac{d}{dt}\,\hat{x}(t) = \frac{d}{dt}\,x[\hat{X}(t),\, t].$$

(8.125)

The **wave velocity** $V(t)$ is identified with respect to an external fixed frame of reference (i.e. as seen by an external observer at rest).

Meantime, the **wave 'intrinsic' velocity** $U(t)$ is defined as the velocity of propagation of the wave front relative to the material in the reference configuration. It is expressed as

$$U(t) = \frac{d}{dt}\,\hat{X}(t)$$

(8.126)

where $\hat{X}(t)$, as defined earlier, is the coordinate of the material point in the reference configuration R at which the wave front is to be found at time t.

The material trajectory of the wave front is given here the notation $\Omega(t)$. It is defined as the set of ordered pairs $[\hat{X}(t),\, t]$, $-\infty < t < \infty$.

Coleman, Gurtin and Herrera (1965), following the standard notation used earlier by Truesdell and Toupin (1960), advanced that if a function $f(X,\, t)$ has a jump discontinuity at $X = \hat{X}(t)$, one may define the jump in $f(X,\, t)$, labelled below by $[f]$, across the trajectory of the wave $\Omega(t)$ at time t by

$$[f] = \lim_{X \to \hat{X}(t)^-} f(X,\, t) - \lim_{X \to \hat{X}(t)^+} f(X,\, t).$$

(8.127a)

This expression may also be written in the form

$$[f] = f^- - f^+$$

(8.127b)

where, with the wave intrinsic velocity $U(t) > 0$, f^+ and f^- are the limiting values of the function $f(X,\, t)$ immediately ahead and behind the wave front respectively. The associated **condition of compatibility** to (8.127) is expressed (Truesdell and Toupin, 1960) as

$$[\partial_t f] = -U[\partial_x f].$$

(8.128)

In the present section, the function $f : f(X,\, t)$ is used to designate the kinematical function $x(X,\, t)$ or one of its derivatives.

(b) Balance laws

Mass balance
In one-dimensional motion, the mass balance is expressed with reference to (1.3) as

$$\frac{\rho(X,\, t)}{\rho_0} = \frac{1}{1 - \varepsilon(X,\, t)}$$

(8.129)

where $\rho(X, t)$ is the current mass density and ρ_0 is the mass density in the reference configuration of the material specimen.

Balances of linear momentum and energy

With the exclusion of external body forces, heat conduction and external heat supply, the balances of linear momentum and energy are expressed respectively as

$$\frac{d}{dt} \int_{x_\alpha}^{x_\beta} \rho_0 v(x, t) \, dx = \sigma(x_\beta, t) - \sigma(x_\alpha, t) \tag{8.130}$$

and

$$\frac{d}{dt} \int_{x_\alpha}^{x_\beta} [\tfrac{1}{2}\rho_0 v^2(x, t) + e(x, t)] \, dx = \sigma(x_\beta, t)v(x_\beta, t) - \sigma(x_\alpha, t)v(x_\alpha, t) \tag{8.131}$$

where σ is the one-dimensional stress and e is the internal energy per unit volume.

Clausius–Duhem inequality (second law of thermodynamics)

$$\frac{d}{dt} \int_{x_\alpha}^{x_\beta} \rho_0 S(x, t) \, dx \geq 0 \tag{8.132}$$

where S is the specific entropy per unit mass.

8.8.2 Material response functions

Following Schuler, Nunziato and Walsh (1973), we consider a strain jump ε suddenly applied to a material point which has been unstrained for all past times, i.e.

$$\varepsilon(\tau) = \begin{cases} \varepsilon & \tau = 0 \\ 0 & \tau < 0, \tau > 0. \end{cases} \tag{8.133}$$

The instantaneous stress, denoted by σ_I, corresponding to the strain jump (8.133) is expressed in the following functional format

$$\sigma_I(\varepsilon) = F(\varepsilon) \tag{8.134}$$

where F is a constitutive functional defined under the restrictions imposed on such functional as discussed previously in Chapter 6. The constitutive functional $F(\varepsilon)$ is assumed to be twice continuously differentiable, i.e. $\partial_\varepsilon F[\varepsilon(t - \tau)]$ and $\partial_\varepsilon^2 F[\varepsilon(t - \tau)]$ exist where the partial differentiation is with respect to the present value of strain $\varepsilon(t)$.

The stress-relaxation function corresponding to the strain history (8.133) is designated by $R(\varepsilon; \tau)$. Meantime, the **instantaneous tangent modulus** is designated by $E_I(\varepsilon)$ where

$$E_I(\varepsilon) = \frac{d\sigma_I(\varepsilon)}{d\varepsilon} = R(\varepsilon; 0). \tag{8.135}$$

Similarly, the **instantaneous second-order modulus** $\check{E}_I$ is defined by

$$\check{E}_I(\varepsilon) = \frac{d^2\sigma_I(\varepsilon)}{d\varepsilon^2}. \tag{8.136}$$

On the other hand, the equilibrium response of the material may be expressed as

$$\sigma_E(\varepsilon) = F(\varepsilon_E). \tag{8.137}$$

From this the **equilibrium tangent modulus** is given as

$$E_E(\varepsilon) = \frac{d\sigma_E(\varepsilon)}{d\varepsilon} = R(\varepsilon; \tau). \tag{8.138}$$

Thus, the **equilibrium second-order modulus** $\check{E}_E(\varepsilon)$ is identified by

$$\check{E}_E(\varepsilon) = \frac{d^2\sigma_E(\varepsilon)}{d\varepsilon^2}. \tag{8.139}$$

Schuler, Nunziato and Walsh (1973) have imposed certain curvature conditions on the constitutive functional $F(\varepsilon)$ of (8.134). They advanced that these conditions would hold valid for most of viscoelastic materials. These conditions may be presented as follows.

For all ε on $(0, 1)$ and all τ on $(0, \infty)$,

1.
$$\sigma_I(\varepsilon) > \sigma_E(\varepsilon) > 0,$$

$$E_I(\varepsilon) > E_E(\varepsilon) > 0, \tag{8.140a}$$

$$\check{E}_I(\varepsilon) > 0, \quad \check{E}_E(\varepsilon) > 0;$$

2.
$$R[\varepsilon(t - \tau), \tau] > 0, \quad R'[\varepsilon(t - \tau), \tau] \leq 0,$$

where
$$R'[\varepsilon(t - \tau), \tau] = \frac{\partial}{\partial\tau} R[\varepsilon(t - \tau), t]. \tag{8.140b}$$

The inequalities under (1) imply that the instantaneous and equilibrium stress–strain curves are strictly convex from below and the instantaneous response curve lies everywhere above the equilibrium curve as shown in Fig. 8.24. It is assumed in the latter figure that $\sigma_I(0) = \sigma_E(0) = 0$. Meantime, the inequalities under (2) affirm that, for all strain histories on $(0, 1)$, the stress-relaxation function is positive and a monotonically decreasing function of the elapsed time τ. Gurtin and Herrera (1965) have also discussed inequalities (2) within the context of linear viscoelasticity.

8.8.3 Acceleration waves

The subject of acceleration wave propagation in nonlinear materials with fading time-memory has been considered by, amongst others, Truesdell and Toupin (1960), Varley (1965), Coleman, Gurtin and Herrera (1965), Coleman and Gurtin (1965a–c), Bailey and Chen (1971) and Schuler, Nunziato and Walsh (1973).

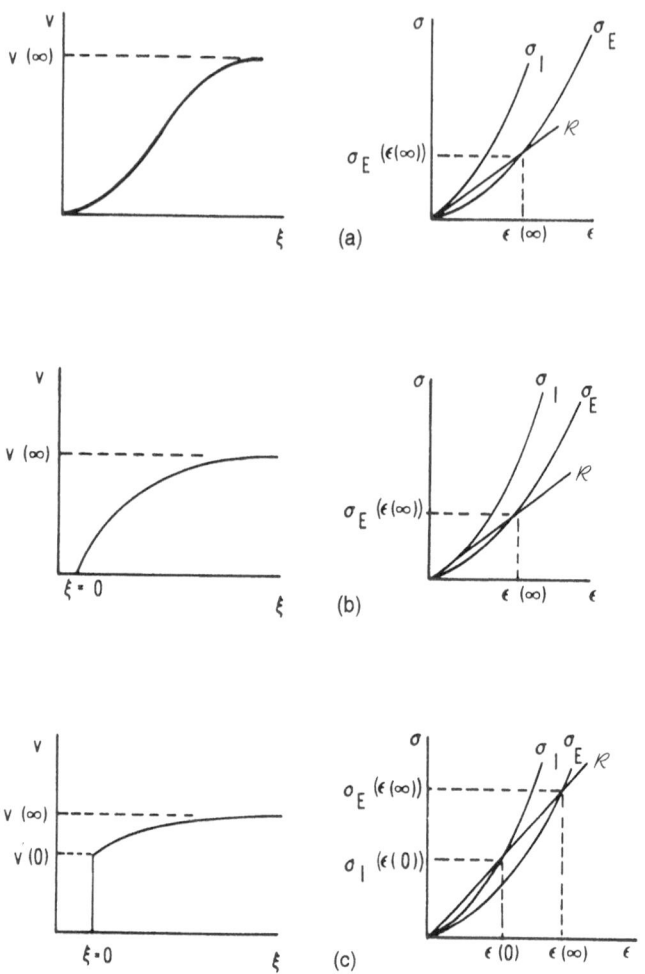

Fig. 8.24 Steady wave solutions (curvature conditions imposed on the instantaneous and equilibrium stress–strain curves). (Reprinted with permission from *Int. J. Solids Struct.*, **9**, Schuler, K. W., Nunziato, J. W. and Walsh, E. K., Recent results in nonlinear viscoelastic wave propagation, copyright (1973), Pergamon Press Ltd.)

Definition Coleman and Gurtin (1965a–c) advanced that if the following two conditions are satisfied, then $\ddot{X}(t)$, $-\infty < t < \infty$, is an acceleration wave.

(A-1) $x(X, \tau)$, $\dot{x}(X, \tau)$ and the deformation gradient F are continuous functions of X and τ jointly for all X and τ, while $\ddot{x}(X, \tau)$, $\partial F(X, \tau)/\partial X$, $\dot{F}(X, \tau)$ have jump discontinuities across the wave material trajectory, but are the continuous in X and τ jointly everywhere else.

(A-2) The past history of the deformation gradient, $F_r^t(X_i)$, is a smooth function of X and t with respect to the norm $\|\cdot\|_r$. $F_r^t(X, \cdot)$ is the restriction on the history of the deformation gradient $F(t - \tau)$ to its domain of definition $(0, \infty)$.

The condition (A-2) limits the wildness of the past history for the material with memory.

Coleman and Gurtin (1965a–c) considered, within the general linear theory of simple materials with fading memory, the case when an acceleration wave which since $t = 0$ has been propagating into a region which had been previously at rest in a fixed homogeneous configuration R. For this case, it was remarked by these authors that the hypothesis (A-2) follows from hypothesis (A-1) for all $X > \hat{X}(0)$ and $t > 0$. In other words, whenever the acceleration front is entering a homogeneous medium at rest, the hypothesis (A-1) would generally suffice to ensure that $F_r^t(X, \cdot)$ is a smooth function of X and t with respect to the norm $\|\cdot\|_r$. Thus, following the condition (A-1) above, the compatibility condition (8.128) and taking $f(X, t) = \varepsilon(X, t)$, one can write at $X = \hat{X}(t)$ that

$$[\dot{v}] = U[\dot{\varepsilon}] = U^2[\partial_x \varepsilon]. \tag{8.141}$$

Following (8.127), the stress and the internal energy must also be discontinuous at $X = \hat{X}(t)$. Thus, with reference to (8.130), the balance of linear momentum asserts that

$$[\sigma] = \rho_0 U[v], \quad [\partial_x \sigma] = -\rho_0[\dot{v}]. \tag{8.142}$$

Also, the balance of energy (8.131) implies that

$$-U[e + \tfrac{1}{2}\rho_0 v^2] = [\sigma v], \quad [\dot{e}] = [\sigma \dot{\varepsilon}] \tag{8.143}$$

(Schuler, Nunziato and Walsh, 1973).

The jump in the particle acceleration is conventionally taken as the amplitude of the wave front. Denoting the latter at any instant of time by $a(t)$, then

$$a(t) = [\dot{v}](t). \tag{8.144}$$

Based on purely kinematical considerations, Coleman and Gurtin (1965a–c) affirmed that the amplitude of an acceleration wave obeys the following relationship:

$$2\frac{da}{dt} - \frac{a}{U}\frac{dU}{dE} = [\ddot{x}] - U^2\frac{\partial \dot{F}}{\partial X} \tag{8.145}$$

where, as introduced earlier, $U = U(t)$ is the intrinsic wave velocity and E_I is the instantaneous tangent modulus. As an alternate expression to (8.145), Coleman and Gurtin (1965a–c) advanced the following equation for the amplitude of an acceleration wave using condition (A-1) and the balance of momentum equation (8.142):

$$2\frac{da}{dt} = \frac{a}{U}\frac{dU}{dt} + \frac{1}{\rho_0}\left[\frac{\partial^2 \sigma}{\partial t\, \partial X}\right] - U^2\left[\frac{\partial \dot{F}}{\partial X}\right]. \tag{8.146}$$

Meantime, Coleman and Gurtin and Herrera (1965) indicated that the intrinsic velocity U of an acceleration wave satisfies the equation

$$U^2 = R(0)/\rho_0. \tag{8.147}$$

In this equation, $R(0)$ is the instantaneous tangent modulus corresponding to the history $F^{(t)}(\hat{X}(t), \cdot)$, i.e.

$$R(0) = \frac{d\sigma_1(\varepsilon)}{d\varepsilon}\bigg|_{\varepsilon=0}. \tag{8.148}$$

Thus, equation (8.147) implies that the intrinsic velocity of an acceleration wave $U = U_0$ is a constant given by

$$\rho_0 U_0^2 = R(0). \tag{8.149}$$

Following the above conditions, Schuler, Nunziato and Walsh (1973), following Coleman and Gurtin (1965a–c), affirmed that the amplitude of an acceleration wave is given by

$$\frac{da}{dt} = -\beta a + \frac{\beta}{\gamma} a^2 \tag{8.150}$$

where β and γ are constants given by

$$\beta = -\frac{R'(0)}{2(E_1)_0} = \text{constant}, \quad \gamma = \frac{R'(0)U_0}{(\check{E}_1)_0} = \text{constant} \tag{8.151}$$

and

$$(\check{E}_1)_0 = \frac{d^2\sigma_1(\varepsilon)}{d\varepsilon^2}\bigg|_{\varepsilon=0}. \tag{8.152}$$

The solution of (8.150) can be written as

$$a(t) = \frac{\gamma}{(\gamma/a(0) - 1)\exp(\beta t) + 1}. \tag{8.153}$$

For a given material, β and γ are constants.

Assuming that the hypothesis of the above theorem holds and supposing that

$$R(0) > 0, \quad R'(0) < 0, \quad (E_1)_0 \neq 0, \tag{8.154}$$

one concludes, with reference to (8.153), following Coleman and Gurtin (1965a–c), the following.

1. If either
   (a) $|a(0)|$ is less than $|\gamma|$, or
   (b) $\operatorname{sgn} a(0) = \operatorname{sgn}(\check{E}_1)_0$
   then $a(t) \to 0$ monotonically as $t \to \infty$.
2. If $a(0) = \gamma$, then $a(t) = a(0)$.

3. If both

  (a) $|a(0)|$ is greater than $|\gamma|$, and
  (b) sgn $a(0) = -\text{sgn }(\check{E}_1)_0$

  then $|a(t)| \to \infty$ monotonically and in a finite time t_∞ given by

$$t_\infty = -\frac{1}{\beta}\ln\left[1 - \frac{\gamma}{a(0)}\right]. \tag{8.155a}$$

It is apparent, from the above discussion, that $|\gamma|$ plays the role of the 'critical amplitude' of the input acceleration. The latter may be denoted by a_c which is expressed with reference to equation (8.151) as

$$a_c = |\gamma| = \left|\frac{R'(0)U_0}{(\check{E}_1)_0}\right| = \text{constant}. \tag{8.155b}$$

Thus (8.155) implies, assuming (8.154), the following.

1. If the amplitude of the input acceleration is sufficiently small ($<$ critical amplitude) or if the amplitude has the same sign as the instantaneous second-order modulus, then an acceleration wave obeying (8.151) is gradually damped out. In this, the internal dissipation of the material is expected to be the governing factor in the mode of wave motion.
2. If, however, the amplitude of the input accleration is greater than the critical amplitude and has its sign opposite to that of the instantaneous second-order modulus, then the wave would achieve an infinite amplitude in a finite time, i.e. a shock wave may be produced. In this, the nonlinearity of the instantaneous stress–strain curve would be the controlling factor (Schuler, Nunziato and Walsh, 1973).

    As noted by Coleman and Gurtin (1965a–c), the presence of internal damping, manifested by a strictly negative value of $R'(0)$, does not always imply that a singular surface moving into a homogeneous region must be damped out.

In the linear theory of simple materials with fading memory, the stress-relaxation function $R(\tau)$, with

$$R'(\tau) = \frac{dR(\tau)}{d\tau},$$

is a material function independent of the strain history $\varepsilon(t - \tau)$. In the physical application of this theory, it is generally expected that (Gurtin and Herrera, 1964)

$$R(0) > 0, \quad R'(0) \leq 0.$$

Coleman, Gurtin and Herrera (1965) ruled out, however, the possibility when $R(0) = 0$ or $R'(0) > 0$ in the applicability of relation (8.151). Meantime, Coleman and Gurtin (1965a–c) considered the applicability of (8.151) in the following two cases.

1. $R'(0) = 0$. In this case, it is advanced that the time dependency of the amplitude $a(t)$ of the acceleration wave is expressed as

$$a(t) = \frac{a(0)}{1 + \zeta a(0)t}, \quad \zeta = \frac{\check{E}_1}{2UR(0)}, \tag{8.156}$$

where $\check{E}_1$ is the instantaneous second-order modulus. Two situations may be considered here:

(a) if $\check{E}_1 \neq 0$, then (8.156) implies, since $\rho_0 > 0$, that, if $a(0)$ has the same sign as $\check{E}_1$, $|a(t)| \to 0$ monotonically in a finite time;
(b) if $\check{E}_1 = 0$, then (8.153) would reduce to

$$a(t) = a(0) \exp\left[\frac{R'(0)t}{2R(0)}\right] \tag{8.157}$$

which may be generally valid for a large class of linear viscoelastic materials.

A special class of materials with $R'(0) = 0$ is the class of perfectly elastic materials for which (8.155) is known to be applicable (Thomas, 1957; Green, 1964; Coleman and Gurtin, 1965a–c).

2. $R'(0) < 0$. In this case, it follows from (8.151) that the amplitude of the acceleration wave $a(t) \to 0$ as $t \to \infty$ regardless of the sign of $a(0)$.

8.8.4 Shock waves

The subject of shock wave propagation in nonlinear materials with fading time-memory has been considered by Duvall and Alverson (1963), Coleman and Gurtin (1965a–c), Coleman, Gurtin and Herrera (1965), Chen and Gurtin (1970, 1972a) and Huilgol (1973), amongst others.

Coleman, Gurtin and Herrera (1965) asserted that the following two conditions must be satisfied for the wave $\hat{X}(t)$, $-\infty < t < \infty$, to be called a shock wave in a material with memory.

(S-1) The coordinate function $x(X, t)$ is continuous in both X and t jointly while the deformation gradient $F(X, t) = \partial x(X, t)/\partial X$ and the time derivative of the coordinate $\dot{x}(X, t) = \partial x(X, t)/\partial t$ have jump discontinuities across the wave material trajectory $\Omega(t)$ but are continuous in X and t jointly everywhere else.
(S-2) The past history of the material is not too wild. For this purpose, it is assumed that the past history of the deformation gradient $F_r^t(X, \cdot)$ is a smooth function of X and t with respect to the norm $\|\cdot\|_r$. In this, $F_r^t(X, \cdot)$ is the restriction on the history of the deformation gradient $F(t - \tau)$ to its domain of definition $(0, \infty)$.

Thus, following condition (S-1), the compatibility condition (8.128) together with $f(X, t) = u(X, t)$ affirm that, at $X = \hat{X}(t)$,

$$[v] = -U[\varepsilon] \tag{8.158}$$

where U is the intrinsic velocity of the shock wave. In view of (8.158), either the jump in the particle velocity $[v]$ or the jump in the strain $[\varepsilon]$ may be taken as a measure of the amplitude of the shock. Meantime, equations (8.142) and (8.143) concerning, respectively, the balance of momentum and the balance of energy are also valid for the case of shock waves.

Coleman, Gurtin and Herrera (1965) showed that the intrinsic velocity U of a shock wave satisfies the relation

$$U^2 = \frac{E_{\mathrm{I}}[F]}{\rho_0} \tag{8.159}$$

where $E_{\mathrm{I}}[F]$ is the instantaneous secant modulus (8.135) corresponding to the history just before the arrival of the shock and a jump of amount $[F]$ where F is the deformation gradient. Consider now the case of a compressive wave propagating into a region at rest and unstrained for all past times (Chen and Gurtin, 1970; Schuler, Nunziato and Walsh, 1973), i.e. for $X > \hat{X}(t)$, $\varepsilon(t - \tau) = 0$, $0 \le \tau < \infty$ and

$$[\varepsilon] = \varepsilon^- > 0, \quad [\partial_x \varepsilon] = (\partial_x \varepsilon)^-.$$

Thus, the corresponding stress jump is expressed in view of the definition of the instantaneous stress σ_{I} as

$$[\sigma] = \sigma_{\mathrm{I}}(\varepsilon^-).$$

This implies, in view of (8.158) and (8.159), that the intrinsic velocity can be expressed by

$$U^2 = \frac{\sigma_{\mathrm{I}}(\varepsilon^-)}{\rho_0 \varepsilon^-} = \frac{E_{\mathrm{I}}(\varepsilon^-)}{\rho_0} \tag{8.160}$$

which, in view of the second inequality of the convexity conditions (8.140a), implies that

$$\frac{\rho_0 U^2}{E_{\mathrm{I}}(\varepsilon^-)} < 1. \tag{8.161}$$

The inequality (8.161) affirms (Schuler, Nunziato and Walsh, 1973) that the shock velocity is subsonic with respect to the material behind the wave front. From (8.160) it can be seen that the shock velocity depends on the strain amplitude ε^-. Furthermore, one can write with reference to (8.160) that

$$\frac{dU}{dt} = \frac{(1 - \hat{\mu})E_{\mathrm{I}}(\varepsilon^-)}{2\rho_0 U \varepsilon^-} \frac{d\varepsilon^-}{dt} \tag{8.162}$$

where

$$\hat{\mu} = \frac{\rho_0 U^2}{E_{\mathrm{I}}(\varepsilon^-)} < 1 \tag{8.163}$$

given earlier by (8.161).

In view of (8.160), one concludes that the time rate of change of the shock velocity U is proportional to that of the amplitude of the strain behind the front ε^-. Following this and using the assumed characteristics of the deformation gradient F, Chen and Gurtin (1970) derived the following '**shock amplitude equation**':

$$\frac{d\varepsilon^-}{dt} = U \frac{1-\hat{\mu}}{1+3\hat{\mu}} [\hat{\lambda} - (\partial_x \varepsilon)^-] \tag{8.164}$$

where

$$\hat{\lambda} = \frac{R'(\varepsilon^-; 0)\varepsilon^-}{UE_i(\varepsilon^-)(1-\hat{\mu})}. \tag{8.165}$$

In (8.165), $R'(\varepsilon^-; 0)$ is the initial slope of the stress-relaxation function corresponding to the jump strain input (8.133). It is evident, in view of (8.140) and (8.165), that $\hat{\lambda} \le 0$.

Thus, with reference to the shock amplitude expression (8.164), one may conclude that the growth or decay behaviour of the shock wave front would depend on the strain gradient immediately behind the front (Schuler, Nunziato and Walsh, 1973). That is,

1. if $\hat{\lambda} < (\partial_x \varepsilon)^-$, $\dfrac{d\varepsilon^-}{dt} < 0$,

2. if $\hat{\lambda} = (\partial_x \varepsilon)^-$, $\dfrac{d\varepsilon^-}{dt} = 0$, $\qquad\qquad$ (8.166)

3. if $\hat{\lambda} > (\partial_x \varepsilon)^-$, $\dfrac{d\varepsilon^-}{dt} > 0$,

In view of the above, Schuler, Nunziato and Walsh (1973) referred to $\hat{\lambda}$ as 'the critical strain gradient'. These authors expressed the shock amplitude equation (8.165) in terms of particle velocity as

$$\frac{dv^-}{dt} = \frac{1-\hat{\mu}^2}{1+3\hat{\mu}} [(\dot{v})^- - U^2|\hat{\lambda}|]. \tag{8.167}$$

In this equation, $(\dot{v})^-$ is the particle acceleration immediately behind the shock front and $U^2|\hat{\lambda}|$ is the '**critical acceleration**'. It is evident from (8.167) that

1. the front grows if $(\dot{v})^- > U^2|\hat{\lambda}|$,
2. is steady if $(\dot{v})^- = U^2|\hat{\lambda}|$, $\qquad\qquad$ (8.168)
3. decays if $(\dot{v})^- < U^2|\hat{\lambda}|$.

Further, it can be shown (Chen and Gurtin, 1970) that equation (8.167) reduces for the case of weak shock waves to the following simple expression:

$$\frac{dv^-}{dt} = -\beta v^- \tag{8.169}$$

where β is a constant given by

$$\beta = -\frac{R'(0;0)}{2(E_1)_0} \tag{8.170a}$$

in which

$$(E_1)_0 = \frac{d\sigma_1(\varepsilon)}{d\varepsilon}. \tag{8.170b}$$

The solution of the differential equation (8.169) is

$$v^-(t) = v^-(0) \exp(-\beta t) \tag{8.171}$$

which asserts that the amplitude of a weak shock wave decays exponentially to zero as $t \to \infty$. Such a response is identical to that predicted by the linear theory of viscoelasticity (e.g. Lee and Kanter, 1953; Chu, 1962; Coleman and Gurtin, 1965a–c; Valanis, 1965).

8.8.5 Thermodynamic influences

(a) Acceleration waves

It is noted by Schuler, Nunziato and Walsh (1973) that thermal effects have no influence on the propagation of acceleration waves in nonconducting materials with memory. Accordingly, on the assumption that a particular material is a thermal nonconductor (which could be reasonable for a large class of polymeric solids), the study of the propagation of acceleration waves in such a material would provide no information about the thermodynamic influence on its mechanical response. Thus, for a thermally nonconducting material, if the relaxtion function and second-order modulus are taken at a fixed entropy, then the velocity of every acceleration wave would satisfy (8.147). Furthermore, the amplitude of an acceleration wave entering a region at rest, unstrained and at uniform temperature, would satisfy (8.153) with the material constants appearing in this equation being given by (8.151) and (8.152).

In the case of conducting materials, however, thermodynamic influences on the propagation of acceleration waves in viscoelastic materials are pronounced. In this, the reader is referred, for instance, to Coleman and Gurtin (1966).

(b) Shock waves

Thermodynamic effects on the propagation of shock waves in nonconducting materials have been considered by Coleman and Gurtin (1966) and Chen and Gurtin (1972b). Meantime, studies on shock wave propagation in materials with memory which conduct heat have been carried out, for instance, by Achenbach, Vogel and Herrmann (1966) and by Dunwoody (1972).

8.9 AN ILLUSTRATIVE EXAMPLE ON THE APPLICATION OF VISCOELASTIC
WAVE PROPAGATION (DETERMINATION OF THE STRESS-RELAXATION
FUNCTION OF A VISCOELASTIC MATERIAL FROM ACOUSTIC DISPERSION
DATA)

We present, in this section, the work of Nunziato and Sutherland (1973) for the
determination of the stress-relaxation function for the solid polymer PMMA using
acoustic dispersion data on this viscoelastic material. This work is based on an earlier
paper by Sutherland and Lingle (1972) concerning the possibility of determining the
phase velocity and attenuation of acoustic waves in polymers over a reasonably wide
range of frequency using the time–temperature superposition concept introduced
earlier in section 5.5. The presented analysis is of interest and offers a possibility of
determining the stress relaxation function for a polymer using acoustic dispersion
data for a reasonably wide range. It was demonstrated by Nunziato and Sutherland
(1973) that the relaxation function for PMMA as determined by the latter method
compares favourably with the relaxation function deduced from experimental observations on steady shock waves in PMMA. This result is shown to be valid for a
time range of 10^{-5}–10^8 μs.

8.9.1 Acoustical representation of the stress-relaxation function

We consider the linear stress-relaxation constitutive law in the form

$$\sigma(x,\ t) = R(0)\ \frac{\partial u(x,\ t)}{\partial x} + \int_0^\infty R'(\tau)\ \frac{\partial u(x,\ t-\tau)}{\partial x}\ d\tau \tag{8.172}$$

where σ is the one-dimensional uniaxial stress, $R(\tau)$ is the corresponding stress
relaxation function and $R'(\tau) = dR(\tau)/d\tau$. Incorporating the dynamical field equation
for a travelling sinusoidal wave, one has the solution

$$u(X,\ t) = u_0\ \text{Re}(\exp[-(\alpha + i\omega/c)X]\ \exp(i\omega t)] \tag{8.173}$$

where, in general, for viscoelastic materials, the phase velocity $c > 0$ and the
attenuation $\alpha > 0$ are functions of the frequency $\omega > 0$.

The solution (8.173) is admissible if and only if (Hunter, 1960; Nunziato and
Sutherland, 1973)

$$\rho_0 c^2(\omega) = |R(0) + \bar{R}'(\omega)|\ \sec^2[\tfrac{1}{2}\delta(\omega)] \tag{8.174a}$$

and

$$\alpha(\omega) = [\omega/c(\omega)]\ \tan[\tfrac{1}{2}\delta(\omega)] \tag{8.174b}$$

where ρ_0 is the reference density. The phase angle $0 \leq \delta \leq \pi/2$ is defined by

$$\tan\ \delta(\omega) = \frac{\text{Im}[R(0) + \bar{R}'(\omega)]}{\text{Re}[R(0) + \bar{R}'(\omega)]} \tag{8.175}$$

and $\bar{R}'(\omega)$ is the Fourier transform

$$\bar{R}'(\omega) = \int_0^{\infty} R'(\tau) \exp(i\omega\tau)\,d\tau. \tag{8.176}$$

If the functions $c(\omega)$ and $\alpha(\omega)$ are known, one can then determine the phase angle $\delta(\omega)$ and the stress-relaxation function $R(\tau)$ by using equations (8.174)–(8.176). One possible representation of these functions is

$$\delta(\omega) = 2\tan^{-1}|\alpha(\omega)c(\omega)/\omega| \tag{8.177}$$

$$R(\tau) = \frac{2}{\pi}\int_0^{\tau}\int_0^{\infty}\left(\frac{\rho_0 c^2(\omega)}{\sec\delta(\omega)\{[\alpha(\omega)c(\omega)/\omega]^2 - 1\}} - \rho_0 c_{\infty}^2\right)\cos(\omega\tau)\,d\omega\,d\tau$$

$$+ \rho_0 c_{\infty}^2 \tag{8.178}$$

in which c_{∞} is the high frequency limit of the phase velocity, i.e.

$$c_{\infty} = \lim_{\omega\to\infty} c(\omega). \tag{8.179}$$

In deducing (8.178), Nunziato and Sutherland (1973) used the standard cosine inversion of a Fourier transform and noted that

$$R(0) = \rho_0 c_{\infty}^2. \tag{8.180}$$

8.9.2 Determination of the stress-relaxation function from experimental acoustic data and using the time–temperature shift concept

In the process of determining experimentally the phase velocity and attenuation for PMMA over an extended range of frequency, Nunziato and Sutherland (1973) used, as a first step, the time–temperature superposition technique. The latter is introduced in section 5.5. It is based (Ferry, 1961; Schapery, 1974) primarily on the observation that, below the glass transition temperature of rubber-like material, a temperature change adjusts the time scale of the stress-relaxation function by a factor a_T. That is,

$$R(\tau)|_{T=T_1} = R\left(\frac{\tau}{a_T}\right)\Bigg|_{T=T_0} \tag{8.181}$$

where $T > 0$ indicates the temperature. Sutherland and Lingle (1972) have employed the relation (8.181) to include the effects of the time–temperature superposition concept on the expressions for the phase velocity and attenuation, namely

$$c(\omega)|_{T=T_1} = c(\omega a_T)|_{T=T_0},$$

$$\alpha(\omega)|_{T=T_1} = a_T\alpha(\omega a_T)|_{T=T_0}. \tag{8.182}$$

Consequently, if the wave propagation velocity and attenuation are experimentally measured as a function of temperature at some frequency, then they can be determined for all frequencies within the frequency range that corresponds to the applicability

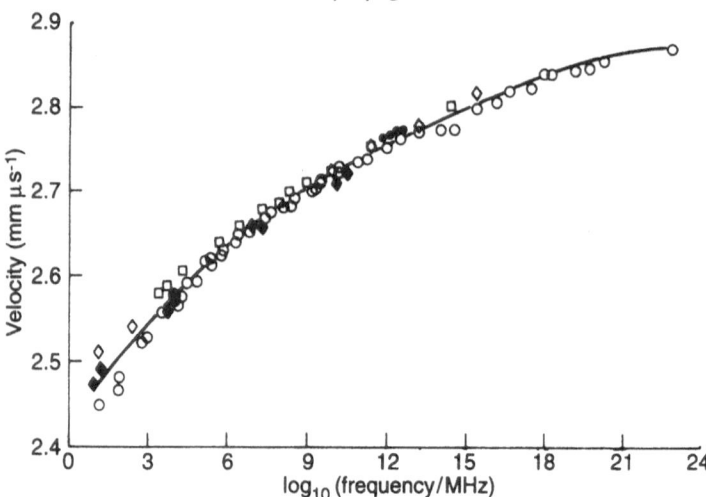

Fig. 8.25 Velocity data for PMMA: ●, Asay, Lamberson and Guenther (1969); □, Lingle (unreferenced data); ◇, Romberger (1970); ◆, Asay *et al.* (unreferenced data); ○, this work. (Source: Sutherland, H. J. and Lingle, R. (1972) An acoustic characterization of polymethyl methacrylate and three epoxy formulations. *J. Appl. Phys.*, **43**(10), 4022–6 (American Institute of Physics). Reprinted with permission.)

of (8.182). Using this technique, Sutherland and Lingle (1972) have determined the values of the functions $c(\omega)$ and $\alpha(\omega)$ for the solid polymer PMMA over 20 decades of frequency. Their results for room temperature are shown in Figs. 8.25 and 8.26 together with similar experimental data due to Asay, Lamberson and Guenther (1969) and Romberger (1970).

To evaluate the stress-relaxation function $R(\tau)$ from the above-mentioned acoustic data, Nunziato and Sutherland (1973) evaluated the integrand

$$F(\omega) = \frac{\rho_0 c^2(\omega)}{\sec \delta(\omega)\{[\alpha(\omega)c(\omega)/\omega]^2 - 1\}} - \rho_0 c_\infty^2 \qquad (8.183)$$

of equation (8.178) pointwise.

Representing the integrand (8.183) with the following series

$$F(\omega) = \sum_{i=1}^{20} f_i \exp\left(-\frac{\omega}{W_i}\right), \qquad (8.184)$$

the equation (8.178) may be integrated to yield

$$R(\tau) = \frac{2}{\pi} \sum_{i=1}^{20} f_i \tan^{-1}(\omega_i \tau) + \rho_0 c_\infty^2 \qquad (8.185)$$

whereby the high frequency limit c_∞ was evaluated by Nunziato and Sutherland (1973), following a method suggested by Walsh (1971), to be 2.87 mm μs^{-1}. The relaxation function $R(\tau)$ (8.185), as determined by Nunziato and Sutherland (1973) using the above procedure is shown in Fig. 8.27 for the solid polymer PMMA.

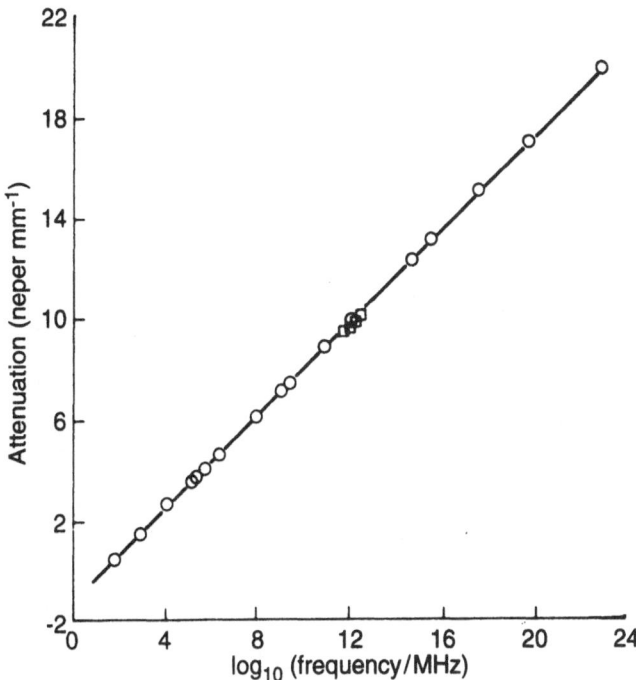

Fig. 8.26 Attenuation data for PMMA: □, Asay, Lamberson and Guenther (1969); ○, this work. (Source: Sutherland, H. J. and Lingle, R. (1972) An acoustic characterization of polymethyl methacrylate and three epoxy formulations. *J. Appl. Phys.*, **43**(10), 4022–6 (American Institute of Physics). Reprinted with permission.)

8.9.3 Determination of the stress-relaxation function from shock wave data

Nunziato and Sutherland (1973) considered plate impact experiments to study the one-dimensional dynamic responses of PMMA. They considered the characteristic time scale for such experiments to be 10^{-2}–1 μs. We denote, over this time scale, the relaxation function by $\hat{R}(\tau)$ where $\hat{R}(0)$ is equivalent to the value of the relaxation function at 10^{-2} μs and $\hat{R}(\infty)$ corresponds to the relaxation function at 1 μs. Using, then, the results (8.173) of the previous subsection, Nunziato and Sutherland (1973) obtained the stress-relaxation function $\hat{R}(\tau)$ shown in Fig. 8.28.

From this figure,

$$\hat{R}(0) = 90.1 \text{ kbar}$$

and   (8.186)

$$\hat{R}(\infty) = 88.0 \text{ kbar.}$$

Meantime, the characteristic relaxation time λ is evaluated by considering the relation

$$\hat{R}(\tau)|_{\tau=\lambda} = \frac{R_0 - R_\infty}{e} + R_\infty \qquad (8.187a)$$

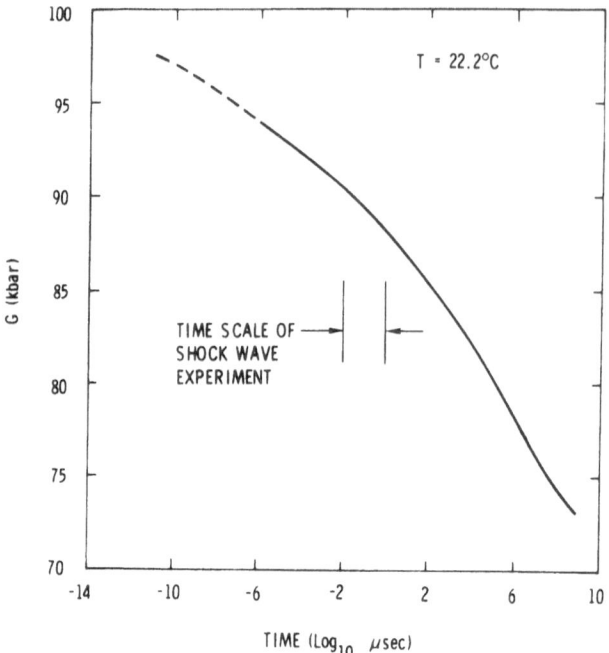

Fig. 8.27 Longitudinal stress relaxation function for PMMA. (Source: Nunziato, J. W. and Sutherland, H. J. (1973) Acoustical determination of stress relaxation functions in polymers. *J. Appl. Phys.*, **44**(1), 184–7 (American Institute of Physics). Reprinted with permission.)

in which e is the Naperian base. From (8.186) and (8.187a), one concludes from Fig. 8.28 that the characteristic relaxation time is

$$\lambda = 0.22 \ \mu\text{s}. \tag{8.187b}$$

Schuler (1970) considered the propagation of steady one-dimensional shock waves in PMMA. The relaxation behaviour of PMMA was characterized by a nonlinear constitutive relation of the form

$$\sigma(x, t) = \sigma_{\text{E}}(\varepsilon) + \int_0^\infty R(\varepsilon; \tau) \left[\frac{1 + \varepsilon(t - \tau)}{1 + \varepsilon(t)} - 1 \right] d\tau \tag{8.188}$$

where $\varepsilon = \partial u/\partial x$ is the longitudinal strain, $\sigma_{\text{E}}(\varepsilon)$ is the equilibrium response function and $R(\varepsilon; \tau)$ is the generalized stress-relaxation function. The latter is assumed to have the form

$$R(\varepsilon; \tau) = R(\varepsilon; 0) \exp(-\tau/\lambda). \tag{8.189}$$

Evaluating (8.188) for the strain jump

$$\varepsilon(t - \tau) = \varepsilon, \qquad \tau = 0,$$

$$= 0, \qquad \tau > 0,$$

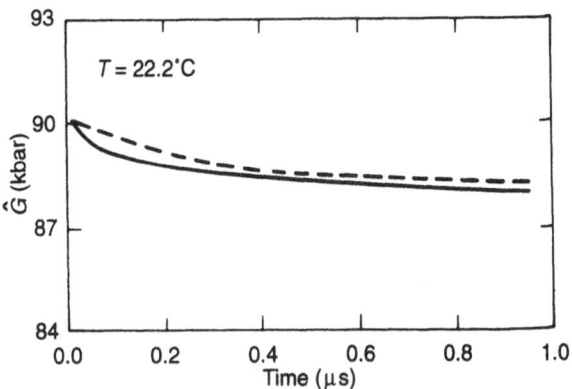

Fig. 8.28 Stress-relaxation function for PMMA appropriate for shock wave experiments: – – –, Schuler (1970, unpublished data); ———, this work. (Source: Nunziato, J. W. and Sutherland, H. J. (1973) Acoustical determination of stress relaxation functions in polymers. *J. Appl. Phys.*, **44**(1), 184–7 (American Institute of Physics). Reprinted with permission.)

it follows from (8.189) that

$$R(\varepsilon; 0) = -(1 + \varepsilon)^2[\sigma_I(\varepsilon) - \sigma_E(\varepsilon)]/\lambda\varepsilon(2 + \varepsilon). \qquad (8.190)$$

From an analysis of steady waves (e.g. Greenberg, 1968; Schuler, 1970; Schuler, Nunziato and Walsh, 1973), one has

$$\sigma_I(\varepsilon) = \rho_0 U_I^2(\varepsilon)\varepsilon$$

and $\qquad (8.191)$

$$\sigma_E(\varepsilon) = \rho_0 U_E^2(\varepsilon)\varepsilon$$

where $U_I(\varepsilon)$ and $U_E(\varepsilon)$ are least-square polynomial functions of the steady shock velocity as a function of strain ε at the shock front and at the tail of the wave (Nunziato and Sutherland, 1973). For small strains, the nonlinear constitutive relation (8.188) would reduce to the corresponding constitutive relation in the linear case, equation (8.172), with (Nunziato and Walsh, 1973)

$$\hat{R}(\tau) = \frac{d\sigma_I(0)}{d\varepsilon} + 2\int_0^\tau R(0; \tau) \, d\tau. \qquad (8.192)$$

Combining, now (8.189) and (8.192), it follows that

$$\frac{d\sigma_I(0)}{d\varepsilon} = \rho_0 U_I^2(0) \qquad (8.193)$$

and consequently

$$\hat{R}(\tau) = (R_0 - R_\infty) \exp(-\tau/\lambda) + R_\infty \qquad (8.194)$$

where

$$R_0 = \rho_0 U_I^2(0) \quad \text{and} \quad R_\infty = \rho_0 U_E^2(\infty).$$

It is noted that $U_I(0)$ and $U_E(0)$ are the zero-strain intercepts of the $U_I(\varepsilon)$ and $U_E(\varepsilon)$ curves. Using data reported by Schuler (reported in Nunziato and Sutherland (1973)) and Barker and Hollenbach (1970), Nunziato and Sutherland (1973) concluded that

$$R_0 = 90.2 \text{ kbars}$$

and   (8.195a)

$$R_\infty = 88.2 \text{ kbars}.$$

Schuler (1970), by fitting the observed steady wave profiles, found that

$$\lambda = 0.25 \ \mu s \qquad (8.195b)$$

which is comparable with the value given earlier by (8.187b). With the above data, the relaxation function (8.194) is given in Fig. 8.28 (after Nunziato and Sutherland, 1973). As demonstrated, there is reasonable agreement with the relaxation function deduced from acoustic dispersion data.

REFERENCES

Achenbach, J. D. and Chao, C. C. (1962) A three-parameter viscoelastic model particularly suited for dynamic problems. *J. Mech. Phys. Solids*, **10**, 245–52.

Achenbach, J. D., Vogel, S. M. and Herrmann, G. (1966) On stress waves in viscoelastic media conducting heat, in *Irreversible Aspects of Continuum Mechanics and Transfer of Physical Characteristics in Moving Fluids* (eds H. Parkus and L. I. Sedov), Springer, Berlin, pp. 1–15.

Ahrens, T. J. and Duvall, G. E. (1966) Stress relaxation behind shock waves in rocks. *J. Geophys. Res.*, **71**, 4349–60.

Alfrey, T., Jr (1948) *Mechanical Behaviour of High Polymers*, Interscience, New York.

Asay, J. R., Lamberson, D. L. and Guenther, A. H. (1969) Pressure and temperature dependence of velocities in polymethyl methacrylate. *J. Appl. Phys.*, **40**, 1768–83.

Bailey, P. and Chen, P. J. (1971) On the local and global behaviour of acceleration waves. *Arch. Ration. Mech. Anal.*, **41**, 121 (addendum: Asymptotic behaviour, *Arch. Ration. Mech. Anal.*, **44**, 212 (1972)).

Ballon, J. W. and Smith, J. C. (1949) Dynamic measurements of polymer physical properties. *J. Appl. Phys.*, **20**, 493–502.

Barker, L. M. (1968) The fine structure of compressive and release wave shapes in aluminium measured by the velocity interferometer technique, in *Behaviour of Dense Media Under High Dynamic Pressures*, Gordon and Breach, London, pp. 483–505.

Barker, L. M. and Hollenbach, R. E. (1964) System for measuring the dynamic properties of materials. *Rev. Sci. Instrum.*, **35**(6), 742–6.

Barker, L. M. and Hollenbach, R. E. (1965) Interferometer technique for measuring the dynamic mechanical properties of materials. *Rev. Sci. Instrum.*, **36**, 1617–20.

Barker, L. M. and Hollenbach, R. E. (1970) Shock wave studies of PMMA, fused silica and sapphire. *J. Appl. Phys.*, **41**, 4208–26.

Barton, C. S., Volterra, E. G. and Citron, S. J. (1958) On elastic impacts of spheres on long rods, in Proc. 3rd US Natl Cong. on Applied Mechanics, pp. 89–94.

Becker, E. C. H. and Carl, H. (1962) Transient-loading technique for mechanical impedance measurement, in *Experimental Techniques in Shock and Vibration* (ed. W. J. Worley), ASME, New York, pp. 1–10.

Bland, D. R. (1960) *The Theory of Linear Viscoelasticity*, Pergamon, New York.

Bradfield, G. (1951) Internal friction of solids. *Nature (London)*, **167**, 1021–3.

Chao, C. and Achenbach, J. D. (1964) A simple viscoelastic analogy for stress waves, in *Stress Waves in Anelastic Solids* (eds. H. Kolsky and W. Prager), IUTAM Symp., Brown University, Providence, RI, April 3–5, 1963, Springer, Berlin, pp. 222–38.

Chen, P. J. and Gurtin, M. E. (1970) On the growth of one-dimensional shock waves in materials with memory. *Arch. Ration. Mech. Anal.*, **36**, 33–46.

Chen, P. J. and Gurtin, M. E. (1972a) On the use of experimental results concerning steady shock waves to predict the acceleration wave response of nonlinear viscoelastic materials. *J. Appl. Mech.*, **39**, 295–6.

Chen, P. J. and Gurtin, M. E. (1972b) Thermodynamic influences on the growth of one-dimensional shock waves in materials with memory. *Z. Angew. Math. Phys.*, **23**, 69–79.

Chen, P. J., Gurtin, M. E. and Walsh, E. K. (1970) Shock amplitude variation in polymethyl methacrylate for fixed values of the strain gradient. *J. Appl. Phys.*, **41**, 3557–8.

Chou, P. C. (1968) Introduction to wave propagation in composite materials, in *Composite Materials Workshop* (eds. S. W. Tsai, J. C. Halpin and N. J. Pagano), Technomic, Stamford, CT, pp. 193–216.

Chree, C. (1889) The equations of an isotropic elastic solid in polar and cylindrical coordinates, their solutions and applications. *Trans. Cambridge Philos. Soc. Math. Phys. Sci.*, **6**, 115–7.

Chu, B. T. (1962) *Stress Waves in Isotropic Viscoelastic Materials*, Division of Engineering, Brown University, Providence, RI.

Coleman, B. D. (1964) Thermodynamics, strain impulses and viscoelasticity. *Arch. Ration. Mech. Anal.*, **17**, 230–54.

Coleman, B. D. and Gurtin, M. E. (1965a) Waves in materials with memory. II. On the growth and decay of one-dimensional acceleration waves. *Arch. Ration. Mech. Anal.*, **19**, 239–65.

Coleman, B. D. and Gurtin, M. E. (1965b) Waves in materials with memory. III. Thermodynamic influences on the growth and decay of acceleration waves. *Arch. Ration. Mech. Anal.*, **19**, 266–98.

Coleman, B. D. and Gurtin, M. E. (1965c) Waves in materials with memory. IV. Thermodynamics and the velocity of general acceleration waves. *Arch. Ration. Mech. Anal.*, **19**(5), 317–38.

Coleman, B. D. and Gurtin, M. E. (1966) Thermodynamics and one-dimensional shock waves in materials with memory. *Proc. R. Soc. London, Ser. A*, **292**, 562–74.

Coleman, B. D., Gurtin, M. E. and Herrera, R. I. (1965) Waves in materials with memory. I. The velocity of one-dimensional shock and acceleration waves. *Arch. Ration. Mech. Anal.*, **19**, 1–19.

Coleman, B. D., Greenberg, J. M. and Gurtin, M. E. (1966) Waves in materials with memory. V. On the amplitude of acceleration waves and mild discontinuities. *Arch. Ration. Mech. Anal.*, **22**, 333–54.

Cunningham, J. R. and Ivey, D. G. (1956) Dynamic properties of various rubbers at high frequencies. *J. Appl. Phys.*, **27**, 967–74.

Dally, J. W. (1968) A dynamic photoelastic study of a doubly loaded half-plane. *Dev. Mech.*, **4**, 649–64.

Dally, J. W. and Lewis, D. (1968) A photoelastic analysis of propagation of Rayleigh waves past a step change in elevation. *Bull. Seismd. Soc. Am.*, **58**, 539–63.

Dally, J. W. and Riley, W. F. (1965) *Experimental Stress Analysis*, McGraw-Hill, New York.

Dally, J. W. and Riley, W. F. (1967) Initial studies in three- dimensional dynamic photoelasticity. *J. Appl. Mech.*, **34**, 405–10.

Dally, J. W. and Thau, S. A. (1967) Observations of stress wave propagation in a half-plane with boundary loading. *Int. J. Solids Struct.*, **3**, 293–307.

Dally, J. W., Durrelli, A. J. and Riley, W. F. (1960) Photoelastic study of stress wave propagation in large plates. *Proc. Soc. Exp. Stress Anal.*, **17**, 33–50.

Davies, R. M. (1948) A critical study of the Hopkinson pressure bar. *Philos. Trans. R. Soc. London, Ser. A*, **240**, 375–457.

Dohrenwend, C. O., Drucker, D. C. and Moore, P. (1944) Transverse impact transients. *Exp. Stress Anal.*, **1**, 1–10.

Dove, R. C. and Adams, P. H. (1964) *Experimental Stress Analysis and Motion Measurement*, Merril, Columbus, OH.

Dunwoody, J. (1972) One-dimensional shock waves in heat conducting materials with memory. I. Thermodynamics. *Arch. Ration. Mech. Anal.*, **47**, 117–48.

Duvall, G. E. and Alverson, R. C. (1963) Fundamental research. *Tech. Summary Rep. 4*, Stanford Research Institute, Menlo Park, CA.

Evans, J. F., Hadley, C. F., Eisler, J. D. and Silverman, D. (1954) A three-dimensional seismic wave model with both electrical and visual observation of waves. *Geophysics*, **19**, 120–36.

Ferry, J. D. (1961) *Viscoelastic Properties of Polymers*, Wiley, New York.

Ferry, J. D. and Williams, M. L. (1952) Second approximation methods for determining the relaxation time spectrum of a viscoelastic material. *J. Colloid Sci.*, **7**, 347–53.

Fisher, H. C. (1954) Stress pulse in bar with neck or swell. *Appl. Sci. Res. A*, **4**, 317–28.

Frederick, J. R. (1965) *Ultrasonic Engineering*, Wiley, New York.

Glauz, R. D. and Lee, E. H. (1954) Transient wave analysis in a linear time-dependent material. *J. Appl. Phys.*, **25**, 947–53.

Goldsmith, W., Polivka, M. and Yang, T. (1966) Dynamic behaviour of concrete. *Exp. Mech.*, **23**, 65–79.

Goodier, J. N., Jahsman, W. E. and Ripperger, E. A. (1959) An experimental surface-wave method for recording force–time curves in elastic impacts. *J. Appl. Mech.*, **26**, 3–7.

Gorsky, W. S. (1936) On the transitions in the CuAu alloy III. On the influence of strain on the equilibrium in the ordered lattice of CuAl. *Phys. Z. Sowjet*, **6**, 77–81.

Graff, K. F. (1975) *Wave Motion in Elastic Solids*, Dover Publications, New York.

Graffi, D. (1982) Mathematical models and waves in linear viscoelasticity, in *Wave Propagation in Viscoelastic Media*, Vol. 52 (ed. F. Mainardi), Pitman, Boston, MA, PR, 1–27.

Green, W. A. (1964) The growth of plane discontinuities propagating into a homogeneously deformed elastic material. *Arch. Ration. Mech. Anal.*, **16**, 79–89.

Greenberg, J. M. (1967) The existence of steady shock waves in nonlinear materials with memory. *Arch. Ration. Mech. Anal.*, **24**, 1–21.

Greenberg, J. M. (1968) Existence of steady waves for a class of nonlinear dissipative materials. *Q. Appl. Math.*, **26**, 27–34.

Gurtin, M. E. and Herrera, I. (1964) A correspondence principle for viscoelastic wave propagation. *Q. Appl. Math.*, **22**, 360–4.

Gurtin, M. E. and Herrera, I. (1965) On dissipation inequalities and linear viscoelasticity. *Q. Appl. Math.*, **23**, 235–45.

Hetenyi, M. (ed.) (1950) *Handbook of Experimental Stress Analysis*, Wiley, New York.

Hill, R. (1962) Acceleration waves in solids. *J. Mech. Phys. Solids*, **10**, 1–16.

Hillier, K. W. (1949) A method of measuring some dynamic elastic constants and its application to the study of high polymers. *Proc. Phys. Soc.*, **52**(2), 701–13.

Hillier, K. W. (1960) A review of the progress in the measurement of dynamic elastic properties, in Int. Symp. on Stress Wave Propagation in Materials (ed. N. Davids), Interscience, London, pp. 183–98.

Hillier, K. W. and Kolsky, H. (1949) An investigation of the dynamic elastic properties of some high polymers. *Proc. Phys. Soc. B*, **62**, 111–21.

Hsieh, D. Y. and Kolsky, H. (1958) An experimental study of pulse propagation in elastic cylinders. *Proc. Phys. Soc.*, **71**, 608–12.

Hudson, G. E. (1943) Dispersion of elastic waves in solid circular cylinders. *Phys. Rev.*, **63**, 46–51.

Huilgol, R. R. (1973) Growth of plane shock waves in materials with memory. *Int. J. Eng. Sci.*, **11**, 75–86.

Hunter, S. C. (1960) Viscoelastic waves, in *Progress in Solid Mechanics*, Vol. I. (eds. I. N. Sneddon and R. Hill), North-Holland, Amsterdam, pp. 1–57.

Ivey, D. G., Mrowea, B. A. and Guth, E. (1949) Propagation of ultrasonic bulk waves in high polymers. *J. Appl. Phys.*, **20**, 486–92.

Janke, E. and Emde, F. (1945) *Tables of Functions*, Dover Publications, New York.

Keast, D. N. (1967) *Measurements in Mechanical Dynamics*, McGraw-Hill, New York.

Kolsky, H. (1954a) Attenuation of short mechanical pulses by high polymers, in Proc. 2nd Int. Congr. on Rheology, Butterworths, London, pp. 79–84.

Kolsky, H. (1954b) The propagation of longitudinal elastic waves along cylindrical bars. *Philos. Mag.*, **45**, 712–26.

Kolsky, H. (1956) The propagation of stress pulses in viscoelastic solids. *Philos. Mag., Ser. 8*, **1**, 693–710.

Kolsky, H. (1958) The propagation of stress waves in viscoelastic solids. *Appl. Mech. Rev.*, **11**, 465–8.

Kolsky, H. (1960) Viscoelastic waves, in Int. Symp. on Stress Wave Propagation in Materials (ed. N. Davids), Interscience, London, pp. 59–90.

Kolsky, H. (1963) *Stress Waves in Solids*, Dover Publications, New York.

Lamb, H. (1904) On the propagation of tremors over the surface of an elastic solid. *Philos. Trans. R. Soc. London, Ser. A*, **203**, 1–42.

Lamb, H. (1917) On waves in an elastic plate. *Proc. R. Soc. London, Ser. A*, **93**, 114–28.

Lee, E. H. and Kanter. I. (1953) Wave propagation in finite rods of viscoelastic materials. *J. Appl. Phys.*, **24**(9), 1115–22.

Lee, E. H. and Morrison, J. A. (1956) A comparison of the propagation of longitudinal waves in rods of viscoelastic materials. *J. Polym. Sci.*, **19**, 93–110.

Lindholm, U. S. (1964) Some experiments with the split Hopkinson pressure bar. *J. Mech. Phys. Solids*, **12**, 317–35.

Magrab, E. B. and Blomquist, D. S. (1971) *The Measurement of Time-Varying Phenomena*, Wiley–Interscience, New York.

Mainardi, F. and Nervosi, R. (1980) Transient-waves in finite viscoelastic rods. *Lett. Nuovo Cim.*, **29**, 443–7.

Mainardi, F. and Turchetti, G. (1975) Wave front expansions for transient viscoelastic waves. *Mech. Res. Commun.*, **2**, 107–12.

Mainardi, F. and Turchetti, G. (1979) Positive constraints and approximation methods in linear viscoelasticity. *Lett. Nuovo Cim.*, **26**, 38–40.

Malvern, L. E. (1951) Plastic wave propagation in a bar of material exhibiting a strain rate effect. *Q. Appl. Math.*, **8**, 405–11.

Mason, W. P. and McSkimin, H. J. (1947) Attenuation and scattering of high frequency sound waves in metals and glasses. *J. Acoust. Soc. Am.*, **19**, 464–73.

Medick, M. A. (1961) On classical plate theory and wave propagation. *J. Appl. Mech.*, **28**, 223–8.

Morrison, J. A. (1956) Wave propagation in rods of Voigt material and viscoelastic materials with three-parameter models. *Q. Appl. Math.*, **14**, 153–169.

Morse, P. and Feshbach, H. (1953) *Methods of Theoretical Physics*, Vols I and II, McGraw-Hill, New York.

Nashif, A. D., Jones, D. I. G. and Henderson, J. P. (1965) *Vibration Damping*, Wiley, New York.

Nunziato, J. W. and Sutherland, H. J. (1973) Acoustical determination of stress relaxation functions in polymers. *J. Appl. Phys.*, **44**(1), 184–7.

Nunziato, J. W. and Walsh, E. K. (1973) Propagation of steady shock waves in nonlinear thermoviscoelastic solids. *J. Mech. Phys. Solids*, **21**, 317–35.

Oliver, J. (1957) Elastic wave dispersion in a cylindrical rod by a wide-band short duration pulse technique. *J. Acoust. Soc. Am.*, **29**, 189–94.

Orowan, E. (1934) Zür Kristall Plastizität. III. Über den Mechanismus des gleitvorganges. Z. Phys., **89**, 634–59.

Pindera, J. T. (1986) New research perspectives opened by isodyne and strain gradient photoelasticity, in Proc. Int. Symp. on Photoelasticity, Tokyo, pp. 193–202.

Polanyi, M. (1934) Über eine Art Gitterstörung, die einen Kristall plastisch machen Könnte. Z. Phys., **89**, 660–4.

Prescott, J. (1942) Elastic waves and vibrations of thin rods. Philos. Mag., **33**, 703–54.

Press, F. and Oliver, J. (1955) Model study of air-coupled surface waves. J. Acoust. Soc. Am., **27**, 45–6.

Rayleigh, J. W. S. (1887) On waves propagated along the plane surface of an elastic solid. Proc. London Math. Soc., **17**, 4–11.

Rayleigh, J. W. S. (1894) Theory of Sound, Dover Publications, New York.

Reinhardt, H. W. and Dally, J. W. (1970) Some characteristics of Rayleigh wave interaction with surface flaws. Mater. Eval., **28**, 213–20.

Riley, W. F. and Dally, J. W. (1966) A photoelastic analysis of stress wave propagation in a layered model. Geophysics, **31**, 881–9.

Ripperger, E. A. (1953) The propagation of pulses in cylindrical bars. An experimental study, in Proc. 1st Midwest Conf. on Solid Mechanics, pp. 29–39.

Rogers, C. A., Barker, D. K. and Jaeger, L. A. (1988) Introduction to smart materials and structures, in Proc. Smart Materials, Structures and Mathematical Issues Workshop, Virginia Polytechnic Institute and State University, Blacksburg, VA, September 15–16, 1992, pp. 17–28.

Romberger, A. B. (1970) MS Thesis, Pennsylvania State University.

Schapery, R. A. (1974) In Viscoelastic Behaviour and Analysis of Composite Materials, Vol. 2 (ed. G. Sendeckj), Academic Press, New York, pp. 86–168.

Schuler, K. W. (1970) Propagation of steady shock waves in polymethyl methacrylate. J. Mech. Phys. Solids, **18**, 277–93.

Schuler, K. W. and Walsh, E. K. (1971) Critical-induced acceleration for shock propagation in polymethyl methacrylate. J. Appl. Mech., **38**, 641–5.

Schuler, K. W., Nunziato, J. W. and Walsh, E. K. (1973) Recent results in nonlinear viscoelastic wave propagation. Int. J. Solid Struct., **91**, 1237–81.

Snoek, J. E. (1941) Effect of small quantities of carbon and nitrogen on the elastic and plastic properties of iron. Physica, **8**, 711–33.

Sokolnikoff, I. S. (1956) Mathematical Theory of Elasticity, 2nd edn, McGraw-Hill, New York.

Srinivasan, S. I. and Haddad, Y. M. (1992) Intelligent materials – an overview. Internal Report, Department of Mechanical Engineering, University of Ottawa, Ottawa.

Sutherland, H. J. and Lingle, R. (1972) An acoustic characterization of polymethyl methacrylate and three epoxy formulations. J. Appl. Phys., **43**(10), 4022–6.

Tanagi, T. (1990) A concept of intelligent material, in US–Japan Workshop on Smart/Intelligent Materials and Systems (eds C. A. Rogers, C. Andrew and A. Masuo), March 19–23, 1990, Honolulu, HI, pp. 3–10.

Tanary, S. and Haddad, Y. M. (1988) Characterization of adhesively bonded joints using acousto-ultrasonics. Final rep. Contract 31946-6-0012/01-ST, Department of Mechanical Engineering, University of Ottawa, Ottawa.

Tatel, H. E. (1954) Note on the nature of a seismogram II. J. Geophys. Res., **59**, 289–94.

Thau, S. A. and Dally, J. W. (1969) Subsurface characteristics of the Rayleigh wave. Int. J. Eng. Sci., **7**, 37–52.

Thomas, T. Y. (1957) The growth and decay of sonic discontinuities in ideal gases. J. Math. Mech., **6**, 455–69.

Thomas, T. Y. (1961) Plastic Flow and Fracture in Solids, Academic Press, New York.

Timoshenko, S. P. (1921) On the correction for shear of the differential equation for transverse vibrations of prismatic bars. Philos. Mag., Ser. 6, **41**, 744–6.

Tobolsky, A., Powell, R. E. and Eyring, H. (1943) The Chemistry of Large Molecules, Interscience, New York.

Truesdell, C. and Toupin, R. A. (1960) The classical field theories, in *Handbuch der Physik*, Vol. III/1 (ed. S. Flügge), Springer, Berlin.

Tschoegl, N. W. (1989) *The Phenomenological Theory of Linear Viscoelastic Behaviour*, Springer, Berlin.

Valanis, K. C. (1965) Propagation and attenuation of waves in linear viscoelastic solids. *J. Math. Phys.*, **44**(3), 227–39.

Vary, A. (1988) The acousto-ultrasonic approach, in *Acousto- Ultrasonics: Theory and Application* (ed. J. C. Duke, Jr), Plenum, New York, pp. 1–21.

Varley, E. (1965) Acceleration fronts in viscoelastic materials. *Arch. Ration. Mech. Anal.*, **19**, 215–25.

Varley, E. and Cumberbatch, E. (1965) Nonlinear theory of wavefront propagation. *J. Inst. Math. Appl.*, **1** (June), 101–12.

Viktorov, I. A. (1967) *Rayleigh and Lamb Waves: Physical Theory and Applications*, Plenum, New York.

Volterra, V. (1931) *Theory of Functionals*, Dover Publications, New York.

Walsh, E. K. (1971) The decay of stress waves in one-dimensional polymer rods. *Trans. Soc. Rheol.*, **15**(2), 345–53.

Worely, W. J. (ed.) (1962) *Experimental Techniques in Shock and Vibration*, ASME, New York.

Zener, C. (1948) *Elasticity and Anelasticity of Metals*, University Press, Chicago, IL.

FURTHER READING

Abbott, B. W. and Cornish, R. H. (1965) A stress wave technique for determining the tensile strength of brittle materials. *Exp. Mech.*, **22**, 148–53.

Achenbach, J. D. and Reddy, D. P. (1967) Note on wave propagation in linearly viscoelastic media. *Z. Angew. Math. Phys.*, **18**, 141–4.

Arenz, R. J. (1964) Uniaxial wave propagation in realistic viscoelastic materials. *J. Appl. Mech.*, **86** (March), 17–21.

Arenz, R. J. (1965) Two-dimensional wave propagation in realistic viscoelastic materials. *J. Appl. Mech.*, **32**(2), 303–14.

Baker, W. E. and Dove, R. C. (1962) Measurements of internal strains in a bar subjected to longitudinal impact. *Exp. Mech.*, **19**, 307–11.

Barberan, J. and Herrera, I. (1966) Uniqueness theorems and speed of propagation of signals in viscoelastic materials. *Arch. Ration. Mech. Anal.*, **23**, 173–90.

Berry, D. S. (1958) A note on stress pulses in viscoelastic rods. *Philos. Mag.*, **8**, 100–2.

Berry, D. S. and Hunter, S. C. (1956) The propagation of dynamic stresses in viscoelastic rods. *J. Mech. Phys. Solids*, **4**, 72–95.

Chu, B. T. (1962) Stress waves in isotropic linear viscoelastic materials (part one). *J. Méc.*, **1**(1), 439–62.

Chu, B. T. (1965) Response of various material media to high velocity loadings. I. Linear elastic and viscoelastic materials. *J. Mech. Phys. Solids*, **13**, 165–87.

Dunwoody, J. (1966) Longitudinal wave propagation in a rate dependent material. *Int. J. Eng. Sci.*, **4**, 277–87.

Dziecielak, R. (1985) The effect of temperature on the propagation of discontinuity waves in a porous medium with a viscoelastic skeleton. *Stud. Geotech. Mech.*, **7**(2), 17–34.

Engelbrecht, J. (1979) One-dimensional deformation waves in nonlinear viscoelastic materials. *Wave Motion*, **1**, 65–74.

Fisher, G. M. C. and Gurtin, M. E. (1965) Wave propagation in the linear theory of viscoelasticity. *Q. Appl. Math.*, **23**, 257–63.

Frydrychowicz, W. and Singh, M. C. (1986) Similarity representation of wave propagation in a nonlinear viscoelastic rod on a group theoretic basis. *Appl. Math. Model.*, **10**(8), 284–93.

Gopalsamy, K. and Aggarwala, B. D. (1972) Propagation of disturbances from randomly moving sources. *Z. Angew. Math. Mech.*, **52**, 31–5.

Graffi, D. (1952) Sulla teoria dei materiali elasticoviscosi. *Atti Accad. Ligure Sci. Lett.*, **9**, 1–10.

Green, W. A. (1960) Dispersion relations for elastic waves in bars, in *Progress in Solid Mechanics*, Vol. I (eds I. N. Sneddon and R. Hill), North-Holland, Amsterdam, Chap. 5.

Harris, C. M. and Crede, E. (1961) *Shock and Vibration Handbook*, Vols I, II and III, McGraw-Hill, New York.

Hatfield, P. (1950) Propagation of low frequency ultrasonic waves in rubbers and rubber-like polymers. *Br. J. Appl. Phys.*, **1**, 252–6.

Hrusa, W. J. and Renardy, M. (1985) On wave propagation in linear viscoelasticity. *Q. Appl. Math.*, **43**(2), 237–53.

Hunter, S. C. (1961) Tentative equations for the propagation of stress, strain and temperature fields in viscoelastic solids. *J. Mech. Phys. Solids*, **9**, 39–51.

Jeffrey, A. (1978) Nonlinear wave propagation. *Z. Angew. Math. Mech.*, **58**, T38–T56.

Jeffrey, A. and Taniuti, T. (1964) *Nonlinear Wave Propagation*, Academic Press, New York.

Knauss, W. G. (1968) Uniaxial wave propagation in a viscoelastic material using measured material properties. *J. Appl. Mech.*, **35**(3), 449–53.

Kolsky, H. (1949) An investigation of the mechanical properties of materials at very high rates of loading. *Proc. Phys. Soc., B*, **62**, 676–700.

Kolsky, H. (1960) Experimental wave-propagation in solids, in *Structural Mechanics* (eds J. N. Goodier and N. Hoff), Pergamon, Oxford, pp. 233–62.

Kolsky, H. (1965) Experimental studies in stress wave propagation, in Proc. Vth US Natl. Congr. on Applied Mechanics, pp. 21–36.

Kolsky, H. and Prager, W. (eds) (1964) *Stress Waves in Anelastic Solids*, IUTAM Symposium, Brown University, Providence, RI, April 3–5, 1963, Springer, Berlin, pp. 1–341.

Langhaar, H. L. (1962) *Energy Methods in Applied Mechanics*, Wiley, New York.

Lifshitz, J. M. and Kolsky, H. (1965) The propagation of spherical divergent stress pulses in linear viscoelastic solids. *J. Mech. Phys. Solids*, **13**, 361–76.

Lindsay, R. B. (1960) *Mechanical Radiation*, McGraw-Hill, New York.

Lockett, F. J. (1962) The reflection and refraction of waves at an interface between viscoelastic materials. *J. Mech. Phys. Solids*, **10**, 53–64.

Love, A. E. H. (1944) *A Treatise on the Mathematical Theory of Elasticity*, Dover Publications, New York.

Mahalanabis, R. K. and Mandal, B. (1986) Propagation of thermomagneto-viscoelastic waves in a half-space of Voigt-type material. *Indian J. Technol.*, **24** (September), 565–7.

Meyer, M. L. (1964) On spherical near fields and far fields in elastic and viscoelastic solids. *J. Mech. Phys. Solids*, **12**, 77–111.

Miklowjiz, J. (1964) Pulse propagation in a viscoelastic solid with geometric dispersion, in *Stress Waves in Anelastic Solids*, Springer, Berlin, pp. 255–76.

Norris, J. M. (1967) Propagation of a stress pulse in a viscoelastic rod. *Exp. Mech.*, **7**(7), 297–301.

Nunziato, J. W. and Walsh, E. K. (1973) Amplitude behaviour of shock waves in a thermoviscoelastic solid. *Int. J. Solids Struct.*, **9**, 1373–83.

Petrof, R. C. and Gratch, S. (1964) Wave propagation in a viscoelastic material with temperature-dependent properties and thermomechanical coupling. *J. Appl. Mech.*, **31**(3), 423–9.

Renardy, M. (1982) Some remarks on the propagation and non-propagation of discontinuities in linearly viscoelastic liquids. *Rheol. Acta*, **21**, 251–4.

Ricker, N. H. (1977) *Transient Waves in Viscoelastic Media*, Elsevier, Amsterdam.

Rubin, J. R. (1954) Propagation of longitudinal deformation waves in a prestressed rod of material exhibiting a strain-rate effect. *J. Appl. Phys.*, **25**, 528–36.

Sackman, J. L. and Kaya, I. (1968) On the propagation of transient pulses in linearly viscoelastic media. *J. Mech. Phys. Solids*, **16**, 349–56.

Sips, R. (1951) Propagation phenomena in elastic viscous media. *J. Polym. Sci.*, **6**, 285–93.

Skalak, R. (1957) Longitudinal impact of a semi-infinite circular elastic bar. *J. Appl. Mech.*, **34**, 59–64.

Stoneley, R. (1924) Elastic waves at the surface of separation of two solids. *Proc. R. Soc. London, Ser. A*, **106**, 416–28.

Sultanov, K. S. (1984) Longitudinal wave propagation in a viscoelastic semispace including an absorbing layer. *J. Appl. Mech. Tech. Phys.*, **25**(5), 790–5.

Timoshenko, S. P. (1928) *Vibration Problems in Engineering*, Van Nostrand, NJ.

Tsai, Y. M. and Kolsky, H. (1968) Surface wave propagation for linear viscoelastic solids. *J. Mech. Phys. Solids*, **16**, 99–109.

Volterra, E. (1955) A one-dimensional theory of wave propagation in elastic rods based on the 'method of internal constraints'. *Ing. Arch.*, **23**, 410.

Watson, G. N. (1960) *A Treatise on the Theory of Bessel Functions*, Cambridge University Press, New York.

Whitham, G. B. (1974) *Linear and Nonlinear Waves*, Wiley, New York.

Zukas, J. A. (1982) Stress waves in solids, in *Impact Dynamics* (eds J. A. Zukas. T. Nicholas, H. F. Swift, L. B. Greszczuk and D. R. Curran), Wiley, New York, Chap. 1, pp. 1–27.

9

Viscoelastic boundary value problem

9.1 INTRODUCTION

In classical elasticity where the response behaviour of the material is time independent, boundary value problems are conventionally classified as static or dynamic in view of the time dependency of the boundary conditions. Static problems of elasticity are often classified into the following two categories (e.g. Fung, 1965; Gakhof, 1966; Fichera, 1972; Gladwell, 1980).

1. Uniform boundary conditions: in this category, either the external loading (stress vector) or, alternatively, the external displacement are specified everywhere on the external boundary.
2. Mixed boundary conditions: here, the external loading is specified over a part of the boundary while the external displacement is specified over the recurring part.

Because of time independency characteristic of the static elastic problem, the boundary conditions as classified by the above two categories are fixed, i.e. they are time invariants. In dynamic problems, however, a set of initial conditions on both the components of the external loading and the displacement must be specified in the volume V and over the surface S of the continuous body.

In view of the time dependency of the response behaviour in viscoelasticity which is further complicated by the form of the constitutive relations and, hence, the associated boundary conditions, serious attempts to solve viscoelastic boundary value problems have lagged considerably behind those in classical elasticity. It is only in the last four decades that viscoelastic boundary value problems have been actively considered. At the beginning, researchers have given attention to the solution of the simpler viscoelastic problems that have analogues in classical elasticity whereby the viscoelastic solution may be expressed directly in terms of the analogous elastic problem. Research efforts have been advanced since then to tackle more difficult viscoelastic boundary value problems with or without correspondence to the theory of elasticity.

9.2 CLASSIFICATION OF BOUNDARY VALUE PROBLEMS IN VISCOELASTICITY

Since the response behaviour of a viscoelastic material is time dependent, it thus follows that no real static viscoelastic problem exists. However, in a large number of cases, it may be admissible (e.g. Hunter, 1967) to neglect the acceleration terms in the equations of motion. In such case, the viscoelastic boundary value problem is referred to as 'quasi-static' or 'quasi-stationary'. As Hunter (1967), for instance, pointed out, the only 'true static' problems in viscoelasticity are those corresponding to the equilibrium limit of complete stress relaxation.

A 'quasi-static' viscoelastic problem is often classified from the point of view of the time dependency of its boundary regions. In this, the following two categories are often dealt with.

1. For quasi-static problems with fixed (time-independent) boundary conditions, the loading history is assumed to be known for all time over a fixed part of the boundary, while the displacement history is specified for the remaining part. This type of problem is generally solvable using a correspondence with an analogous elastic problem, i.e. by employing the correspondence principle (introduced earlier in Chapter 8). This is essentially due to the possibility of obtaining Laplace (or Fourier) transforms of the boundary conditions as illustrated below in Section 9.6 (e.g. Hunter, 1960, 1967; Lee, 1960; Schapery, 1955, 1962, 1974).

2. Quasi-static problems with mixed boundary regions which are time dependent are not generally susceptible to solution by the correspondence principle as it may be impossible to obtain appropriate transforms of the boundary conditions. Examples of such type of problems may include contact problems where the load on the indentor is varying or the indentor is moving into the viscoelastic material specimen with an indentation of varying geometry (for instance, Hunter, 1968; Graham, 1969; Graham and Williams, 1972; Atkinson and Coleman, 1977; Aboudi, 1979; Nachman and Walton, 1978; Sabin, 1987).

Much less research work has been carried out on inertial and dynamic viscoelastic boundary value problems. In this domain, a large portion of the research has concentrated primarily on viscoelastic wave propagation problems that involve only one space variable (Chapter 8). Chao and Achenbach (1964) and Gurtin and Herrera (1964), among others, considered the use of the correspondence principle for the solution of viscoelastic wave propagation problems of this type. In general, however, viscoelastic waves may propagate in three dimensions with different magnitudes of attenuation and dispersion (e.g. Lockett, 1962). In this case, an associated boundary value problem may not be solvable via a dynamic correspondence principle (Hunter, 1967).

Research efforts to solve thermoviscoelastic boundary value problems have often been distracted by the fact that mechanical properties of viscoelastic materials are sensitive to temperature variations. This is complicated further by the heat generated in the viscoelastic material during deformation. The formulation of the governing equations has thus been proven to be difficult. Morland and Lee (1960), for instance, have considered the case of a thermorheologically simple solid (e.g. Schwarzl and Staverman, 1952; Hunter, 1961) in the absence of internally generated heat and thermodynamic coupling effects. Morland and Lee (1960) applied, then, the resulting equations to the quasi-static problem of an incompressible long cylinder subject to radial temperature gradient and internal pressure. Muki and Sternberg (1961) have dealt with the thermal stresses in viscoelastic materials with temperature-dependent properties and considered transient stress problems in plane slabs and spheres subject to temperature variation. Rogers and Lee (1962) have considered the solution of the quasi-static thermoviscoelastic problem of a sphere with an internally ablating cavity. Sternberg and Gurtin (1963, 1964) considered the uniquness of the theory of thermorheologically simple ablating viscoelastic solids.

A classification of boundary value problems in viscoelasticity is presented in Fig. 9.1. For comprehensive studies of the subject matter, the reader is referred further to Read (1950), Lee, Radok and Woodward (1959), Sternberg (1964), Predeleanu (1965), Rogers (1965), Lee (1966) and Golden and Graham (1988), among others.

9.3 FORMULATION OF THE VISCOELASTIC BOUNDARY VALUE PROBLEM

In compliance with the principles of continuum mechanics, the motion of a viscoelastic (continuum) body is generally governed by the laws of conservation of mass and momentum, the stress–strain constitutive relations, the boundary conditions and the

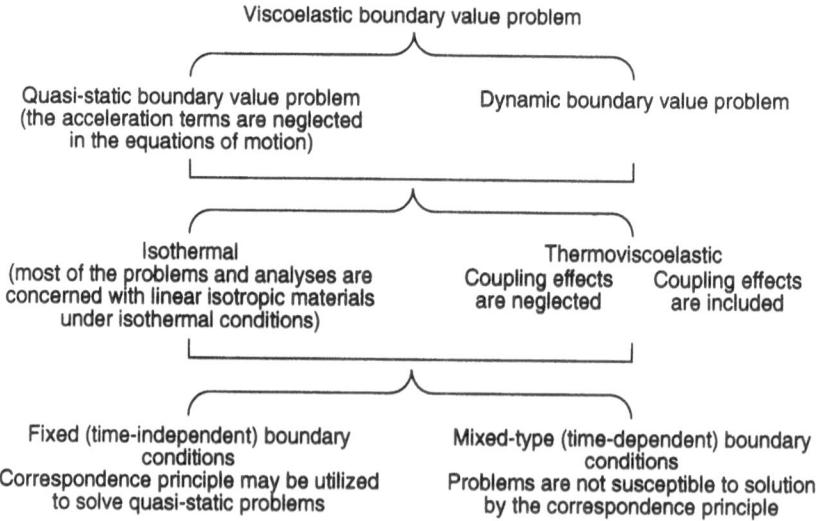

Fig. 9.1 Classification of viscoelastic boundary value problems.

initial conditions. As demonstrated in the remainder of this chapter, the formulation of this set of governing conditions is determined by the type of the boundary value problem considered.

9.4 ISOTHERMAL, LINEAR VISCOELASTIC BOUNDARY VALUE PROBLEM

In this class of boundary value problem, all the geometrical assumptions of infinitesimal elasticity theory are implied. These would usually include the assumptions of small deformations and small strains, the boundary conditions applied to undisturbed surfaces and the neglect of any convective terms in the acceleration. In this class of viscoelastic boundary value problem, only the viscoelastic stress–strain relations would differ from the linear elastic constitutive equations. All other governing conditions would follow directly from linear elasticity with proper inclusion of the time dependency of the pertaining variables. The governing set of conditions for an isothermal, linear viscoelastic boundary value problem are as follows.

9.4.1 Initial conditions

We assume that the body is initially undisturbed. In other words, it is initially stress free and in mechanical equilibrium. Thus, the initial conditions are

$$u_i(t) = 0, \quad \varepsilon_{ij}(t) = 0, \quad \sigma_{ij}(t) = 0, \quad -\infty < t < 0, \tag{9.1}$$

where u_i designate the components of the displacement vector in a rectangular Cartesian coordinate system.

9.4.2 Boundary conditions

The boundary B of the body is considered to be composed of two parts B_σ and B_u. That is,

$$B = B_\sigma + B_u$$

where B_σ denotes the part of the boundary of the body over which the components of the stress σ are prescribed and B_u indicates the remaining part of the boundary over which the components of the displacement u are specified. The boundary conditions may be assigned in the form of magnitudes of

- the traction vector components T_i over B_σ such that

$$\sigma_{ij}(\mathbf{x}, t)n_j = T_i(\mathbf{x}, t), \quad \mathbf{x} \text{ on } B_\sigma, \tag{9.2a}$$

  where n_j are the components of the outward unit normal to B_σ, or
- the displacement vector components U_i over B_u as

$$u_i(\mathbf{x}, t) = U_i(\mathbf{x}, t), \quad \mathbf{x} \text{ on } B_u. \tag{9.2b}$$

The boundary conditions (9.2) are assumed to be fixed; that is, both the traction vector components T_i and the displacement vector components U_i are considered to be prescribed for all t.

9.4.3 Balance of linear momentum

One of the following two situations may be considered:

• for a quasi-static problem, the equilibrium equation is

$$\sigma_{ij,j} + \chi_i = 0; \tag{9.3}$$

• for a dynamic problem, the equation of motion is

$$\sigma_{ij,j} + \chi_i = \rho \frac{\partial^2 u_i}{\partial t^2}. \tag{9.4}$$

In (9.3) and (9.4) χ_i are the body force components per unit volume.

9.4.4 Linear strain–displacement relations

$$\varepsilon_{ij}(t) = \tfrac{1}{2}[u_{i,j}(t) + u_{j,i}(t)] \tag{9.5}$$

in which a comma indicates partial differentiation with respect to the coordinates x_i of the material particle.

9.4.5 Stress–strain relations

General linear constitutive equations for a viscoelastic material with an arbitrary degree of anisotropy may be expressed in the form of Boltzmann superposition integral.

• For the creep case,

$$\varepsilon_{ij}(t) = \int_0^t F_{ijkl}(t - \tau) \frac{\partial \sigma_{kl}(\tau)}{\partial \tau} \, d\tau \tag{9.6}$$

where $F_{ijkl}(t - \tau)$ are the components of the creep function.
• For the relaxation case

$$\sigma_{ij}(t) = \int_0^t R_{ijkl}(t - \tau) \frac{\partial \varepsilon_{kl}(\tau)}{\partial \tau} \, d\tau \tag{9.7}$$

where $R_{ijkl}(t - \tau)$ are the components of the relaxation function.

As mentioned in section 4.4, the constitutive relation for an isotropic material can be reduced to two pairs of operators, one for the stress–strain deviatoric constitutive relation which covers shear response and one for average hydrostatic tension and dilatation. In this case, a differential operator law in the following form may be used (Lee, 1960):

$$P_1(D)\sigma'_{ij}(t) = Q_1(D)\varepsilon'_{ij}(t)$$
$$P_2(D)\sigma_{kk}(t) = Q_2(D)\varepsilon_{kk}(t)$$

(9.8)

where σ'_{ij}, ε'_{ij} are the stress and strain deviators defined, respectively, by

$$\sigma'_{ij} = \sigma_{ij} - \tfrac{1}{3}\delta_{ij}\sigma_{kk}, \quad \sigma'_{ii} = 0,$$
$$\varepsilon'_{ij} = \varepsilon_{ij} - \tfrac{1}{3}\delta_{ij}\varepsilon_{kk}, \quad \varepsilon'_{ii} = 0.$$

(9.9)

In equation (9.8), P_1, P_2, Q_1 and Q_2 are polynomials of the time derivative operator $D = \partial/\partial t$.

Alternatively, the stress–strain relations (9.8) may be used in either of the following constitutive forms:

- in the creep case,

$$\varepsilon'_{ij} = \sigma'_{ij}(t) * dF_1(t), \quad \varepsilon_{kk}(t) = \sigma_{kk}(t) * dF_2(t)$$

(9.10)

where $F_1(t)$ and $F_2(t)$ are the creep functions in pure shear and pure dilatation, respectively (section 4.4);
- in the relaxation use,

$$\sigma'_{ij}(t) = \varepsilon'_{ij}(t) * dR_1(t), \quad \varepsilon_{kk}(t) = \sigma_{kk}(t) * dR_2(t)$$

(9.11)

where $R_1(t)$ and $R_2(t)$ are the relaxation functions in pure shear and pure dilatation, respectively.

For an isotropic viscoelastic material, an approximate form of the constitutive relation in the relaxation case may be expressed (Hunter, 1967) as

$$\sigma_{ij}(t) = \delta_{ij} \int_0^t \lambda(t - \tau) \frac{\partial \varepsilon_{kk}(\tau)}{\partial \tau}\, d\tau + 2 \int_0^t \mu(t - \tau) \frac{\partial \varepsilon'_{ij}(\tau)}{\partial \tau}\, d\tau$$

(9.12)

where $\lambda(t)$ and $\mu(t)$ are appropriate relaxation functions.

In terms of deviatoric and dilatational components, the isotropic constitutive equations in the relaxation case can be further written as

$$\sigma'_{ij}(t) = 2 \int_0^t \mu(t - \tau) \frac{\partial \varepsilon'_{ij}(\tau)}{\partial \tau}\, d\tau$$

and

(9.13)

$$\sigma_{ii}(t) = 3 \int_0^t k(t - \tau) \frac{\partial \varepsilon_{jj}(\tau)}{\partial \tau}\, d\tau$$

where $\sigma'_{ij}(t)$ and $\varepsilon'_{ij}(t)$ are, respectively, the deviatoric stress and the deviatoric strain components and $\mu(t)$ and $k(t)$ are the relaxation functions in pure shear and pure dilatation, respectively.

In the isothermal linear boundary value problem, the three balance of linear momentum equations (9.3) or (9.4), the six strain–displacement relations (9.5) and the six stress–strain constitutive equations, e.g. (9.13), constitute a set of fifteen field equations for the fifteen dependent variables u_i, ε_{ij} and σ_{ij} under the prescribed boundary conditions $T_i(\mathbf{x}, t)$ and $U_i(\mathbf{x}, t)$, (9.2), and the assumed initial conditions (9.1).

9.5 VISCOELASTIC UNIQUENESS THEOREMS: UNIQUENESS OF SOLUTION

An important question concerning the solution of a boundary value problem in continuum mechanics is whether the formulated problem has a solution and whether the solution is unique or not (Fung, 1965). On physical grounds, this question may be dealt with by reference to the thermodynamics of the problem involved. On mathematical grounds, however, this question must be answered by the theory of partial differential equations. A satisfactory solution to the problem in hand must comply with both the laws of physics and the principles of mathematics. In solving boundary value problems of static equilibrium within classical elasticity, for example, one may proceed in the following sequence: (i) one solves the equations of equilibrium for the stresses σ_{ij}; (ii) the constitutive response equations are then solved for the strains ε_{ij} by using the stress components σ_{ij} obtained from (i). Here, an infinite set of solutions may be found. However, the unique solution would be singled out by employing, for instance, the conditions of compatibility (equation (1.80)).

The existence and uniqueness of solution theorems in classical elasticity have been extended by Gurtin and Sternberg (1962) to the class of linear boundary value problems in viscoelasticity. This was carried out in light of an earlier work by Volterra (1909).

For the most direct case of isothermal, isotropic, linear viscoelastic boundary value problem under a quasi-static condition, Christensen (1971), following Gurtin and Sternberg (1962), presented a uniqueness condition of solution. This condition may be stated, in view of the set of governing equations mentioned in section 9.4, as follows.

Theorem 9.1 Uniqueness theorem

The isotropic, quasi-static, viscoelastic boundary value problem governed by the initial conditions (9.1), the boundary conditions (9.2), the equations of equilibrium (9.3), the strain–displacement relations (9.5) and the stress–strain equations (9.13) possesses a unique solution provided that the initial values of the relaxation functions appearing in the constitutive equations (9.13) satisfy the conditions

$$\mu(0) > 0 \quad \text{and} \quad k(0) > 0. \tag{9.14}$$

For a proof of the uniqueness theorem stated above, the reader is referred to Christensen (1971). Other versions of uniqueness theorems for the above class of boundary value problems are given by Onat and Breuer (1963), Edelstein and Gurtin (1964), Odeh and Tadjbakhsh (1965), Barberan and Herrera (1966) and Lubliner and Sackman (1967), amongst others.

9.6 CORRESPONDENCE PRINCIPLE: THE ELASTIC–VISCOELASTIC ANALOGY

For a large number of technical viscoelastic problems, it is possible to relate mathematically the solution of a linear, viscoelastic boundary value problem to an analogous problem of an elastic body of the same geometry and under the same initial and boundary conditions. This is carried out by transforming the governing equations of the viscoelastic problem to be mathematically equivalent to those governing a corresponding elastic problem. In this, both Laplace and Fourier transforms are often used. Accordingly, one would be able to employ the tools of the theory of elasticity to solve different boundary value problems in linear viscoelasticity.

The above analogy is referred to as the 'correspondence principle'. It implies that elastic analysis procedures may be utilized to derive transformed viscoelastic solutions (for instance, Lee, 1955; Morland and Lee, 1960; Schapery, 1967). Lee (1955) demonstrated the correspondence principle for isotropic media at constant temperature. Meantime, Morland and Lee (1960) considered the application of the correspondence principle for isotropic materials with temperature variations. Biot (1958) argued that the correspondence principle may be also applied to anisotropic materials because of the symmetry of the relaxation modulus tensor, i.e. $R_{ijkl}(t) = R_{klij}(t)$.

9.6.1 Laplace-transformed isothermal, linear viscoelastic boundary value problem

(a) Initial conditions

The body is assumed to be initially undisturbed. Thus, the initial conditions (9.1) will hold.

(b) Boundary conditions

The Laplace-transformed forms of the boundary conditions (9.2a) and (9.2b) are, respectively,

$$\bar{\sigma}_{ij}(\mathbf{x}, s)n_j = \bar{T}_i(\mathbf{x}, s), \quad \mathbf{x} \text{ on } B_\sigma, \tag{9.15a}$$

and

$$\bar{u}_i(\mathbf{x}, s) = \bar{U}_i(\mathbf{x}, s), \quad \mathbf{x} \text{ on } B_u, \tag{9.15b}$$

where s is the Laplace transform variable and the overbar designates the Laplace transform of the variable, i.e.

$$\bar{T}_i(\mathbf{x}, s) = \int_0^\infty T_i(\mathbf{x}, t) \exp(-st) \, dt \tag{9.16a}$$

and

$$\bar{U}_i(\mathbf{x}, s) = \int_0^\infty U_i(\mathbf{x}, t) \exp(-st)\, dt \qquad (9.16b)$$

(Appendix C).

(c) Balance of linear momentum

The quasi-static case is dealt with here. Recalling (9.3), multiplying it by $\exp(-st)$ and integrating over $-\infty < t < \infty$, the Laplace transform of the equilibrium equation is

$$\bar{\sigma}_{ij,j} + \bar{\chi}_i = 0. \qquad (9.17)$$

(d) Linear strain–displacement relations

The Laplace-transformed strain–displacement relation (9.5) is

$$\bar{\varepsilon}_{ij} = \tfrac{1}{2}(\bar{u}_{i,j} + \bar{u}_{j,i}) \qquad (9.18)$$

(e) Stress–strain relations

The constitutive equations (9.6) and (9.7) can be transformed by the rule of convolution integrals (Schapery, 1967) to yield, respectively, the algebraic relations

$$\bar{\varepsilon}_{ij} = \tilde{F}_{ijkl}\bar{\sigma}_{kl} \qquad (9.19)$$

and

$$\bar{\sigma}_{ij} = \tilde{R}_{ijkl}\bar{\varepsilon}_{kl} \qquad (9.20)$$

where $\tilde{F}_{ijkl}$ and $\tilde{R}_{ijkl}$ are the s-multiplied (Laplace) transforms of the creep and relaxation functions, respectively, i.e.

$$\tilde{F}_{ijkl} = s\bar{F}_{ijkl} \qquad (9.21a)$$

and

$$\tilde{R}_{ijkl} = s\bar{R}_{ijkl}. \qquad (9.21b)$$

The quantities $\tilde{F}_{ijkl}$ and $\tilde{R}_{ijkl}$ are interrelated operational functions, i.e.

$$[\tilde{F}_{ijkl}] = [\tilde{R}_{ijkl}]^{-1}, \qquad (9.22)$$

and both are completely symmetric. Thus, in view of the thermodynamic theory, the transformed constitutive equations in terms of these operational functions are identical to those of an elastic body with compliances $\tilde{F}_{ijkl}$ and moduli $\tilde{R}_{ijkl}$ and of the same degree of geometric symmetry (Schapery, 1967).

For the case of an isotropic material, the constitutive equation (9.12) may be used. The Laplace transform of this equation is (Hunter, 1967)

$$\bar{\sigma}_{ij}(s) = \lambda(s)\bar{\varepsilon}_{ii}(s)\delta_{ij} + 2\mu(s)\bar{\varepsilon}'_{ij}(s) \qquad (9.23)$$

where $\bar{\sigma}_{ij}$ and $\bar{\varepsilon}'_{ij}$ are the Laplace transforms of σ_{ij} and ε'_{ij}, respectively. In this equation, the transform moduli $\lambda(s)$ and $\mu(s)$ are defined by

$$\lambda(s) = k(s) - \tfrac{2}{3}\mu(s) \qquad (9.24)$$

where $k(s)$ is the Laplace transform of the relaxation function in pure dilatation, i.e.

$$k(s) = s \int_0^\infty R_1(t) \exp(-st) \, dt \qquad (9.25a)$$

and $\mu(s)$ is the Laplace transform of the relaxation function in pure shear, i.e.

$$\mu(s) = s \int_0^\infty R_2(t) \exp(-st) \, dt. \qquad (9.25b)$$

In the general case of nonhomogeneous material, the field quantities $\bar{\sigma}_{ij}$, $\bar{\varepsilon}_{ii}$ and $\bar{\varepsilon}'_{ij}$ of (9.23) are usually functions of both the transform parameter s and the position vector $\mathbf{x}$. However, the transform moduli $\lambda(s)$ and $\mu(s)$ are functions of the transform variable s only (Hunter, 1967).

The corresponding format to (9.23) in linear elasticity is the constitutive equation

$$\sigma_{ij} = \lambda\varepsilon_{ii}\delta_{ij} + 2\mu\varepsilon'_{ij} \qquad (9.26)$$

where λ and μ are the Lamé constants. Such an analogy reflects the basis of the correspondence principle.

The set of Laplace-transformed relations (9.15), (9.17) and (9.18), together with the transformed constitutive equations (9.19) and (9.20), or alternatively (9.23), constitutes an 'associated' elastic problem corresponding to the original (quasi-static) viscoelastic boundary value problem for the same geometry and subject to surface tractions $\bar{T}_i = \bar{T}_i(\mathbf{x}, s)$, displacements $\bar{U}_i = \bar{U}_i(\mathbf{x}, s)$ and body forces $\bar{\chi}_i = \bar{\chi}_i(\mathbf{x}, s)$. The task then would be to solve this analogous elastic problem (Laplace transformed of the original viscoelastic problem) to determine the transformed components of the stress $\bar{\sigma}_{ij}$ and/or the transformed components of the displacement $\bar{u}_i$ throughout the body. A Laplace inversion procedure would follow afterwards to determine the components of the stress and displacement in the original viscoelastic boundary value problem. The reader is referred, in this context, to Sips (1951), Brull (1953) and Lee (1955, 1960), among others.

Although the presentation above uses the Laplace transform procedure, a similar treatment could be accomplished using Fourier transform (e.g. Read, 1950).

9.6.2 Remarks on the use of the correspondence principle to solve linear viscoelastic boundary value problems

In the course of solving a linear viscoelastic boundary value problem using the correspondence principle, one might consider some simplifications in order to ease the difficulty which might arise in the inversion of the resulting Laplace transforms. For instance (Hunter, 1967), the following could be used.

- In a large number of boundary value problems it may be unnecessary to invert the resulting Laplace transform for all positions on the boundary (i.e. x on B_σ or B_u) if the stress and/or displacement is only required at one particular position.
- In some situations, the integral value of the stress and/or displacement is required to be determined rather than individual values of these variables. In such case, it might be easier if one establishes the relevant integral property before the inversion process.
- The inversion procedure can be simplified significantly if one assumes a constant Poisson's ratio model and particularly if the body forces χ_i are neglected. In this case, if B_σ is considered to be stress free, then, for the same boundary conditions, the resulting displacement field at any given instant of time would be identical to the displacement field of the corresponding elastic problem. A similar example here is when $B_u = 0$ and $B_\sigma = B$, i.e. the traction vector is specified everywhere on the total boundary; then the resulting viscoelastic stress field would be identical with the stress field of the corresponding elastic problem.

9.6.3 Illustrative examples on the use of the correspondence principle to solve linear viscoelastic boundary value problems

Example 9.1 Torsional quasi-static twisting of a linear, viscoelastic cylinder (Hunter, 1967)

This example considers the determination of the time-dependent twisting moment and displacement of a solid cylinder of radius a and length l made of linear viscoelastic material. Let u_r, u_θ, u_z represent the displacement components in cylindrical polar components, θ denote the angle of twist at $z = l$ and M be the twisting moment.

From the theory of elasticity, the displacement field of an elastic solid cylinder under the action of a twisting moment and a stress-free cylindrical surface condition is expressed (e.g. Love, 1944) as

$$u_\theta = \theta r z / l, \quad u_r = u_z = 0. \tag{9.27}$$

For an analogous quasi-static viscoelastic problem with prescribed displacement such as

$$u_\theta = \begin{cases} r_\theta, & z = l, \\ 0, & z = 0, \end{cases}$$

and a stress-free cylindrical surface, the displacement field is given by (9.27) with θ now a time-dependent variable. In this case, the only nonvanishing strain component is

$$\varepsilon_{\theta z} = \tfrac{1}{2}\theta(t)\,\frac{r}{l}. \tag{9.28}$$

The nonvanishing (transformed) stress corresponding to the above strain becomes

$$\bar{\tau}_{\theta z} = \bar{\mu}(s)\bar{\theta}(s)\,\frac{r}{l} \tag{9.29}$$

where $\bar{\mu}(s)$ is the transformed shear modulus.

Thus, the total (transformed) couple required to maintain the (transformed) angle of twist $\bar{\theta}(s)$ can be expressed as

$$\bar{M} = 2\pi \int_0^a \bar{\tau}_{\theta z} r^2 \, dr = \frac{\pi a^4}{2l} \bar{\mu}(s)\bar{\theta}(s) \tag{9.30}$$

which may be inverted in either the relaxation form

$$M(t) = \frac{\pi a^4}{2l} \int_0^t \dot{\theta}(t')G(t-t') \, dt' \tag{9.31a}$$

or the creep form

$$\theta(t) = \left(\frac{\pi a^4}{2l}\right)^{-1} \int_0^t \dot{M}(t')G^{-1}(t-t') \, dt' \tag{9.31b}$$

where $G(t-t')$ is the relaxation function of the material in shear.

Equations (9.31) provide a quasi-static linear viscoelastic solution of the presented problem for prescribed $\theta(t)$ or $M(t)$.

Example 9.2 Impact of a flat circular punch on a linear, viscoelastic half-space (Hunter, 1967)

(a) Elastic solution

According to the theory of elasticity, the solution of the problem of the normal identation of an elastic half-space by a flat-ended rigid-circular punch of radius a gives the pressure distribution (Boussinesq, 1885)

$$p(r) = \begin{cases} \dfrac{4\mu ad}{1-v}(a^2-r^2)^{-1/2}, & r < a, \\ 0 & r > a, \end{cases} \tag{9.32}$$

where d is the depth of the penetration and v is Poisson's ratio of the elastic half-space. Further, the total load is given by

$$F = 2\pi \int_0^a pr \, dr = \frac{8\pi a}{1-r}\mu d. \tag{9.33}$$

(b) Viscoelastic solution

For the linear viscoelastic case, equation (9.33) becomes

$$\bar{F} = \frac{8\pi a}{1 - v} \, \bar{u}(s)\bar{d}, \quad v = \text{constant}, \tag{9.34}$$

where v is Poisson's ratio of the viscoelastic half-space. Accordingly, given $F(t)$ or $d(t)$, equation (9.34), when inverted, gives d or F respectively.

9.6.4 Extension of Example 9.2 to include impact

For the impact problem, the load F is expressed by Newton's second law of motion:

$$F = -m\ddot{d}$$

where m is the mass of the indentor and $\ddot{d}$ is its acceleration. On taking the Laplace transform of the above expression, the (transformed) force is written as

$$\bar{F} = -m(s^2\bar{d} - v) \tag{9.35}$$

where v is the initial impact velocity at the initial conditions $d = 0$ at $t = 0$.
  Solving (9.34) and (9.35) for $\ddot{d}$ gives

$$\bar{d} = v\left[s^2 + \frac{8\pi a}{m(1 - v)} \mu(s) \right]^{-1}. \tag{9.36}$$

With some physical approximation (Hunter, 1967), the inversion of (9.36) gives

$$d = \frac{v}{\omega}\left(1 + \frac{\tan \delta}{\pi} \right) \exp[-\tfrac{1}{2}(\omega \tan \delta)t]\,\sin(\omega t - \tan \delta)$$

$$+ \frac{v \tan \delta}{2\omega} \exp\left(-\frac{2\omega t}{\pi} \right) \tag{9.37a}$$

where ω is the solution of

$$\frac{\omega(1 - v)}{8\pi a}\,\omega^2 = \mu_1(\omega) \tag{9.37b}$$

where $\mu_1(\omega)$ is the real part of the complex shear modulus and $\tan \delta = \mu_2(\omega)/\mu_1(\omega)$. For the values of ω where the viscoelasticity is significant, $\mu_2(\omega) \ll \mu_1(\omega)$ and $\tan \delta \ll 1$. Further, the impact terminates at a time given by $\ddot{d} = 0$ with solution

$$\omega t = d - \frac{1}{2}\left[1 + \left(\frac{2}{\pi e} \right)^2 \right] \tan \delta + O(\tan^2 \delta)$$

when the indentor velocity is $-v(1 - \gamma \tan \delta)$ where

$$\gamma = \tfrac{1}{2}\pi - (1 + e^{-2})/\pi = 1.205.$$

This results in a coefficient of restitution e given by

$$e = 1 - 1.2 \tan \delta \qquad (9.38)$$

Thus, the energy absorbed by the solid is

$$E = 1.2 \, mv^2 \tan \delta \qquad (9.39)$$

(Hunter, 1967).

9.7 SOLUTION OF QUASI-STATIC VISCOELASTIC PROBLEMS IN THE ABSENCE OF THE CORRESPONDENCE PRINCIPLE: MIXED BOUNDARY VALUE PROBLEMS

In section 9.4, the set of conditions governing an isothermal, linear viscoelastic boundary value problem has been introduced. In section 9.6, the correspondence principle was presented to solve a boundary value problem of this type, subject to the condition that B_σ and B_u are independent of time where these are the parts of the boundary upon which stress vector components and displacement components, respectively, are specified. This is necessitated by the requirement that the assumed boundary conditions at a point are time invariant so that the integral transform methods would be applicable. Consequently, an elastic–viscoelastic correspondence principle does not exist when the parts of the boundary B_σ and B_u are functions of time, i.e. when the boundary conditions at the particular point in question may involve with the passage of time both stress and displacement vectors. A representative boundary value problem of the latter type is the time-dependent indentation of a viscoelastic half-space by a curved rigid indentor. In this case, as the indentor is loaded and depression into the viscoelastic half-space is progressing, there are some points on the boundary of the indentation region that, at first, may have traction-free boundary conditions, but later could have displacement followed by stress boundary conditions. In other words, a portion of the boundary is the boundary B_u part of the time and is the boundary B_σ at other times, so that the half-space would conform to the geometry of the indentor in the contact region. Studies concerning this problem were presented, for instance, by Lee and Radok (1960), Hunter (1960), Graham (1965, 1967), Calvit (1967) and Ting (1966, 1968). Other examples of mixed boundary value problems are, for instance, those involving rolling of rigid bodies over a viscoelastic half-space (e.g. Hunter, 1961; Morland, 1962, 1967) and ablation problems in which a phase change could cause the boundaries of a viscoelastic medium to change size and shape. An example of this problem is the case of a spinning rocket's filling burning internally. A similar problem of an internally ablating sphere was considered by Rogers and Lee (1962). Other examples of boundary value problems where integral transform methods are invalid are nonisothermal problems in which the mechanical properties are assumed to be temperature dependent. A number of boundary value problems of the latter types have been solved, but it appears that no systematic methods of solution are available.

Example 9.3 Deformation of a uniform viscoelastic beam by a curved rigid indentor (Christensen, 1971)

The schematics of the problem are shown in Fig. 9.2. As indicated, $P(t)$ is the force applied to the rigid indentor, $a(t)$ is half-length of the contact region (considered a basic unknown of the problem) and $d(t)$ is the displacement (vertical) of the indentor. The indentor is assumed to have a cubic profile expressed by

$$y = d(t) - c|x|^3 \tag{9.40}$$

where c is a given constant.

In this problem, classical beam theory with simply supported end conditions is assumed. Inertia effects are neglected. Contact is considered to begin at $t = 0$.

Based on the above assumptions, elasticity theory gives

$$EI \frac{d^4 w}{dx^4} = q(x) \tag{9.41}$$

where I is the moment of inertia of the cross-section of the beam, w is the transverse displacement and $q(x)$ is the lateral load.

Meantime, viscoelastic beam theory gives

$$I \int_0^t R(t - \tau) \frac{\partial}{\partial \tau} \left[\frac{\partial^4 w(x, \tau)}{\partial x^4} \right] d\tau = q(x, \tau) \tag{9.42}$$

in which $R(t - \tau)$ is the uniaxial relation function.

- In the contact region, $x < a(t)$, the deflection of the beam must conform to the geometry of the indentor; then, with reference to equation (9.40),

$$w(x, t) = d(t) - cx^3, \quad x < a(t) \quad \text{and} \quad t \geq 0. \tag{9.43}$$

- Outside the contact region, $x > a(t)$, the lateral load vanishes and equation (9.42) is satisfied by

$$w(x, t) = C_1(t) + C_2(t)x + C_3(t)x^2 + C_4(t)x^3, \quad x > a(t) \quad \text{and} \quad t \geq 0, \tag{9.44}$$

where $C_1(t)$, $C_2(t)$, $C_3(t)$ and $C_4(t)$ are functions of time required to be determined.

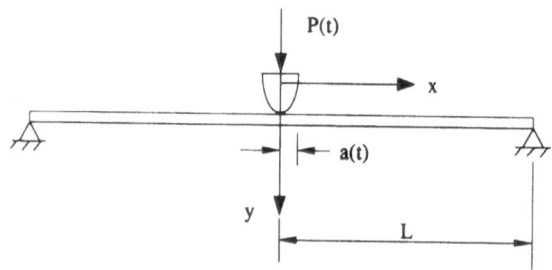

Fig. 9.2 Deformation of a uniform beam by a curved rigid indentor.

Implying the condition that the shear resultants on the ends of the beam balance the applied load $P(t)$ gives, in view of (9.44),

$$12I \int_0^t R(t - \tau) \frac{dC_4(\tau)}{d\tau} \, d\tau = P(t). \tag{9.45}$$

- The end conditions are

$$at - x = L, \quad w = \frac{\partial^2 w}{\partial x^2} = 0$$

which can be specified by (Fig. 9.2)

$$C_1(t) + C_2(t)L + C_3(t)L^2 + C_4(t)L^3 = 0 \tag{9.46}$$

and

$$2C_3(t) + 6C_4(t)L = 0. \tag{9.47}$$

- The continuity conditions at the edge of the contact region $x = a$ imply that w, $\partial w/\partial x$ and $\partial^2 w/\partial x^2$ are continuous. Accordingly, equations (9.43) and (9.44) give

$$C_1(t) + C_2(t)a(t) + C_3(t)a^2 + C_4(t)a^3(t) = d(t) - ca^3(t), \tag{9.48}$$

$$C_2(t) + 2C_3(t)a(t) + 3C_4(t)a^3(t) = -3ca^2(t) \tag{9.49}$$

and

$$2C_3(t) + 6C_4(t)a(t) = -6ca(t). \tag{9.50}$$

Relations (9.45–9.50) give six nonlinear equations.

1. If the load $P(t)$ is considered to be known, then the six equations would be solved for the six unknowns $C_1(t)$, $C_2(t)$, $C_3(t)$, $C_4(t)$, $a(t)$ and $d(t)$.
2. Alternatively, if the displacement of the indentor $d(t)$ is taken to be specified, then the above-mentioned equations can be solved for the unknowns $C_1(t)$, $C_2(t)$, $C_3(t)$, $C_4(t)$, $a(t)$ and $P(t)$.

These two cases are considered separately by Christensen (1971).

Example 9.4 A spherical indentor on a viscoelastic half-space

As a second example of a viscoelastic boundary value problem with mixed-type boundary conditions, the problem of indentation of a viscoelastic half-space by a rigid spherical indentor is considered. Previous studies on this type of problem were carried out, for instance, by Lee and Radok (1960), Hunter (1960) and Graham (1965). The analysis presented below follows that of Hunter (1967) after Graham (1965). Reference is also made to Christensen (1971).

The indentor is considered to be applied at the origin of a rectangular Cartesian coordinate system (x, y, z) and its motion is vertical in the z direction. The shear stresses over the entire boundary of the half-space are assumed to be identically equal to zero. In the contact region, however, the normal component of the displacement of the boundary is considered to conform to the shape of the indentor. Let R be the radius of the indentor, $\delta(t)$ be the depth of penetration, $a(t)$ be the radius of the contact surface, $z = 0$ be the surface of the (viscoelastic) half-space and $r^2 = x^2 + y^2$.

(a) The Problem

This is to determine the stress distribution under the spherical indentor and the relation between the depth of penetration $\delta(t)$ and the radius of the contact surface $a(t)$ subject to the following boundary conditions:

$$\text{at } x = 0 \begin{cases} u_z = \delta(t) - \dfrac{1}{2R}(x^2 + y^2)H(t), & r \leq a(t), & (9.51a) \\[2mm] \sigma_z = 0, & r > a(t), & (9.51b) \\[2mm] \tau_{rz} = 0, & & (9.51c) \end{cases}$$

where $H(t)$ is the Heaviside step function.

(b) Elastic solution

The starting point of Graham's solution is taken as (Hunter, 1967) the Boussinesq formula for the normal surface displacement of an elastic solid subjected to a normal point load P at x', y', i.e.

$$u_z(x, y, 0) = \frac{(1 - \nu)P}{2\pi\mu} [(x - x')^2 + (y - y')^2]^{-1/2} \tag{9.52}$$

where μ is the elastic shear modulus.

(c) Viscoelastic solution

Generalizing (9.52) to the case of a viscoelastic half-space subjected to a time variable distributed load $P(x, y, t)$ gives under the assumption of a constant Poisson's ratio (Hunter, 1967)

$$u_z(x, y, 0, t) = \frac{1 - \nu}{2\pi} G^{-1}(t) * d \iint_{\Omega_m} P(x', y', t')[(x - x')^2 + (y - y')^2]^{-1/2} \, dx' \, dy' \tag{9.53}$$

where $G^{-1}(t)$ is the creep function in shear and where the following notation of Gurtin and Sternberg (1962) is used:

$$a * d\beta = \int_{-\infty}^{t} a(t - t')\, d\dot{\beta}(t') = \int_{-\infty}^{t} a(t - t')\dot{\beta}(t')\, dt'.$$

In (9.53) the double (surface) integral is taken over the maximum range Ω_m enclosing all points x', y' for which $P(x', y', t')$ is nonzero for any time t' in the range $-\infty < t' < t$. With reference to equation (9.53), Hunter (1967) considered the following four situations.

1. For given $P(x, y, t)$, equation (9.53) presents the solution for the normal surface displacement and such problems may be considered within the class of the boundary value problems that can be solved by the correspondence principle (section 9.6).
2. For the indentation problem, 'mixed-type boundary conditions' boundary value problem, equation (9.53) may be considered as an integral equation for P subject to the condition that, for $r < a(t)$, the surface displacement u_z is given by (9.51a), while, for $r > a(t)$, P vanishes.
3. For monotonically increasing $a(t)$, Ω_m is time dependent and can be taken as $\Omega(t) = \pi a^2(t)$. In this case, the orders of space and time integration in (9.53) can be changed to give

$$u_z = \frac{1 - v}{2\pi} \iint\limits_{\Omega(t)} dx'\, dy'[(x - x')^2 + (y - y')^2]^{-1/2} G^{-1}\, dP \tag{9.54}$$

in which

$$G^{-1}\, dP = \eta(x, y, a) \tag{9.55}$$

where η is the unique solution of the corresponding elastic problem whose solution (Boussinesq, 1885) is given by

$$\eta = \frac{4}{\pi(1 - v)R}\,(a^2 - r^2)^{1/2} \tag{9.56a}$$

and

$$\eta = 0 \quad \text{for} \quad r > a \tag{9.56b}$$

so that (9.55) leads to

$$P(x, y, t) = \frac{4}{\pi(1 - v)R} \int_0^t G(t - t')\, \frac{d}{dt'}\, [(a^2(t') - r^2)^{1/2}]\, dt' \tag{9.57}$$

in which, for fixed r, the lower limit of the integral may be taken as t'' where t'' is the unique solution of

$$a(t'') = r.$$

Meantime, the total load on the indentor is given by

$$F(t) = 2\pi \int_0^{a(t)} rP(x, y, t) \, dr$$

which can be evaluated by interchanging the order of the space and time integrations (Hunter, 1960, 1967) to give

$$F(t) = \frac{8}{3R(1 - v)} \int_0^t G(t - t') \frac{d}{dt'} a^3(t') \, dt'. \tag{9.58}$$

Further, it can be shown that the depth of penetration can be expressed by

$$\delta(t) = a^2(t)/R \tag{9.59}$$

which is the same for the corresponding elastic problem.

4. The radius of the contact surface $a(t)$ increases monotonically to a maximum value at $t = t_m$ and then decreases to zero. In this case, the solution given above is valid for $t \leq t_m$. For $t > t_m$, however, the solution fails because it is no longer permissible to replace Ω_m in (9.53) by $\Omega(t)$. To obtain the solution of (9.53) for $t > t_m$, Hunter (1967) introduced the time function $t_1(t)$ defined by the relations

$$\begin{aligned} \text{for} \quad t \leq t_m, \quad & t_1(t) = t, \\ \text{for} \quad t \geq t_m, \quad & a(t_1) = a(t), \quad t_1 < t_m. \end{aligned} \tag{9.60}$$

In other words, t_1 is the time prior to t for which the radius of the contact circle is equal to the current value.

Further studies concerning the viscoelastic contact problem have been dealt with, for instance, by Calvit (1967) and Ting (1968). Graham (1968) and Ting (1968) have outlined restricted classes of viscoelastic contact problems which may be solved directly using the elastic–viscoelastic correspondence principle.

9.8 THERMOVISCOELASTIC BOUNDARY VALUE PROBLEM

The set of conditions that governs a thermoviscoelastic boundary value problem may be stated as follows.

9.8.1 Initial conditions

Assuming the body is initially undisturbed at a base temperature θ_0, then, the initial conditions are

$$u_i(t) = 0, \quad \varepsilon_{ij}(t) = 0, \quad \sigma_{ij}(t) = 0, \quad \theta(t) = 0, \quad -\infty < t < 0, \tag{9.61}$$

where θ denotes the temperature deviation from the base temperature θ_0.

9.8.2 Boundary conditions

In order to account for the temperature effect, the boundary is visualized (Christensen, 1971) to be composed of two regions, i.e. B_θ is that region of the boundary upon which the temperature is prescribed and $B - B_\theta$ is the complimentary region over which the surface is taken to be perfectly insulated against heat flow. The thermal boundary conditions can then be stated as

$$\theta(\mathbf{x}, t) = \hat{\theta}(\mathbf{x}, t), \qquad \mathbf{x} \text{ on } B_\theta, \quad t \geq 0 \tag{9.62}$$

and

$$k_{ij}(\mathbf{x}, t)\theta_{,i} n_j = 0 \qquad \mathbf{x} \text{ on } B - B_\theta, \quad t \geq 0. \tag{9.63}$$

where k_{ij} as a second-order tensor accounts for the mechanical properties of the material.

Combining the thermal boundary conditions (9.62) and (9.63) with the traction and displacement boundary conditions stated earlier for the isothermal problem, (9.2), the set of boundary conditions for the thermoviscoelastic problem is written as follows:

$$
\begin{aligned}
\sigma_{ij}(\mathbf{x}, t)n_j &= T_i(\mathbf{x}, t), & \mathbf{x} \text{ on } B_\sigma; \\
u_i(\mathbf{x}, t) &= U_i(\mathbf{x}, t), & \mathbf{x} \text{ on } B_u; \\
\theta(\mathbf{x}, t) &= \hat{\theta}(\mathbf{x}, t), & \mathbf{x} \text{ on } B_\theta; \\
k_{ij}(\mathbf{x}, t)\theta_{,i} n_j &= 0, & \mathbf{x} \text{ on } B - B_\theta.
\end{aligned}
\tag{9.64}
$$

9.8.3 Balance of linear momentum

The equations of (quasi-static) equilibrium

$$\sigma_{ij,j} + \chi_i = 0 \tag{9.65}$$

or, alternatively, the equations of motion

$$\sigma_{ij,j} + \chi_i = \rho \frac{d^2 u_i}{dt^2} \tag{9.66}$$

can be used.

9.8.4 Strain–displacement relations

$$\varepsilon_{ij}(t) = \tfrac{1}{2}[u_{i,j}(t) + u_{j,i}(t)]. \tag{9.67}$$

9.8.5 Stress–strain relations

(a) Anisotropic materials

The relaxation constitutive relation is expressed as

$$\sigma_{ij}(t) = \int_0^t R_{ijkl}(t - \tau) \frac{\partial \varepsilon_{kl}(\tau)}{\partial \tau} \, d\tau - \int_0^t \psi_{ij}(t - \tau) \frac{\partial \theta(\tau)}{\partial \tau} \, d\tau. \qquad (9.68a)$$

The creep constitutive relation corresponding to (9.68a) is written as

$$\varepsilon_{ij}(t) = \int_0^t F_{ijkl}(t - \tau) \frac{\partial \sigma_{kl}(\tau)}{\partial \tau} \, d\tau - \int_0^t \alpha_{ij}(t - \tau) \frac{\partial \theta(\tau)}{\partial \tau} \, d\tau. \qquad (9.68b)$$

In the case of thermorheologically simple materials, one may employ a stress–strain relation of the form (Schapery, 1964)

$$\sigma_{ij}(t) = \int_0^t R_{ijkl}(\xi - \xi') \frac{\partial \varepsilon_{kl}}{\partial \tau} \, d\tau - \int_0^t \psi_{ij}(\xi - \xi') \frac{\partial \theta(\tau)}{\partial \tau} \, d\tau \qquad (9.69a)$$

or, equivalently,

$$\sigma_{ij}(\xi) = \int_0^\xi R_{ijkl}(\xi - \xi') \frac{\partial \varepsilon_{kl}}{\partial \xi'} \, d\xi' - \int_0^\xi \psi_{ij}(\xi - \xi') \frac{\partial \theta(\tau)}{\partial \tau} \, d\tau \qquad (9.69b)$$

where ξ, introduced in section 5.5, is the so-called **reduced time** defined by the relation

$$d\xi = dt/a_\theta(\theta). \qquad (9.70)$$

Also,

$$\xi = \int_0^t dt/a_\theta(\theta), \quad \xi' = \int_0^\tau dt/a_\theta(\theta), \qquad (9.71)$$

where $\tau \leq t$.

The relaxation function $\psi_{ij}(\xi)$ appearing in (9.69) is assumed to have the following exponential series form

$$\psi_{ij}(\xi) = \sum_m \psi_{ij}^{(m)} \exp(-\xi/\gamma_m) + \psi_{ij}' \qquad (9.72)$$

where the constants $\psi_{ij}^{(m)}$ and ψ_{ij}' define the thermal stress characteristics of the material before loading (Schapery, 1964, 1967) and γ_m are appropriate exponent factors. Meantime, the relaxation moduli in (9.69) are considered (Schapery, 1964) to be given by

$$R_{ijkl}(\xi) = \sum_m R_{ijkl}^{(m)} \exp(-\xi/\gamma_m) + R_{ijkl}'. \qquad (9.73)$$

On the other hand, when the temperature is constant, the relaxation moduli may be taken as

$$R_{ijkl}(t/a_\theta) = \sum_m R_{ijkl}^{(m)} \exp[(-t/\gamma_m)a_\theta] + R_{ijkl} \tag{9.74}$$

which reflects the effect of constant temperature on relaxation (or creep) behaviour, that is, to simply shift the time scale. Accordingly, a_θ is often referred to as 'time shift factor'.

The creep constitutive equation corresponding to (9.69b) is

$$\varepsilon_{ij}(\xi) = \int_0^\xi F_{ijkl}(\xi - \xi') \frac{\partial \sigma_{kl}}{\partial \xi'} d\xi' + \int_0^\xi \alpha_{ij}(\xi - \xi') \frac{\partial \theta}{\partial \xi'} d\xi' \tag{9.75}$$

where the function $\alpha_{ij}(\xi)$ accounts for the strain response in the absence of the stress. It is expressed (Schapery, 1964) by

$$\alpha_{ij}(\xi) = \sum_m \alpha_{ij}^{(m)}[1 - \exp(-\xi/\gamma_m)] + \alpha_{ij} \tag{9.76}$$

where $\alpha_{ij}^{(m)}$ and α_{ij} define the thermal strain characteristics of the material before loading.

(b) Isotropic materials

The relaxation constitutive relations corresponding to (9.68a) are

$$\sigma_{ij}'(t) = \int_0^t R_1(t - \tau) \frac{\partial \varepsilon_{ij}'(\tau)}{\partial \tau} d\tau$$

and $\hspace{10cm}$ (9.77)

$$\sigma_{kk}(t) = \int_0^t R_2(t - \tau) \frac{\partial \varepsilon_{kk}(\tau)}{\partial \tau} d\tau - 3 \int_0^t \psi(t - \tau) \frac{\partial \theta(\tau)}{\partial \tau} d\tau$$

where σ_{ij}' and ε_{ij}' denote, respectively, the deviatoric components of the stress and strain.

The creep constitutive equations corresponding to (9.77) are expressed as

$$\varepsilon_{ij}'(t) = \int_0^t F_1(t - \tau) \frac{\partial \sigma_{ij}'(\tau)}{\partial \tau} d\tau$$

and $\hspace{10cm}$ (9.78)

$$\varepsilon_{kk}(t) = \int_0^t F_2(t - \tau) \frac{\partial \sigma_{kk}(\tau)}{\partial \tau} - 3 \int_0^t \alpha(t - \tau) \frac{\partial \theta(\tau)}{\partial \tau} d\tau.$$

In the case of thermorheologically simple materials, constitutive equations for isotropic materials are expressed (Schapery, 1964) for the relaxation case by

$$\sigma_{ij}(\xi) = 2 \int_0^\xi R(\xi - \xi') \frac{\partial \varepsilon_{ij}}{\partial \xi'} \, d\xi'$$

$$+ \delta_{ij} \int_0^\xi \left[\lambda(\xi - \xi') \frac{\partial \varepsilon}{\partial \xi'} - \psi(\xi - \xi') \frac{\partial \varepsilon}{\partial \xi'} \right] d\xi' \qquad (9.79)$$

where

$$\varepsilon = \varepsilon_{11} + \varepsilon_{22} + \varepsilon_{33}.$$

In (9.79), $R(\xi)$, $\lambda(\xi)$ and $\psi(\xi)$ are relaxation functions which, for thermodynamic reasons (Schapery, 1964), are considered to have the forms

$$R(\xi) = \sum_m R^{(m)} \exp(-\xi/\gamma_m) + R_e,$$

$$\lambda(\xi) = \sum_m \lambda^{(m)} \exp(-\xi/\gamma_m) + \lambda_{\bar{e}}, \qquad (9.80)$$

$$\psi(\xi) = \sum_m \psi^{(m)} \exp(-\xi/\gamma_m) + \psi_e,$$

with constants having the properties

$$\rho_m > 0,$$

$$R^{(m)} \geq 0, R_e \geq 0, \sum_m R(m) + R_e > 0,$$

$$K^{(m)} = \lambda^{(m)} + \tfrac{2}{3} R(m) \geq 0, \qquad (9.81)$$

$$K_e = \lambda_e + \tfrac{2}{3} R_e \geq 0,$$

$$\sum_m K^{(m)} + K_e > 0,$$

where K_e and $K^{(m)}$ define the bulk relaxation modulus

$$K(\xi) = \lambda(\xi) + \tfrac{2}{3} R(\xi) = \sum_m K^{(m)} \exp(-\xi/\gamma_m) + K_e. \qquad (9.82)$$

9.8.6 The heat conduction equation

- For isotropic materials

$$\frac{k_{ij}}{\theta_0} \theta_{,ij} = \frac{\partial}{\partial t} \int_0^t m(t - \tau) \frac{\partial \theta(\tau)}{\partial \tau} \, d\tau + \frac{\partial}{\partial t} \int_0^t \psi_{ij}(t - \tau) \frac{\partial \varepsilon_{ij}(\tau)}{\partial \tau} \, d\tau \qquad (9.83)$$

- and for anisotropic materials

$$\frac{k}{\theta_0}\,\theta_{,ii} = \frac{\partial}{\partial t}\int_0^t m(t-\tau)\,\frac{\partial\theta(\tau)}{\partial\tau}\,d\tau + \frac{\partial}{\partial t}\int_0^t \psi(t-\tau)\,\frac{\partial\varepsilon_{kk}(\tau)}{\partial\tau}\,d\tau \qquad (9.84)$$

where k_{ij} or k, $m(t)$, and $\psi_{ij}(t)$ or $\psi(t)$ are mechanical properties of the material.

In the general anisotropic case, the Laplace-transformed governing equations for the thermoviscoelastic boundary value problem are given by the following.

- The boundary conditions (equations (9.64)) are

$$\bar{\sigma}_{ij}n_j = \bar{T}_i \text{ on } B_\sigma,$$

$$\bar{u}_i = \bar{U}_i \text{ on } B_u,$$

$$\bar{\theta} = \hat{\bar{\theta}} \text{ on } B_\theta \qquad (9.85)$$

and

$$k_{ij}\bar{\theta}_{,i}n_j = 0 \text{ on } B - B_\theta.$$

- For the balance of linear momentum, the equations of quasi-static equilibrium (9.65)

$$\bar{\sigma}_{ij,j} + \bar{\chi}_i = 0 \qquad (9.86)$$

or, alternatively, the equation of motion (9.66)

$$\bar{\sigma}_{ij,j} + \bar{\chi}_i = \rho s^2 \bar{u}_i \qquad (9.87)$$

where s is the Laplace transform variable can be used.
- The strain–displacement relations (9.21) are

$$\bar{\varepsilon}_{ij} = \tfrac{1}{2}(\bar{u}_{i,j} + \bar{u}_{j,i}) \qquad (9.88)$$

- The relaxation constitutive relation (9.68a) is

$$\bar{\sigma}_{ij} = s\bar{R}_{ijkl}\bar{\varepsilon}_{kl} - s\bar{\psi}_{ij}\bar{\theta}. \qquad (9.89)$$

- The heat conduction equation (9.83) is

$$(k_{ij}/\theta_0)\bar{\theta}_{,ij} = s^2\bar{m}\bar{\theta} + s^2\bar{\psi}_{ij}\bar{\varepsilon}_{ij}. \qquad (9.90)$$

The viscoelastic boundary value problem governed by the set of equations (9.85)–(9.90) can be solved in the same manner as in the case of coupled thermoelastic problems. Consequently, the complete solution of the viscoelastic boundary value problem under consideration is obtained by inverting the transformed solution. The procedure here is the same as in the case of treating isothermal linear viscoelastic boundary value problems discussed earlier in section 9.4.

In problems where the coupling term involving ε_{ij} in (9.83) and (9.84) can be neglected, mechanical response problems and thermal response problems may be separated. Thus, after obtaining the temperature distribution, either by solving the heat conduction equation or from experimental results, the mechanical response problem would then be governed by (9.61) and (9.64)–(9.68). Integral transform methods could thus provide a useful tool in solving such problems.

REFERENCES

Aboudi, J. (1979) The dynamic indentation and impact of a viscoelastic half-space by an axisymmetric rigid body. *Comput. Math. Appl. Mech. Eng.*, **20**, 135–50.

Atkinson, C. and Coleman, C. J. (1977) On some steady-state moving boundary problems in the linear theory of viscoelasticity. *J. Inst. Math. Appl.*, **20**, 85–106.

Barberan, J. and Herrera, I. (1966) Uniqueness theorems and speed of propagation of signals in viscoelastic materials. *Arch. Ration. Mech. Anal.*, **23**, 173–90.

Biot, M. A. (1958) Linear thermodynamics and the mechanics of solid,s in Proc. 3rd US Natl Congr. on Applied Mechanics, ASME, New York, pp. 1–18.

Boussinesq, M. J. (1885) In Todhunter, I. and Pearson, K. (1960) *A History of Theory of Elasticity and of the Strength of Materials*, Vol. II, Part 2, Dover Publications, New York, pp. 185–357.

Brull, M. A. (1953) A structural theory incorporating the effect of time-dependent elasticity, in Proc. 1st Midwestern Conf. on Solid Mechanics, pp. 141–7.

Calvit, H. H. (1967) Numerical solution of the problem of impact of a rigid sphere onto a linear viscoelastic half-space and comparison with experiment. *Int. J. Solids Struct.*, **3**, 951–66.

Chao, C. and Achenbach, J. D. (1964) Simple viscoelastic analogy for stress waves, in Proc. Stress Waves in Anelastic Solids Symp., Brown University, April 3–5, 1963, Springer, Berlin, pp. 222–38.

Christensen, R. M. (1971) *Theory of Viscoelasticity*, Academic Press, New York.

Edelstein, W. S. and Gurtin, M. E. (1964) Uniqueness theorems in the linear dynamic theory of anisotropic viscoelastic solids. *Arch. Ration. Mech. Anal.*, **17**, 47–60.

Fichera, G. (1972) Boundary value problems of elasticity with unilateral constraints, in *Encyclopedia of Physics*, Vol. VI a/2, Mechanics of Solids II (ed. C. Truesdell), Springer, Berlin, pp. 391–423.

Fung, Y. C. (1965) *Foundations of Solid Mechanics*, Prentice-Hall, Englewood Cliffs, NJ.

Gakhof, F. D. (1966) *Boundary Value Problems*, Pergamon, Oxford.

Gladwell, G. M. L. (1980) *Contact Problems in the Classical Theory of Elasticity*, Sijthoff and Noordhoff, Alphen aan den Riju.

Golden, J. M. and Graham, G. A. C. (1988) *Boundary Value Problems in Linear Viscoelasticity*, Springer, Berlin.

Graham, G. A. C. (1965) The contact problem in the linear theory of viscoelasticity. *Int. J. Eng. Sci.*, **3**, 27–46.

Graham, G. A. C. (1967) The contact problem in the linear theory of viscoelasticity when the time-dependent contact area has any number of maxima and minima. *Int. J. Eng. Sci.*, **5**, 495–514.

Graham, G. A. C. (1968) The correspondence principle of linear viscoelasticity for mixed boundary value problems involving time-dependent boundary regions. *Q. Appl. Math.*, **26**, 167–74.

Graham, G. A. C. (1969) The solution of mixed boundary value problems that involve time-dependent boundary regions, for viscoelastic materials with one relaxation function. *Acta Mech.*, **8**, 188–204.

Graham, G. A. C. and Williams, F. M. (1972) Boundary value problems for time-dependent regions in aging viscoelasticity. *Util. Math.*, **2**, 291–303.

Gurtin, M. E. and Herrera, I. (1964) A correspondence principle for viscoelastic wave propagation. *Q. Appl. Math.*, **22**, 360–4.

Gurtin, M. E. and Sternberg, E. (1962) On the linear theory of viscoelasticity. *Arch. Ration. Mech. Anal.*, **11**(4), 291–356.

Hunter, S. C. (1960) The Hertz problem for a rigid spherical indentor and a viscoelastic half-space. *J. Mech. Phys. Solids*, **8**, 219–34.

Hunter, S. C. (1961) The rolling contact of a rigid cylinder with a viscoelastic half-space. *J. Appl. Mech.*, **28**, 611–7.

Hunter, S. C. (1967) The solution of boundary value problems in linear viscoelasticity, in Proc. 4th 1965 Symp. on Naval Structural Mechanics (eds A. C. Eringen, H. Liebowitz, S. L. Koh and J. M. Crowley), Pergamon, Oxford, pp. 257–95.

Hunter, S. C. (1968) The motion of a rigid sphere embedded in an adhering elastic or viscoelastic medium. *Proc., Edinburgh Math. Soc., Ser. II*, **16** (Part I), 55–69.

Lee, E. H. (1955) Stress analysis in viscoelastic bodies. *Q. Appl. Math.*, **13**, 183–90.

Lee, E. H. (1960) Viscoelastic stress analysis, in 1st Symp. on Naval Structural Mechanics (eds J. N. Goodier and N. J. Hoff), Pergamon, New York, pp. 456–82.

Lee, E. H. (1966) Some recent developments in linear viscoelastic stress analysis, in Proc. 11th Int. Congr. of Applied Mechanics (ed. H. Gortler), Springer, Berlin, pp. 396–402.

Lee, E. H. and Radok, J. R. M. (1960) The contact problem for viscoelastic bodies. *J. Appl. Mech.*, **27**, 438–44.

Lee, E. H., Radok, J. R. M. and Woodward, W. B. (1959) Stress analysis for linear viscoelastic materials. *Trans. Soc. Rheol.*, **3**, 41–59.

Lockett, F. J. (1962) The reflection and refraction of waves at an interface between viscoelastic materials. *J. Mech. Phys. Solids*, **10**, 53–64.

Love, A. E. H. (1944) *A Treatise on the Mathematical Theory of Elasticity*, Cambridge University Press, Cambridge.

Lubliner, J. and Sackman, J. L. (1967) On uniqueness in general linear viscoelasticity. *Q. Appl. Math.*, **25**, 129–38.

Morland, L. W. (1962) A plane problem of rolling contact in linear viscoelasticity theory. *J. Appl. Mech.*, **29**, 345–58.

Morland, L. W. (1967) Exact solution for rolling contact between viscoelastic cylinders. *Q. J. Appl. Math.*, **20**, 73–106.

Morland, L. W. and Lee, E. H. (1960) Stress analysis for linear viscoelastic materials with temperature variation. *Trans. Soc. Rheol.*, **4**, 233–63.

Muki, R. and Sternberg, E. (1961) On transient thermal stresses in viscoelastic materials with temperature dependent properties. *J. Appl. Mech.*, **28**, 193–207.

Nachman, A. and Walton, J. R. (1978) The sliding of a rigid indentor over a power law viscoelastic layer. *J. Appl. Mech.*, **45**, 111–3.

Odeh, F. and Tadjbakhsh, I. (1965) Uniqueness in the linear theory of viscoelasticity. *Arch. Ration. Mech. Anal.*, **18**, 244–50.

Onat, E. T. and Breuer, S. (1963) On uniqueness in linear viscoelasticity, in *Progress in Applied Mechanics*, The Prager Anniversary Volume (ed. D. C. Drucker), Macmillan, New York, pp. 349–53.

Predeleanu, M. (1965) Stress analysis in bodies with time-dependent properties. *Bull. Math. Soc. Sci. Math., Roum.*, **9**, 115–27.

Read, W. T. (1950) Stress analysis for compressible viscoelastic materials. *J. Appl. Phys.*, **21**, 671–4.

Rogers, T. G. (1965) Viscoelastic stress analysis, in Proc. Princeton University, Conf. on Solid Mechanics, Princeton, NJ, pp. 49–74.

Rogers, T. G. and Lee, E. H. (1962) *Brown University Rep. NORD 18594/6*.

Sabin, G. C. W. (1987) The impact of a rigid axisymmetric indentor on a viscoelastic half-space. *Int. J. Eng. Sci.*, **25**, 235–51.

Schapery, R. A. (1955) A method of viscoelastic stress analysis using elastic solutions. *J. Franklin Inst.*, **279**(4), 268–89.

Schapery, R. A. (1962) Approximate method of transform inversion for viscoelastic stress analysis, in Proc. 4th US Natl Congr. on Applied Mechanics, ASME, New York, pp. 1075–85.

Schapery, R. A. (1964) Application of thermodynamics to thermomechanical, fracture, and birefringant phenomena in viscoelastic media. *J. Appl. Phys.*, **35**(5), 1451–65.

Schapery, R. A. (1967) Stress analysis of viscoelastic composite materials. *J. Compos. Mater.*, **1**, 228–66.

Schapery, R. A. (1974) Viscoelastic behaviour and analysis of composite materials, in *Mechanics of Composite Materials*, Vol. 2 (ed. G. P. Sendeckj), Academic Press, New York, pp. 85–168.

Schwarzl, F. and Staverman, A. J. (1952) Time–temperature dependence of linear viscoelastic behaviour. *J. Appl. Phys.*, **23**(8), 838–43.

Sips, R. (1951) General theory of deformation of viscoelastic substances. *J. Polym. Sci.*, **9**, 191–205.

Sternberg, E. (1964) On the analysis of thermal stresses in viscoelastic solids, in *High Temperature Structures and Materials*, Proc. 3rd Symp. on Naval Structural Mechanics (eds A. M. Freudenthal, B. A. Boles and H. Liebowitz), Pergamon, Oxford, pp. 348–82.

Sternberg, E. and Gurtin, M. E. (1963) Uniqueness in the theory of thermorheologically simple ablating viscoelastic solids, in *Progress in Applied Mechanics*, The Prager Anniversary Volume (ed. D. C. Drucker), Macmillan, New York, pp. 373–84.

Sternberg, E. and Gurtin, M. E. (1964) Further study of thermal stresses in viscoelastic materials with temperature dependent properties, in Proc. IUTAM Symp. on Second Order Effects in Elasticity, Plasticity and Fluid Mechanics, Haifa, pp. 51–76.

Ting, T. C. T. (1966) The contact stresses between a rigid indentor and a viscoelastic half-space. *J. Appl. Mech.*, **33**, 845–54.

Ting, T. C. T. (1968) Contact problems in the linear theory of viscoelasticity. *J. Appl. Mech.*, **35**, 248–54.

Volterra, V. (1909) Sulle equazioni integro-differenziali della teoria dell' elasticita. *Atti Reale Accad. Lincei*, **18**(1), 167; **18**(2), 295.

FURTHER READING

Alblas, J. B. and Kuipers, M. (1970) The contact problem of a rigid cylinder rolling on a thin viscoelastic layer. *Int. J. Eng. Sci.*, **8**, 363–80.

Alfrey, T. (1944) Nonhomogeneous stresses in viscoelastic media. *Q. Appl. Math.*, **2**, 113–9.

Atkinson, C. and Coleman, C. J. (1977) On some steady-state moving boundary problems in the linear theory of viscoelasticity. *J. Inst. Maths. Appl.*, **20**, 85–106.

Battiato, G., Ronca, G. and Varga, C. (1977) Moving loads on a viscoelastic double layer: prediction of recoverable and permanent deformations, in Proc. 4th Int. Conf. on Structural Design of Asphalt Pavements, The University of Michigan, Ann Arbor, MI, pp. 459–60.

Calvit, H. H. (1967) Experiments on rebound of steel spheres from blocks of polymers. *J. Mech. Phys. Solids*, **15**, 141–50.

Comninou, M. (1976) Contact between viscoelastic bodies. *J. Appl. Mech.*, **43**, 630–2.

Edelstein, W. S. (1969a) The cylinder problem in thermoviscoelasticity. *J. Res. Natl. Bur. Stand., Sect. B*, **73**, 31–40.

Edelstein, W. S. (1969b) Ablation and thermal effects in a viscoelastic cylinder. *Acta Mech.*, **8**, 174–82.

Gakhof, F. D. (1966) *Boundary Value Problems*, Pergamon, Oxford.

Gaul, L. (1992) Substructure behaviour of resilient support mounts for single and double stage mounting systems. *Comput. Struct.*, **44**(1/2), 273–8.

Gaul, L., Klein, P. and Kemple, S. (1991) Damping description involving fractional operators. *Mech. Syst. Signal Process.*, **5**(2), 81–2.

Gaul, L., Schanz, M. and Fiedler, C. (1992) Viscoelastic formulations of BEM in time and frequency domain. *Eng. Anal. Bound. Elem.*, **10**, 137–41.

Golden, J. M. and Graham, G. A. C. (1988) *Boundary Value Problems in Linear Viscoelasticity*, Springer, Berlin.

Graham, G. A. C. (1965) On the use of stress functions for solving problems in linear viscoelasticity theory that involve moving boundaries. *Proc. R. Soc. Edinburgh, Sect. A*, **67**, 1–8.

Graham, G. A. C. and Golden, J. M. (1988) The generalized partial correspondence principle in linear viscoelasticity. *Q. Appl. Math.*, **56**(3), 527–38.

Graham, G. A. C. and Sabin, G. C. W. (1973) The correspondence principle of linear viscoelasticity for problems that involve time-dependent regions. *Int. J. Eng. Sci.*, **11**, 123–40.

Graham, G. A. C. and Sabin, G. C. W. (1978) The opening and closing of a growing crack in a linear viscoelastic body that is subject to alternating tensile and compressive loads. *Int. J. Fract.*, **14**, 639–49.

Graham, G. A. C. and Sabin, G. C. W. (1981) Steady-state solutions for a cracked standard linear viscoelastic body. *Mech. Res. Commun.*, **8**, 361–8.

Harvey, R. B. (1975) On the deformation of a viscoelastic cylinder, rolling without slipping. *Q. J. Mech. Appl. Math.*, **28**, 1–24.

Hunter, S. C. (1960) The Hertz problem for a rigid spherical indentor and a viscoelastic half-space. *J. Mech. Phys. Solids*, **8**, 219–34.

Hunter, S. C. (1961) The rolling contact of a rigid cylinder with a viscoelastic half-space. *J. Appl. Mech.*, **28**, 611–7.

Hunter, S. C. (1967) The transient temperature distribution in a semi-infinite viscoelastic rod, subject to longitudinal oscillations. *Int. J. Eng. Sci.*, **5**, 119–43.

Kalker, J. J. (1975) Aspects of contact mechanics, in *The Mechanics of the Contact Between Deformable Media* (eds A. D. de Pater and J. J. Kalker), Delft University Press, pp. 1–25.

Kalker, J. J. (1977) A survey of the mechanics of contact between solid bodies. *Z. Angew. Math. Phys.*, **57**, 13–17.

Koeller, R. C. (1984) Application of fractional calculus to the theory of viscoelasticity. *J. Appl. Mech.*, **51**, 299–307.

Lee, E. H. (1966) Some recent developments in linear viscoelastic stress analysis, in Proc. 11th Congr. of Applied Mechanics (ed. H. Gortler), Springer, Berlin, pp. 396–402.

Lifshitz, J. M. and Kolsky, H. (1964) Some experiments on anelastic rebound. *J. Mech. Phys. Solids*, **12**, 35–43.

Lockett, F. J. (1961) Interpretation of mathematical solutions in viscoelasticity theory illustrated by a dynamic spherical cavity problem. *J. Mech. Phys. Solids*, **9**, 215–29.

Lockett, F. J. and Morland, L. W. (1967) Thermal stresses in a viscoelastic thin-walled tube with temperature-dependent properties. *Int. J. Eng. Sci.*, **5**, 879–98.

Margeston, J. (1971) Rolling contact of a smooth viscoelastic strip between rotating rigid cylinders. *Int. J. Mech. Sci.*, **13**, 207–15.

Margeston, J. (1972) Rolling contact of a rigid cylinder over a smooth elastic or viscoelastic layer. *Acta Mech.*, **13**, 1–9.

McCartney, L. N. (1978) Crack propagation in linear viscoelastic solids: some new results. *Int. J. Fract.*, **14**, 547–54.

Morland, L. W. (1963) Dynamic stress analysis for a viscoelastic half-plane subject to moving surface tractions. *Proc. London Math. Soc.*, **13**, 471–92.

Morland, L. W. (1968) Rolling contact between dissimilar viscoelastic cylinders. *Q. Appl. Math.*, **25**, 363–76.

Pao, Y. H. (1955) Extension of the Hertz theory of impact to the viscoelastic case. *J. Appl. Phys.*, **26**, 1083–8.

Rabotonov, Y. N. (1969) *Creep Problems in Structural Members*, North-Holland, Amsterdam.

Rogers, T. G. (1965) Viscoelastic stress analysis, in Proc. Princeton University Conf. on Solid Mechanics, Princeton, NJ, pp. 49–74.

Sabin, G. C. W. (1975) Some dynamic mixed boundary value problems in linear viscoelasticity. PhD Thesis, University of Windsor, Windsor, Ontario.

Schapery, R. A. (1978) A method for predicting crack growth in nonhomogeneous viscoelastic media. *Int. J. Fract.*, **14**, 293–309.

Schapery, R. A. (1979) On the analysis of crack initiation and growth in nonhomogeneous viscoelastic media, in *Fracture Mechanics*, Proc. Symp. in Applied Mathematics of the AMS and SIAM, Vol. XII (ed. R. Burridge), American Mathematical Society, Providence, RI, pp. 137–52.

Sokolnikoff, I. S. (1956) *Mathematical Theory of Elasticity*, 2nd edn, McGraw-Hill, New York.

Stackgold, I. (1967) *Boundary Value Problems of Mathematical Physics*, Vol. 1, Macmillan, New York.

Ting, T. C. T. (1969) A mixed boundary value problem in viscoelasticity with time-dependent boundary regions, in Proc. 11th Midwestern Mechanics Conf. (eds H. J. Weiss, D. F. Young, W. F. Riley and T. R. Rogge), Iowa University Press, pp. 591–8.

Ting, E. C. (1970) Stress analysis for a nonlinear viscoelastic cylinder with ablating inner surface. *J. Appl. Mech.*, **37**, 44–7.

Willis, J. R. (1967) Crack propagation in viscoelastic media. *J. Mech. Phys. Solids*, **15**, 229–40.

10

Extension of the phenomenological theory of viscoelasticity to include microscopic effects

10.1 INTRODUCTION

A large class of engineering materials have a distinct microstructure and, hence, their response behaviour may be formulated with the inclusion of the effects of such a microstructure. Although most engineering materials may be characterized within this category, the present chaper will consider only one group of such materials, namely randomly structured fibrous systems. The latter may be considered as a good representative of discrete viscoelastic systems. Whilst it might appear that structured viscoelastic materials are significantly different in nature, it is the author's point of view that a general deformation theory could be developed that would be applicable to a large category of such material systems.

It is the main purpose of this chapter to show the development of such a theory within the framework of 'probabilistic micromechanics' (Axelrad, 1970, 1971, 1978, 1984). In this theory, the discrete nature of individual microstructural elements forming the macroscopic material system as well as the interaction effects between these elements are taken, from the onset, into account. Because of the discrete nature of the microstructure, the analysis is based on the concepts of the mathematical theory of probability and statistical micromechanics. Thus, the significant stochastic parameters pertaining to the deformation process are expressed in the formulation of the theory in terms of their statistical distribution functions. Such distribution functions are, in part, experimentally accessible. Although significant research work is still required to be carried out for the development of the probabilistic micro- mechanical approach, it is, by no doubt, a promising approach towards the formulation of the response behaviour of material systems with the inclusion of the real microstructure. For further applications of the latter approach, the reader is referred to Axelrad (1978, 1984), Axelrad and Jaeger (1969a, b), Axelrad, Provan and Basu (1974) and Haddad (1986, 1990).

10.2 STOCHASTIC MICROMECHANICAL APPROACH TO THE RHEOLOGY OF RANDOMLY STRUCTURED MATERIAL SYSTEMS

In the present section, a stochastic microstructural approach to the viscoelastic (rheological) response of randomly structured material systems is developed. This approach recognizes that the microstructure of such a material system consists of randomly arranged microstructural elements which are mutually interacting through a bonding mechanism at particular regions where they cross or are neighbours. Thus, the analysis takes into account the rheological response of individual microelements as well as the effect of interelemental bonding. Because of the inherent randomness of the physical and geometrical characteristics of the microstructure, probabilistic concepts are used. Moreover, the microstructural elements forming the material system possess, in general, time-dependent response characteristics. As a consequence, it is appropriate to consider the significant quantities governing the deformation process as stochastic variables and the deformation process itself is seen in this approach as a stochastic process.

In order to describe the mechanical response of the material system from a microstructural point of view, it is necessary to consider the response of the individual structural elements which on a local scale could differ considerably from an average response if the phenomenological continuum approach was taken. Such local deviations in the response which are usually neglected if one ignores the microstructure are directly related to basic properties of the nonhomogeneous material system. Accordingly, the present analysis begins with a definition of the structural element of the particular material system under consideration and deals with the formulation of its rheological response in a probabilistic sense.

In order to extend the analysis to the practical case of a macroscopic material system, it is necessary to make use of 'intermediate quantities' arising from the consideration of the existence of a statistical ensemble of structural elements within an intermediate domain of the material specimen. Further, it is equally important to find a connection between the microscopic and the macroscopic response formulations. Thus, the analysis aims at the formulation of a set of 'governing response equations' for the structured material system that, in contrast to the classical continuum mechanics formulations, are based on the concepts of statistical theory and probabilistic micromechanics (Axelrad, 1978, 1984; Haddad, 1985). In this context, it has been found useful to employ an operational representation of the various relations. Hence, the notion of a 'material operator' (Axelrad, 1978, 1984) characteristic of the viscoelastic response of an intermediate domain of the material is introduced. This material operator provides the connection between the stress field and the occurring deformations within the intermediate domain under consideration. It contains in its argument those stochastic variables or functions of such variables distinctive of the microstructure within the intermediate domain. In a very reduced and simplified form, such an operator is expressed as

$$\Gamma(t) = \Gamma({}^t\Gamma, {}^i\Gamma, {}^\alpha K, p_1, p_2, \cdots, t) \qquad (10.1)$$

Table 10.1 A comparison between basic concepts of the probabilistic micromechanical approach and the corresponding concepts of classical continuum mechanics

	Classical continuum mechanics	*Probabilistic micromechanics*
Material system	Continuous	Discrete
Local description	Mathematical point	Structural element
Stress and deformation	Continuous	Discontinuous
Analytical approach	• Deterministic	• Stochastic
	• Constitutive theory	• Operational formalism of a structured material system

where $^f\Gamma$ and $^j\Gamma$ are random material operators expressing the response characteristics of elements of different classes of the microstructure (e.g. a fibre segment and a junction area between two overlapping fibres in a fibrous material system), ^xK is a function of one or more geometrical parameters, p_1 and p_2 are geometrical probabilities and t is the time parameter. Other variables that may be included in the argument of the material operator $\Gamma(t)$ could include, for instance, the temperature T and relative

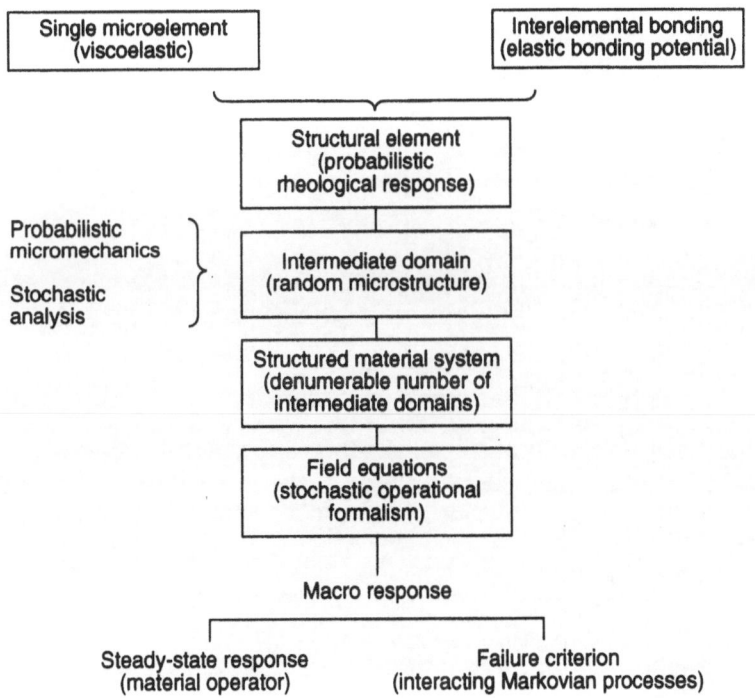

Fig. 10.1 Scope of the stochastic micromechanical approach to the rheological response of a randomly structured (viscoelastic) material system.

humidity ϕ. A comparison between basic concepts of the probabilistic micro-mechanical approach adopted here and the corresponding postulates of the conventional continuum mechanics approach (adopted in the previous chapters of the text) is shown in Table 10.1.

The scope of the stochastic micromechanical approach to the rheological response of a randomly structured material system (of mutually interacting viscoelastic microelements) is demonstrated in Fig. 10.1.

10.3 CASE STUDY: VISCOELASTIC RESPONSE BEHAVIOUR OF A CLASS OF RANDOMLY STRUCTURED FIBROUS SYSTEMS

10.3.1 Theoretical analysis

We consider here the application of the stochastic micromechanical approach, as presented in the previous section, to the rheological behaviour of a class of randomly oriented fibrous systems. In order to illustrate the type of material investigated here, a micrograph (× 170) of a (cellulosic) fibrous system is shown in Fig. 10.2. It represents a material system of bleached sulphite paper. The microstructure of the system is seen to be heterogeneous showing a random arrangement of single fibres

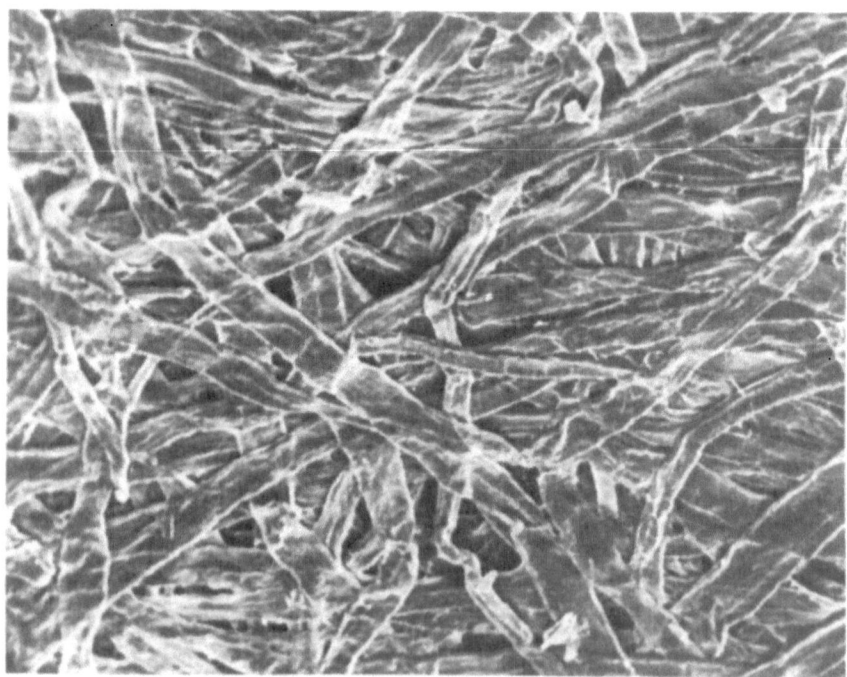

Fig. 10.2 Micrograph (× 170) of a typical, randomly structured fibrous system; beaten sulphite paper, bleached and dried (Haddad, 1975).

which are bonded together at certain junctions. Fibrous systems such as introduced above have been classified with respect to their response behaviour as viscoelastic materials (Brezinski, 1956; Schulz, 1961).

Traditionally, models that are based on continuum theories have been used for the prediction of the response behaviour of fibrous systems. These models, in general, refer to a homogeneous medium ignoring, thereby, the effect of the microstructure. Several attempts, however, have been made to modify the continuum mechanics approach by allowing for microscopic or 'local' quantities to enter into the analysis, but without removing the main restrictions imposed by continuum physics on such formulations. In this sense, 'modified continuum' models have been proposed by Nissan (1959), Onogi and Sasaguri (1961), Sternstein and Nissan (1962) and Van den Akker (1962), amongst others.

The necessity, however, to develop a new approach to the response of fibrous systems that would be based explicitly on microstructural considerations has been frequently discussed in the literature (e.g. Rance, 1948, 1962; Haddad, 1975). Rance (1948) advanced the argument that the nonrecoverable déformation of a paper sheet, for example, is essentially due to the breakage of interfibre bonding that occurs at an increasing rate leading to the final rupture of the macroscopic specimen. Page, Tydeman and Hunt (1962) verified experimentally that this phenomenon could occur under the effect of elastic tension. Corte (1966) and Nissan (1967) reported, in this regard, that hydrogen bonding is the most effective binding mechanism between two adjacent cellulosic fibres.

On the other hand, the failure of fibre segments during elastic tension of paper samples has been reported by Van den Akker *et al.* (1958). Thus, the opinion has been put forward that the mechanical response of a fibrous network depends on the response characteristics of interfibre bonding as well as those of individual fibres. Experimental investigations by McIntosh and Leopold (1962) supported the last statement. Van den Akker (1970) highlighted further the importance of the statistical approach to the problem of the prediction of the macroscopic response of fibrous systems with the inclusion of the microstructure. In this context, Corte and Kallmes (1962) presented a comprehensive study concerning the statistical geometry of a model of fibrous systems.

(a) Probablistic, micromechanical response

A structural element
A structural element (α) is defined as the smallest part of the medium that represents the mechanical and physical characteristics of the microstructure at the 'micro' level. For the particular material system under consideration (Fig. 10.2), a model of the structural element is introduced that includes the contribution of a single fibre segment between two neighbouring junctions, as well as one-half of each of the junction areas associated with the actual bonding between the overlapping fibres (Fig. 10.3).

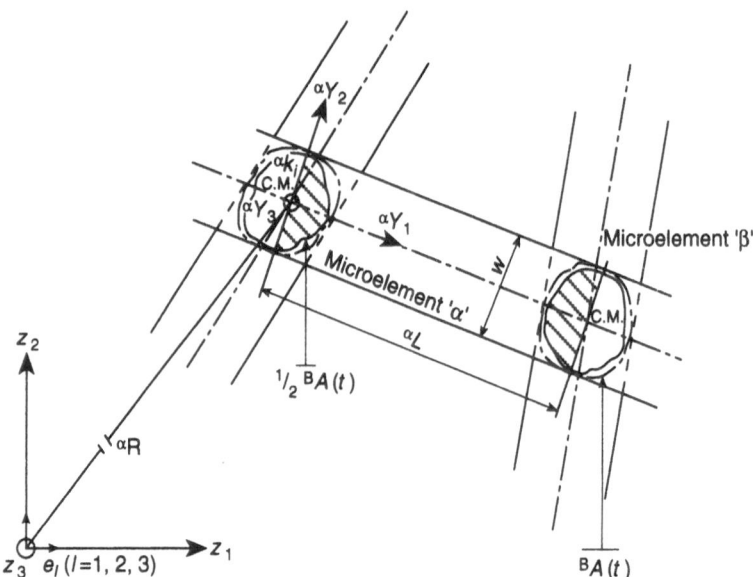

Fig. 10.3 Schematic representation of a structural element (α) of a randomly structured fibrous system.

Throughout the analysis, a superscript (α) to the left of the symbol will refer, in general, to a structural element. However, since a distinction is made between the fibre segment and the two half-junction areas, the quantities referring to such a segment will be denoted by a superscript 'f' while those referring to the bonding interaction within the junction area between two matching fibres are designated by a superscript 'B'.

The kinematics of deformation are considered from the point of view of two coordinate systems, i.e. a local frame of reference $^{\alpha}Y_i$ ($i = 1, 2, 3$) attached to the geometrical centre of the junction between two overlapping fibres and an external coordinate frame Z_i ($I = 1, 2, 3$); Fig. 10.3.

Viscoelastic response of a fibre segment
In view of the fact that we consider, here, the response of 'natural fibres', such as cellulose fibres, which, in most cases, exhibit rheological properties, the viscoelastic response of such fibres will be considered. While, for the simplification of the analysis, the continuum mechanics approach is maintained for the response characterization of the fibre segment, it is understood that the effect of fibre substructural parameters is not considered at this stage of presentation.

Information concerning the microstructure of single fibres of natural cellulose is given by, amongst others, Meredith (1956), Roelofsen (1959), Mark (1967) and Bikales and Segal (1971). Mark (1967) presented a corrolation between the fibril angle in the S_2 layer (the main layer in the single wood pulp fibre) and the tensile strength of the fibre. Schematics of such fibres are shown in Figs. 10.4 and 10.5 after Roelofsen (1959). The importance of synthetic fibres, in particular with respect to papermaking, has been discussed by Battista (1964). The response behaviour of synthetic single fibres has been discussed by Lai and Findley (1968), among others.

We shall consider here, the formulation of the creep behaviour of a single fibre with the inclusion of pertaining experimental data. This is with the understanding that the same approach can be adopted for the relaxation behaviour as we have presented earlier in Chapter 7.

Referring to the creep response equation (7.20), once the values of the unknown constants of this equation have been determined by the procedure mentioned in section 7.2, this equation can be used to represent a 'model equation' for the response behaviour of an actual viscoelastic fibre. For this purpose, an equivalent strain $^f\varepsilon(t)$, in the fibre segment, is assumed which corresponds to the actual strain in such segment that can be determined for the probabilistic theory only from experimental observations, e.g. via a distribution function. Under these circumstances, the model equation (7.20) for the viscoelastic response for the fibre segment can be written as

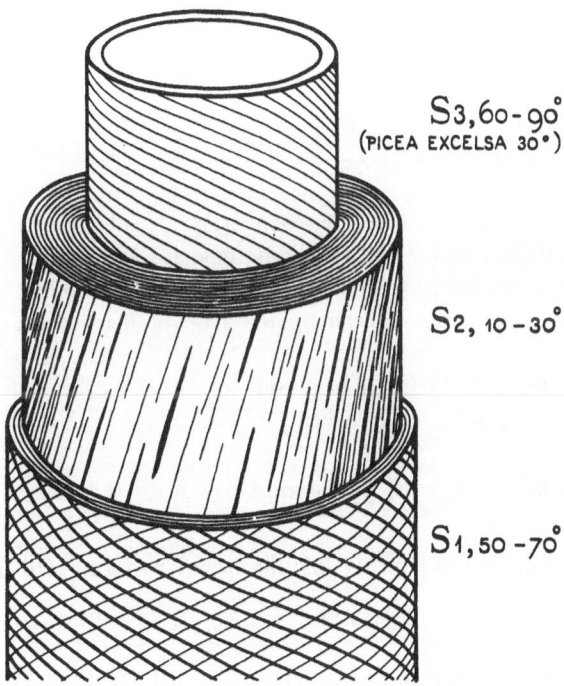

Fig. 10.4 Schema of the orientation of the cellulose microfibrils in the normal three-ply structure. (Source: Roelofsen, P. A. (1959) *The Plant Cell Wall*, © Gebrüder Borntraeger, Berlin-Nikolassee. Reprinted with permission.)

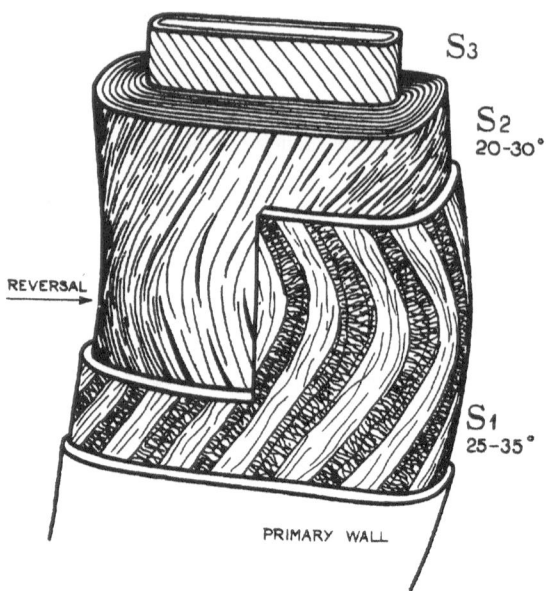

Fig. 10.5 Schema of the structure of the secondary wall of a natural cellulose fibre. (Source: Roelofsen, P. A. (1959) *The Plant Cell Wall,* © Gebrüder Borntraeger, Berlin-Nikolessee. Reprinted with permission.)

follows, where the index for the number of experiment i has been omitted:

$$^f\varepsilon(t) = {}^fE^{-1f}\breve{\xi} + h({}^f\breve{\xi}, b_1, b_2, \cdots) \sum_{I=1}^{N} D_I[\exp(A_I t) - 1] \qquad (I = 1, 2, \cdots, N) \quad (10.2)$$

where $^f\varepsilon(t)$ is the strain response in the fibre segment and $^f\breve{\xi}$ is the input level of the stress. In (10.2), E is the uniaxial elastic modulus and $b_1, b_2, \cdots, D_I$ and A_I ($I = 1, 2, \cdots, N$) are material parameters that can be determined by the inclusion of the experimental creep data concerning the class of fibres under consideration (Chapter 7).

The second term on the right-hand side of (10.2) maybe considered to be represented by an operator $^fL({}^f\breve{\xi}, t)$ where

$$^fL({}^f\breve{\xi}, t)^f\breve{\xi} = h({}^f\breve{\xi}, b_1, b_2, \cdots) \sum_{I=1}^{N} D_I[\exp(A_I t) - 1] \qquad (I = 1, 2, \cdots, N) \quad (10.3)$$

so that, by including the first term of (10.2), the operational form of (10.2) becomes

$$^f\varepsilon(t) = [{}^fE^{-1} + {}^fL({}^f\breve{\xi}, t)]. \qquad (10.4)$$

It is noted, with reference to (10.3) and (10.4), that the form of the operator $^fL({}^f\breve{\xi}, t)$ depends on the form of the function $h(\cdot)$ which represents, from the discussion in Chapter 7, the nonlinear behaviour of the fibre material. In the present case of natural cellulose fibres, for instance, the function $h(\cdot)$ may be assumed to have the following simple form:

$$h(\cdot) = \exp(b^f\overset{\vee}{\xi}) - 1. \tag{10.5}$$

Expanding asymptotically the exponential term in the form of $h(\cdot)$ given by (10.5) and retaining only the first two terms of the resulting expansion, expression (10.3) can be written as

$$^fL(t)^f\overset{\vee}{\xi} = b^f\overset{\vee}{\xi} \sum_{I=1}^{N} D_I[\exp(A_I t) - 1] \qquad (I = 1, 2, \cdots, N) \tag{10.6}$$

whereby the operator $^fL(\cdot)$ is now a function of time only.

While the above formulation has been illustrated by a procedure applied to the specific case of cellulosic fibres, it is valid for any other class of viscoelastic fibres provided that the proper material characteristics and the form of experimental curves representing the viscoelastic behaviour of the particular class of material are observed. Hence, in view of the foregoing, a constitutive relation in operational form can be written, in general, as

$$^f\varepsilon(t) = [^fE^{-1} + {}^fL(t)]^f\xi(t). \tag{10.7}$$

Further, if one considers, from a system theoretical point of view, that the microstress in the fibre segment is the stimulus and the occurring microstrain is the corresponding response, then equation (10.7) can be written in a more compact operational form as

$$^f\varepsilon(t) = {}^f\Gamma(t)^f\xi(t) \tag{10.8}$$

where $^f\Gamma(t)$ is a transform operator. This transform operator will be included subsequently as part of another material operator $^M\Gamma(t)$ valid for an intermediate domain of the macroscopic material specimen.

The simplification and systematization achieved with the operator formulation is rather important, since, in many cases, it can be used to achieve a response formalism with the inclusion of the microstructure of a material, such as fibrous network, which otherwise cannot be achieved directly (e.g. Axelrad and Basu, 1974).

In view of equations (10.2), (10.5) and (10.6), the expression for the transform operator $^f\Gamma(t)$ appearing in (10.8) is written as

$$^f\Gamma(t) = \left\{ {}^fE^{-1} + b \sum_{I=1}^{N} D_I[\exp(A_I t) - 1] \right\} \qquad (I = 1, 2, \cdots, N) \tag{10.9}$$

where the material parameters appearing in the above expression can be determined (as discussed in Chapter 7) with the inclusion of the viscoelastic experimental data concerning the particular fibre material under consideration.

Interfibre bonding

In any mathematical approach to the response behaviour of fibrous systems that would be based explicitly on microstructural considerations, it is of utmost importance to include in the formulation the effect due to bonding between two overlapping fibres. It is readily noticed from Fig. 10.6 showing a micrograph (× 1120) of unbeaten, low yield sulphite network that a junction area between two adjacent fibres shows a distinct discontinuous of fibrils in this area. The fibrils are bonded either partially or wholly as a result of interactions between them at the molecular level.

Corte (1966) and Nissan (1967) dealt comprehensively with the nature of bonding in a cellulosic structure. These authors, among others, claim that hydrogen bonding is the most effective binding mechanism between the fibrils on the surfaces of two adjacent fibres. A hydrogen bond is usually formed by the hydroxyl groups (OH) of the cellulose molecules (e.g. Corte, 1966; Pauling, 1960; Paradowski, 1991). Here, a hydrogen-containing OH group on the surface of a fibre may bond to a corresponding receptor, an oxygen atom on the surface of the matching fibre, forming an intermolecular hydrogen bond, within a junction area between the two fibres. Hemicellulose and portions of lignin molecules, usually encountered in pulp fibres, are also capable of forming this type of bond (Nissan, 1967). Available X-ray data on cellulosic systems indicate that cellulose in natural fibres crystallizes partially or wholly within the fibrils in a manner in which unit cells are formed and which repeat themselves as a common chain. A model of the repeating unit of cellulose (Jacobson, Wunderlich and Lipscomb, 1961) is shown in Fig. 10.7. This, as shown in the figure, is composed of two so-called β-D glucose residues that are linked by an oxygen bridge to adjacent residues and are rotated with respect to one another about a screw axis to form continuous chain segments.

A model (Meyer and Misch, 1937), of the unit cell of cellulose is shown in Fig. 10.8. It has a monoclinic character in which the length of the cellulose repeating unit

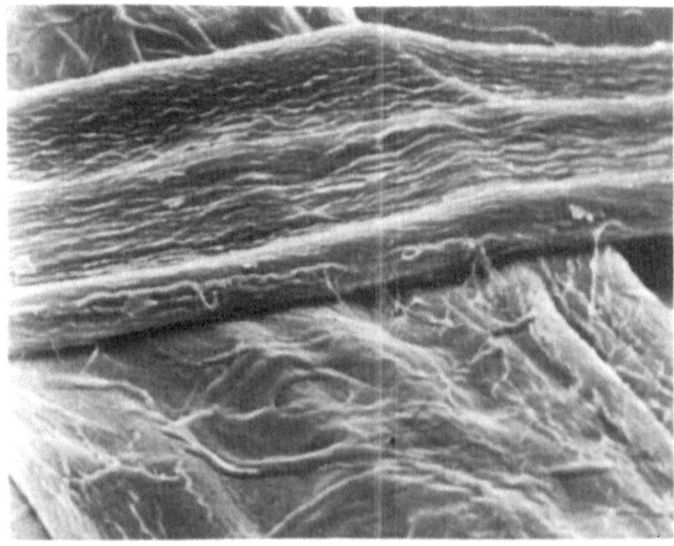

Fig. 10.6 Micrograph (× 1120) of unbeaten low-yield sulphite network (Haddad, 1975).

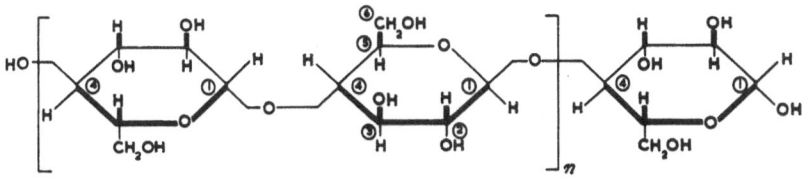

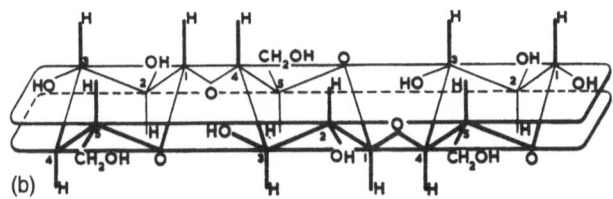

(b)

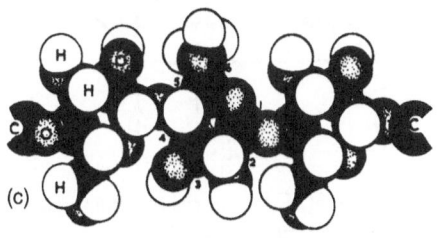

(c)

Fig. 10.7 (a) Structural formula of cellulose molecule; (b) three-dimensional structure of cellulose molecule; (c) atom model of cellulose molecule. (Source: Jacobson, R. A., Wunderlich, J. A. and Lipscomb, W. N. (1961) The crystal and molecular structure of cellulose. *Acta Crystallogr.*, **14**, 598–607 (International Union of Crystallography). Reprinted with permission.)

is perpendicular to two other edges which include an angle β. The three edges b, a and c of the unit cell define the crystallographic directions (010), (100) and (001), respectively. Gardener and Blackwell (1974) have shown that the rigid-body least-squares refinement technique (Arnott and Wonacott, 1966) could be used to distinguish between the various possible structures of the cellulosic unit cell, e.g. relative chain polarities and elucidation of the intra- and intermolecular hydrogen bonding. Structural models have been also proposed to describe the way in which the fibrils are built up of molecular chains (e.g. Nissan, 1967; Tonnesen and Ellefsen, 1971; Manley and Inoue, 1965).

In view of the above-indicated Meyer and Misch model of the unit cell, one may define a unit cell area $^{\alpha\beta}\mu$ in the junction as

$$^{\alpha\beta}\mu = b \times a = 10.30 \text{ Å} \times 8.35 \text{ Å}. \tag{10.10}$$

In order to visualize the notion of matching between two fibres (α, β), it is proposed in the present analysis to follow the geometrical theory of coincidence site lattices

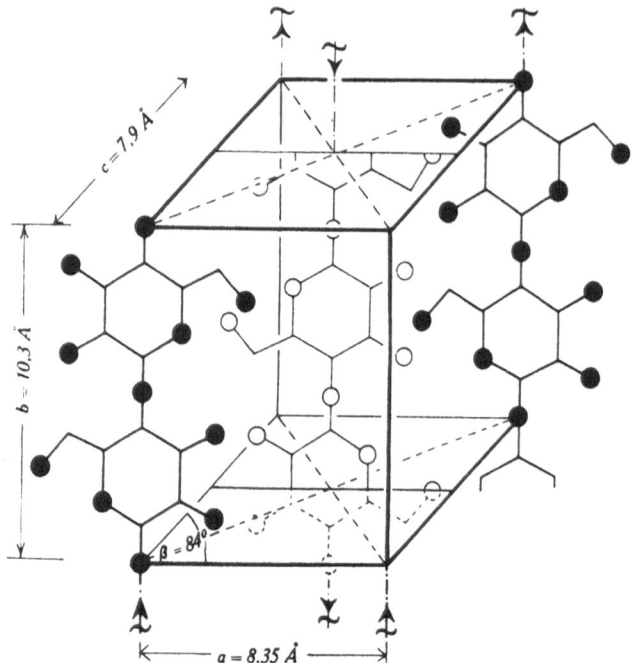

Fig. 10.8 Schematic representation of the unit cell of natural cellulose. (Source: Meyer, K. H. and Misch, L. (1937) Positions des atomes dans le nouveau modèle spatial de la cellulose. *Helv. Chem. Acta*, **20**, 232–44. Reprinted with permission.)

(Bollmann, 1970) in such a manner that the two surfaces of a junction, between two matching fibres, are modelled in terms of two interpenetrating lattices with areas $^{\alpha}A$ and $^{\beta}A$. A three-dimensional picture of the situation, visualized to hold in the present case, is shown in Fig. 10.9. In this figure, each lattice is considered to be formed by cellulosic unit cells in such a manner that the length of the repeating unit it taken in the direction of the appropriate fibre axis. The lattice $^{\alpha\beta}A$ on the surface of one fibre may be assessed by a polarized light technique (Page, Tydeman and Hunt, 1962). The number of unit cells M within $^{\alpha\beta}A$ may be taken as $M = {}^{\alpha\beta}A/{}^{\alpha\beta}\mu$ where $^{\alpha\beta}\mu$ is the unit cell area as defined by equation (10.10).

In Fig. 10.9, a free OH group on the surface of a fibre (α) may bond with a corresponding receptor of the oxygen atoms contained on the surface of the overlapping fibre (β). The lattice area, $^{\alpha\beta}A$, is said to be totally bonded if all its available hydroxyl groups have formed proper interfibre bonds. With reference to Fig. 10.8, for the case of natural cellulose, the number of OH groups on one side of the junction within the unit area $^{\alpha\beta}\mu$ is 6.

It is apparent, however, that not all the hydroxyl groups on the surface of one fibre (α) are free to form a bond with an adjacent fibre (β). This may be attributed to different reasons reviewed previously by Haddad (1975, 1980).

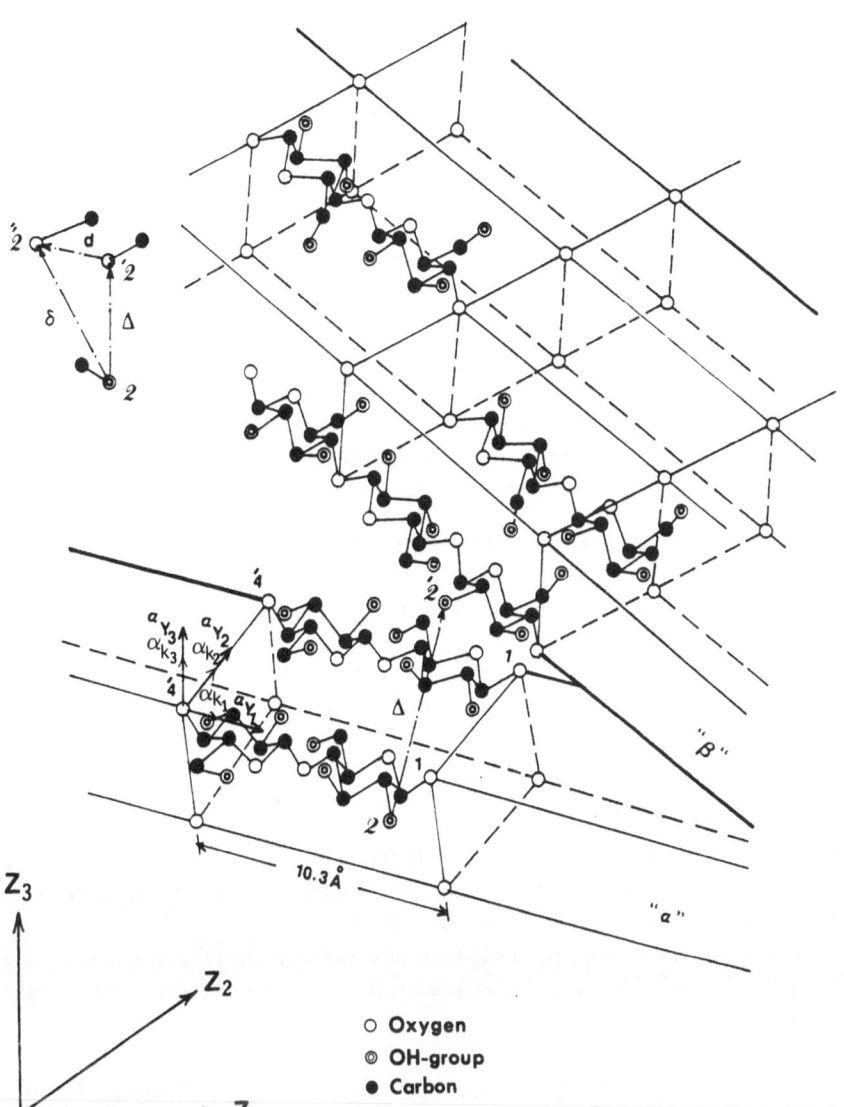

Fig. 10.9 Fibre–fibre interface in a natural cellulosic structure (three-dimensional) (Haddad, 1975).

As a consequence, a restriction on interfibre bonding must be taken into account so that one can assume that only an actual bonded region within the junction area will prevail. This region will, henceforth, be referred to as an 'actual bonded area' ^{B}A within the 'junction area' $^{\alpha\beta}A$. This actual bonded area can be considered as a percentage of $^{\alpha\beta}A$ in such a way that $^{B}A = \zeta^{\alpha\beta}A$, where ζ is a bonding ratio, $0 \leq \zeta \leq 1$, accessible only from proper experimental observations. Table 10.2 shows some values of the bonding ratio as proposed by different authors.

Table 10.2 The actual bonded area $^B A$ as compared with the junction area $^{\alpha\beta} A$ between two overlapping fibres. Natural cellulose fibrous systems

Reference	Bonding ratio $(\zeta = {}^B A/{}^{\alpha\beta} A)$	Matching fibres
Page and Tydeman (1960)	1.0	Spruce–sulphite pulp finish
Jayme and Hunger (1962)	0.1–0.2	Wood tracheids
Nissan and Sternstein (1964)	0.001–0.008	(Theoretical reasoning)

With reference to Fig. 10.9, the distance vector Δ between, for instance, two matching points 2 and 2′ is considered to be the basic kinematic parameter of a hydrogen bond in a junction area between two overlapping natural cellulosic fibres. The counterpart of this vector in the deformed state is denoted by δ and the microdeformation in an individual bond can, thus, be read as

$$\mathbf{d}(t) = \delta(t) - \Delta. \tag{10.11}$$

From classical considerations, one of the usual forms of a binding potential in the one-dimensional case is represented by a 'Morse function' (Morse, 1929). The analytical form of this function, which will represent the three-dimensional case in accordance with the atomistic arrangement of the bonding indicated in Fig. 10.9, can be written as

$$\psi = \psi_0[\exp(-2\kappa|\mathbf{d}(t)|) - 2\exp(-\kappa|\mathbf{d}(t)|)] \tag{10.12}$$

where ψ_0 is the equilibrium value of the Morse potential at the value of $|\mathbf{d}(t)| = 0$ and κ is the Morse constant. The values of ψ_0 and κ are obtainable from spectroscopic data (Corte and Schaschek, 1955; Sokolov, 1959).

Considering the binding potential form of expression (10.12) and differentiating this potential with respect to the displacement, a discrete interaction force in the bond can be obtained.

$$\mathbf{f}(t) = -\frac{\partial\psi}{\partial|\mathbf{d}(t)|}\,\mathbf{k} = 2\kappa\psi_0[\exp(-2\kappa|\mathbf{d}(t)|) - \exp(-\kappa|\mathbf{d}(t)|)]\mathbf{k} \tag{10.13}$$

where $\mathbf{k}$ is a unit base vector associated with the local coordinate frame attached to the centre of mass of the junction (Fig. 10.9; Fig. 10.3 should also be referred to).

Postulating that a bond stress in the continuum sense exists, one may express this stress as

$$^B\xi(t) = \frac{{}^{\alpha\beta}\mathbf{n}}{{}^B A(t)}\sum_{j=1}^{{}^B N}\mathbf{f}_j(t) \tag{10.14}$$

where $^{\alpha\beta}\mathbf{n}$ is the unit normal to the junction area and $j = 1, 2, \cdots, {}^B N$ is the number of interfibre hydrogen bonds in the junction. Combining (10.13) and (10.14), the stress–deformation relation for the interfibre bonding may be written as

$$^B\zeta(t) = \frac{2\kappa\psi_0}{^BA(t)} {}^{\alpha\beta}\mathbf{n} \sum_{j=1}^{^BN} [\exp(-2\kappa|\mathbf{d}_j(t)|) - \exp(-\kappa|\mathbf{d}_j(t)|)]\mathbf{k}. \qquad (10.15)$$

In a manner similar to the operational formulation of the response of a single fibre, presented earlier in this section, one may express the bonding response equation (10.15) in an operational form. For this reason, a transform operator $^B\Gamma^{-1}$ for the bonding interaction is introduced such that

$$^B\zeta(t) = {}^B\Gamma^{-1} \; {}^B\varepsilon(t) \qquad (10.16a)$$

where $^B\varepsilon(t)$ is the configurational bonding microstrain in the junction area such that

$$^B\varepsilon(t) = {}^\alpha\mathbf{V} \sum_{j=1}^{^BN} \mathbf{d}_j(t)$$

where $^\alpha\mathbf{V}$ is the gradient operator on the bonding displacement expressed, as illustrated in Fig. 10.3, by $^\alpha\mathbf{V} = \partial/\partial^\alpha Y$.

In order, however, to derive from (10.15) the operational form (10.16a), one may expand asymptotically the exponential terms in (10.15) and retain only the first two terms of each resulting series. As a consequence, one may write for the operator $^B\Gamma$ (equation (10.16a)), the following expression:

$$^B\Gamma^{-1} = \frac{-2\kappa^2\psi_0{}^{\alpha\beta}\mathbf{n}kk^{-1}}{^BA(t)} {}^\alpha\mathbf{V}^{-1}. \qquad (10.16b)$$

By the inversion of (10.16), the operational response equation for the bonding interaction, corresponding so that for the fibre segment (equation (10.8)), can be written as

$$^B\varepsilon(t) = {}^B\Gamma \; {}^B\zeta(t) \qquad (10.17)$$

The adoption of an atomistic approach to the interfibre bonding, as dealt with in the foregoing, is a key point in the present model. This is supported, as indicated earlier, by the fact that the interfibre hydrogen bonding determines primarily the strength of the junction between two matching cellulose fibres.

Probabilistic response of a structural element α
With reference to Fig. 10.10, the following geometrical probabilities are introduced with respect to a scanning line S–S that intersects the surface of a macroscopic specimen:

- p is the probability that the scanning line S–S intersects, at a certain point of the specimen, an individual fibre segment;

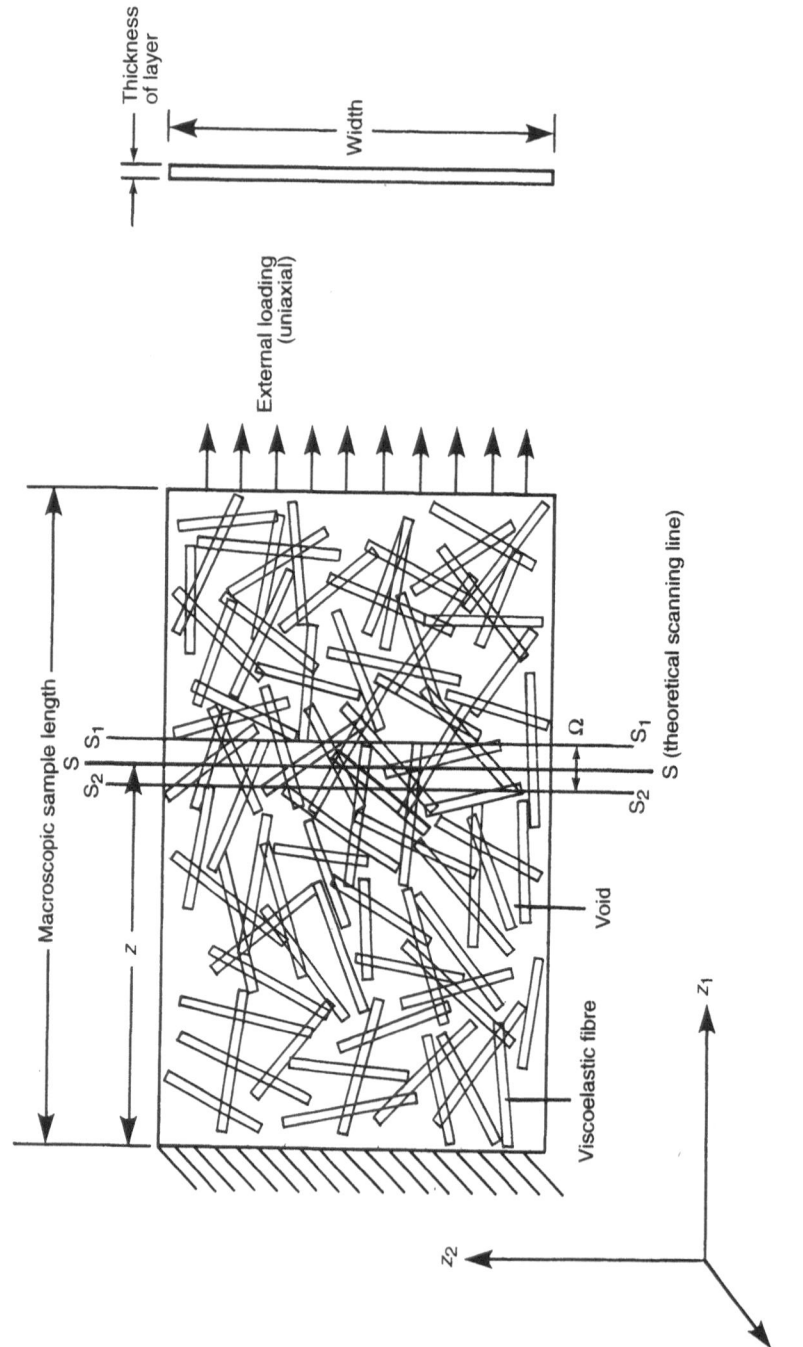

Fig. 10.10 A model of a two-dimensional fibrous network with randomly oriented microstructure.

- q is the probability that S–S intersects, at the same point, a junction between two fibres.

Thus, at the considered point, one may express the microstrain, in a structural element α, in a probabilistic manner as

$$^\alpha\varepsilon(t) = p\,{}^f\varepsilon(t) + q\,{}^B\varepsilon(t) \tag{10.18}$$

where $0 \leq p, q \leq 1$ and $p + q = 1$. In this question, the microstrain $^B\varepsilon(t)$ is considered to be totally due to the bonding displacement. That is, the fibre displacement at the point in question is taken to zero. The probability q appearing in (10.18) can be determined in terms of the physical and geometrical characteristics of the fibrous microstructure (Haddad, 1987).

Substituting for the microstrains $^f\varepsilon(t)$ and $^B\varepsilon(t)$ in (10.18) in terms of the corresponding microstresses from (10.8) and (10.17), respectively, it follows that

$$^\alpha\varepsilon(t) = (1 - q)\ {}^f\Gamma(t)\ {}^f\xi(t) + q\ {}^B\Gamma\ {}^B\xi(t). \tag{10.19}$$

However, in order to formulate an operational relation that will serve eventually in establishing a response relation for the structural element, it is necessary to employ a relationship between the fibre microstress and the microstress in the junction. This relation may be expressed as (Haddad, 1984),

$$^B\xi(t) = {}^\alpha K(t)\ {}^f\xi(t) \tag{10.20a}$$

where

$$^\alpha K(t) = \frac{{}^f a}{{}^B A(t)}\ [1 - {}^f\lambda(t)] \tag{10.20b}$$

where $^f a$ is the fibre cross-sectional area and $^f\lambda(t)$ is a time-dependent parameter associated with the equilibrium of the junction between two overlapping fibres α and β. In Haddad (1984), the parameter $^f\lambda(t)$ is determined in terms of the orientation of the fibrous elements forming a particular junction.

Now, substituting for $^B\xi(t)$ from (10.20a) into (10.19), the latter equation can be written as

$$^\alpha\xi(t) = [(1 - q)\ {}^f\Gamma(t) + q\ {}^B\Gamma\ {}^\alpha K(t)]\ {}^f\xi(t). \tag{10.21}$$

We introduce the transform operator $^{\alpha f}\Gamma(t)$, associated with the mechanical response of the structural element, such that it corresponds to the expression between square brackets in (10.21), i.e.

$$^{\alpha f}\Gamma(t) = (1 - q)\ {}^f\Gamma(t) + q\ {}^B\Gamma^\alpha K(t). \tag{10.22}$$

Thus, the response equation (10.21) can be written in the following operational form:

$$^\alpha\varepsilon(t) = {}^{\alpha f}\Gamma(t)\ {}^f\xi(t). \tag{10.23}$$

Further, it is possible, by the inversion of (10.23), to determine the microstress ${}^f\xi(t)$ from the microstrain ${}^\alpha\varepsilon(t)$ as

$$ {}^f\xi(t) = {}^{\alpha f}\Gamma^{-1}(t)\ {}^\alpha\varepsilon(t). \tag{10.24}$$

In a similar manner, given (10.20), one can express (10.23) as a function of the bonding microstress ${}^B\xi(t)$ as

$$ {}^\alpha\varepsilon(t) = {}^{\alpha B}\Gamma(t)\ {}^B\xi(t) \tag{10.25}$$

where

$$ {}^{\alpha B}\Gamma(t) = {}^{\alpha f}\Gamma(t)\ {}^\alpha K^{-1}(t). \tag{10.26}$$

(b) Transition to the macroscopic response behaviour

Since the fibrous network that occupies a given physical domain is regarded in the present analysis as a discrete medium, a transition from the discrete description to the macroscopic one must be attempted. In this context, the concept of the intermediate domain, or mesodomain (Axelrad, 1978), is employed. The latter is specified by the requirements mentioned by Axelrad (1978, 1984). It is the smallest region of the medium on the boundary of which the macroscopic observables are still valued, but, on the other hand, is large enough to contain a statistical number of structural elements. This permits statistical principles to be introduced in the analysis. It is further postulated that within the macroregion of the system, the mesodomains are denumerable and nonintersecting such that

$$ \bigcup_{M=1}^{Q} {}^M V = V $$

and

$$ {}^{M_1}V \cap {}^{M_2}V = \varnothing \qquad (M_1 \neq M_2) $$

where Q designates the total number of mesodomains within a macrovolume V. In the case of a two-dimensional fibrous system under loading in the Z_1 direction, for instance (Fig. 10.10), a mesodomain M ($M = 1, 2, \cdots, Q$) may be specified by the region bounded by the two theoretical scanning lines S_1–S_1 and S_2–S_2 which are perpendicular to the direction of loading. The width of this domain is determined, following the above, in relation to the actual dimensions of the structural element α such that, within M, $\alpha = 1, 2, \cdots, {}^M N$, where ${}^M N$ is very large. In the case of a cellulosic network, such as shown in Fig. 10.2, the width of the mesodomain may be taken of the order of 0.5–2 mm.

Following the concepts of the micromechanical theory of structured media (Axelrad, 1978, 1984), all microscopic field quantities within the intermediate domain are considered to be stochastic functions of primitive random variables. Thus, the components of the microstress, for instance, are seen as stochastic functions ${}^\alpha\xi(r, t)$

that can be regarded as a family of random variables $^{\alpha}\xi_t(\mathbf{r})$ within the intermediate domain depending on the time parameter t, or a family of curves $^{f}\xi_r(t)$ depending on the structural element position vector $^{\alpha}\mathbf{r}$.

Letting $^{M}P\{\cdot\}$ denote the probability distribution of a random variable within an intermediate domain M, then, in view of the structural element response equation (10.23), the probability distribution of the microstrain within the intermediate domain may be expressed as

$$^{M}P\{^{\alpha}\varepsilon(t)\} = {}^{M}P\{^{\alpha f}\Gamma(t)\}\ {}^{M}P\{^{f}\xi(t)\}. \tag{10.27}$$

At the same time, with reference to (10.24), one can express the probabilistic distribution of the fibre microstress as

$$^{M}P\{^{f}\xi(t)\} = {}^{M}P\{^{\alpha f}\Gamma^{-1}(t)\}\ {}^{M}P\{^{\alpha}\varepsilon(t)\}. \tag{10.28}$$

Further, following (10.25), the distribution of the bonding microstress becomes

$$^{M}P\{^{B}\xi(t)\} = {}^{M}P\{^{\alpha B}\Gamma^{-1}(t)\}\ {}^{M}P\{^{\alpha}\xi(t)\}. \tag{10.29}$$

In addition, the distribution of the bonding microstress is associated, in view of (10.20), with the distribution of the fibre stress via the relation

$$^{M}P\{^{B}\xi(t)\} = {}^{M}P\{^{\alpha}K(t)\}\ {}^{M}P\{^{f}\xi(t)\}. \tag{10.30}$$

Having established the time evolution of the internal deformation process via equation (10.21), a local failure criterion $^{\alpha}S(t)$ of the microstructure, within the intermediate domain, may be conjectured (Haddad, 1986) by setting

$$^{\alpha}S(t) = 1 - \int_{^{\alpha}\varepsilon_{maximum}}^{\infty} d^{M}P\{^{\alpha}\varepsilon(t)\}. \tag{10.31}$$

In this connection, $^{\alpha}S(t)$ may be interpreted, in a probabilistic sense, to be associated with the failure of fibrous elements or with the failure of interfibre bonding within the intermediate domain through the relation

$$^{\alpha}S(t) = (1 - q)\ ^{f}S(t) + q^{B}S(t) \tag{10.32}$$

where the probability q is used.

In (10.32), $^{f}S(t)$ and $^{B}S(t)$ may be expressed, respectively, in terms of the statistical distributions $^{M}P\{^{f}\xi(t)\}$ and $^{M}P\{^{B}\xi(t)\}$ in an analogous manner to (10.31) as

$$^{f}S(t) = 1 - \int_{^{f}\varepsilon_{maximum}}^{\infty} d^{M}P\{^{f}\varepsilon(t)\} \tag{10.33}$$

and

$$^{B}S(t) = 1 - \int_{^{B}\varepsilon_{maximum}}^{\infty} d^{M}P\{^{B}\varepsilon(t)\}. \tag{10.34}$$

This is with the understanding that the two deformation processes ($^{f}\varepsilon(t)$, $^{B}\varepsilon(t)$; $t_1 \leq t \leq t_2$) in the intermediate domain are interrelated via equations (10.26) and (10.30). In the context of the above formulations, it may be mentioned that the distribution functions of the significant parameters for the class of material considered can be determined by means of X-ray diffraction techniques, holographic interferometry and electron microscopy (e.g. Axelrad and Kalousek, 1971; Kalousek, 1973).

(c) Extension of the model to include breakage of interfibre bonding

In the present model, a junction area between two adjoining fibres is seen to consist of two regions: a cohesive zone, in which the adjoining fibres act as completely bonded and debonded (free) zone in which interfibre bonding has ceased to exist. The existence of such a free zone could have been initiated by a debonding process due to, for instance, the increase of local stress. From this point of view (Axelrad, 1984), the free zone may initiate an interfibre debonding process towards the cohesive zone. Further, one may assume that the interfibre debonding process occurs in a rather cooperative manner, i.e. bonds can dissociate and reform within the same mechanical state. Thus, it may be visualized that the breakage of interfibre bonds will occur in such a manner that energy is released activating bond formation within the same or neighbouring junction areas. Hence, we consider a process such that the number of interfibre bonds, within the fibrous network, can experience positive as well as negative jumps. Thus, in general, a time-dependent nonhomogeneous birth-and-death model of a process (e.g. Bharucha-Reid, 1960) is seen to be applicable.

If at time t, the material specimen is in the state Σ ($\Sigma = 1, 2, \cdots$) corresponding to a number of existing interfibre bonds $n(t):n_{\Sigma}$, one considers that both the intensities of positive and negative transitions to be time dependent. The latter are designated, respectively, in the following analysis by $\lambda(t)$ and $\mu(t)$. Accordingly,

1. the probability of transition from the state Σ to $\Sigma + 1$ in the interval $(t, t + \Delta t)$ is $\lambda(t) \Delta t + O(\Delta t)$,
2. the probability of transition from the state Σ to $\Sigma - 1$ in the interval $(t, t + \Delta t)$ is $\mu(t) \Delta t + O(\Delta t)$,
3. the probability of a transition to a state other than a neighbouring state is $O(\Delta t)$,
4. the probability of no change is $1 - [\lambda(t) + \mu(t)] + O(\Delta t)$ and
5. the state $\Sigma = 0$ is an absorbing state corresponding to the breakage of all interfibre bonds within the material specimen.

The above assumptions lead to the relation

$$P_{\Sigma}(t + \Delta t) = \lambda(t)P_{\Sigma-1}(t) \, \Delta t + \left\{1 - [\lambda(t) + \mu(t)] \, \Delta t\right\}P_{\Sigma}(t)$$
$$+ \mu(t)P_{\Sigma+1}(t)\Delta t + O(\Delta t) \tag{10.35}$$

where P_{Σ} is the probability that the material system is in the state Σ as defined above. Equation (10.35) leads in the limit to the following differential equation:

$$\frac{dP_{\Sigma}(t)}{dt} = \lambda(t)P_{\Sigma-1}(t) - [\lambda(t) + \mu(t)]P_{\Sigma}(t) + \mu(t)P_{\Sigma+1}(t) \tag{10.36a}$$

which holds for $\Sigma = 1, 2, \cdots$. For $\Sigma = 0$, however, one has

$$\frac{dP_0(t)}{dt} = \mu(t)P_1(t). \tag{10.36b}$$

The solution of equations (10.36) can be obtained with the aid of generating functions. Hence,

$$P_\Sigma(t) = [1 - \zeta(t)][1 - Y(t)][Y(t)]^{\Sigma-1}, \quad \Sigma = 1, 2, \cdots, \tag{10.37a}$$

and

$$P_0(t) = \zeta(t) \tag{10.37b}$$

where

$$\zeta(t) = 1 - \frac{\exp[-\gamma(t)]}{\Omega(t)}$$

$$Y(t) = 1 - \frac{1}{\Omega(t)}$$

$$\gamma(t) = \int_0^t [\mu(\tau) - \lambda(t)] \, d\tau$$

and

$$\Omega(t) = \exp[-\gamma(t)]\left\{1 + \int_0^t \mu(t) \exp[\gamma(t)] \, d\tau\right\}. \tag{10.37c}$$

The probability of total intergranular bond dissociation is, then, given with reference to (10.37) by

$$P_0(t) = \frac{\int_0^t \mu(t) \exp[\gamma(\tau)] \, dt}{1 + \int_0^t \mu(\tau) \exp[\gamma(\tau)] \, dt}. \tag{10.38}$$

10.3.2 Numerical illustration

In the present section, the theoretical analysis proposed in the previous section is applied to the case of a natural cellulose fibrous network. The aim here is to incorporate available experimental data concerning the microstructure into the proposed formulation. Hence, in accordance with the introduced model, the internal stress distribution and the response behaviour of an intermediate domain of the material may be predicted. In the context of the application of the theoretical analysis, the following remarks are first made.

1. In practice, the experimental values concerning the response of single fibres, in general, are obtainable, in most cases, from a uniaxial test situation; thus one may only proceed to evaluate numerically the material parameters characterizing the creep behaviour of cellulose fibres for this particular case. Consequently, one may assume, for the purpose of this section, that the microdeformation in the fibre segment as well as the bond deformation in the junction area occur only in the direction of the fibre segment axis, i.e. in the direction of the $^{\alpha}Y_1$ axis of the local coordinate system.

2. Because of the random orientation of the fibrous elements, these elements may experience tension or compression at different instants of time. Hence, it is necessary to assume that the rheological response of a fibre segment in compression is governed by the same constitutive law in tension. This hypothesis is considered to be valid only for the case of small strain (e.g. Hill, 1967).

3. A review of the literature on the determination of the distribution of orientation of the fibrous elements revealed that no experimental work has been carried out to determine such a distribution for the general three-dimensional case. However, the distribution of orientation of the fibrous elements is accessible for the two-dimensional case (e.g. Kallmes, 1969). Consequently, in the context of the application of the theoretical analysis, one must assume further that

$$^{\alpha}K_3, \quad ^{\alpha}k_3 \equiv e_3$$

where $^{\alpha}K_3$, $^{\alpha}k_3$ are the local unit vectors (undeformed and deformed states, respectively) that are associated with the local coordinate frame $^{\alpha}Y_i$ ($i = 1, 2, 3$), attached to the junction area, and e_3 is the unit vector associated with the external frame of reference Z_I ($I = 1, 2, 3$) (Fig. 10.3).

(a) Local response behaviour

Viscoelastic response of a fibre segment
We consider here, as an example, the case of dry summerwood fibres of a longleaf pine holocellulose pulp (conditioned at 50% RH and 23 °C). The numerical evaluation of the creep response of these fibres has been considered in section 7.4.1, where the numerical values of the parameters characterizing the model equation (7.6) or (7.26), with the inclusion of (7.36), were determined for a range of stress input between 29.1 and 49.1 dyn μm^2 and an extent of time up to 10^5 s. These numerical values are given in Tables 7.3 and 7.4 for an order of optimization $N = 3, 4, \cdots, 8$.

Interfibre bonding
The response behaviour of interfibre bonding has been determined for the present case of natural cellulose by using equations (10.16). In this, the numerical values of the parameters characterizing the material operator $^B\Gamma(t)$, equation (10.17) are taken as $\psi_0 = 3.14 \times 10^5$ dyn Å (Corte and Schaschek, 1955) and $\kappa = 2$ Å^{-1} (Sokolov, 1959).

Table 10.3 Geometrical and physical characteristics of a structural element (α) of natural cellulose

Fibre width, w	36 μm
Fibre thickness, *s*	3.6 μm
Fibre cross-sectional area, $^f a = ws$	129.60 μm^2
Actual bonded area of the junction, $^B A$	2000 μm^2
Parameter, $^f \lambda$ (associated with the equilibrium of the junction)	0.50
Probability, *q*	0.30
Bonding equilibrium potential, ψ_0	3.14 $\times$ 10^5 dyn Å
Morse constant, κ	2 Å^{-1}
Rheological response of a fibre segment*	
Order of optimization, *N*	5
Parameter *b*	0.0123 $\times$ 10^8 dyn μm^{-2}
Parameters D_I ($I = 1, 2, \ldots, 5$)	-0.8435×10^8 dyn μm^{-2}
	-9.8879×10^8 dyn μm^{-2}
	-39.3752×10^8 dyn μm^{-2}
	-104.2946×10^8 dyn μm^{-2}
	-231.0216×10^8 dyn μm^{-2}

* Creep of dry symmerwood fibres of longleaf pine holocellulose pulp after conditioning at 50% RH and 23 °C (Table 7.4).

Probabilistic viscoelastic response of a structural element (α)
The response behaviour of a structural element (α) is considered here for the case of natural cellulose. The geometrical and physical characteristics of this element are given in Table 10.3.

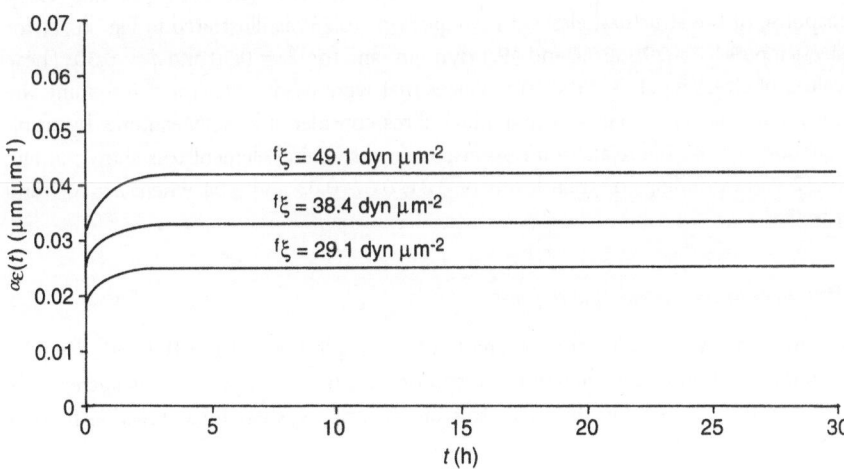

Fig. 10.11 Creep response of a structural element (α), $^f \lambda = 0.50$ and $q = 0.30$; equation (10.23).

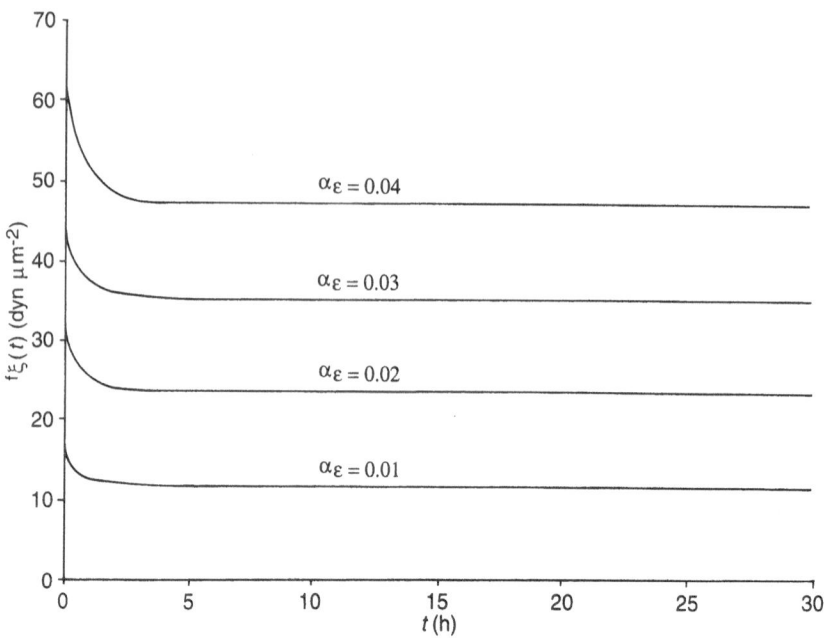

Fig. 10.12 Relaxation response of a structural element (α), ${}^f\lambda = 0.50$ and $q = 0.30$; equation (10.24).

Proceeding from the information derived earlier in the context of the response behaviour of the fibre segment and that of the interfibre bonding, the numerical values of the material operator ${}^{\alpha f}\Gamma(t)$, expressed by equation (10.22), for the structural element have been evaluated for different values of the probability q and also for different values of the parameter ${}^f\lambda$, equations (10.20). The corresponding creep response of the structural element (α), equation (10.23), is illustrated in Fig. 10.11 for stress inputs ${}^f\xi = 29.1, 38.4$ and 49.1 dyn μm^2 and for ${}^f\lambda = 0.50$ and $q = 0.30$. These values of stress inputs are the same values that were used earlier for determining the creep response of the class of individual fibres considered here. Meantime, by using equation (10.24), the relaxation response of the structural element α is shown in Fig. 10.12 corresponding to strain levels of 0.01, 0.02, 0.03 and 0.04 when ${}^f\lambda = 0.5$ and $q = 0.3$.

Transition to the macroscopic response

In order to give an illustrative example of the application of the theoretical model to the prediction of the rheological response of a (macroscopic) fibrous system, the parameters characterizing the microstructure of such a system are assumed as follows (equation (10.22)).

1. The value of the probability q is assumed to be constant throughout the intermediate domain and is equal to 0.30.

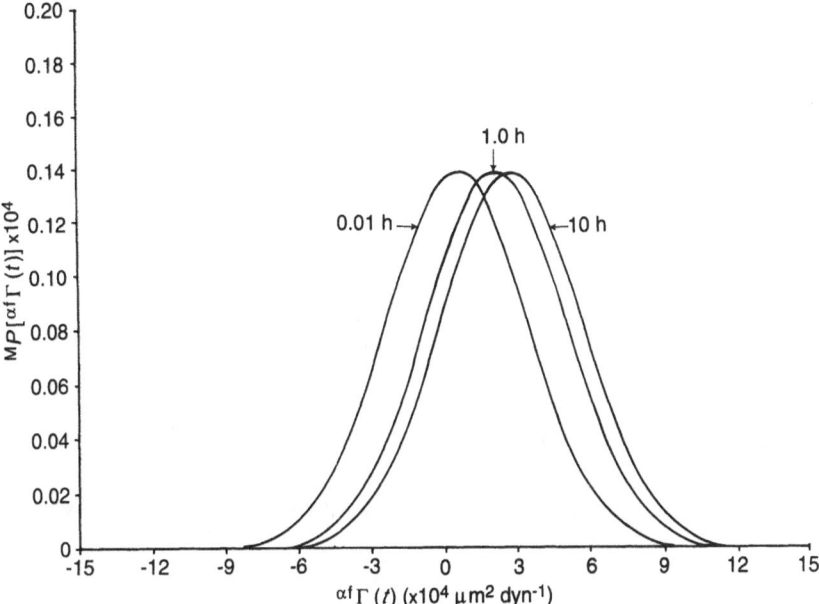

Fig. 10.13 Statistical distribution of the material operator $^{af}\Gamma(t)$ of a structural element.

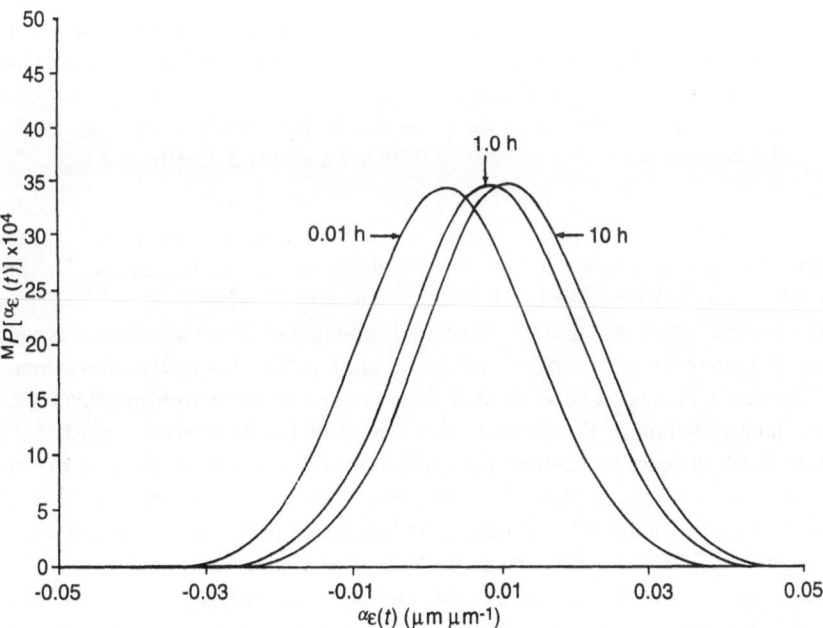

Fig. 10.14 Statistical distribution of microstrain $^{\alpha}\varepsilon(t)$.

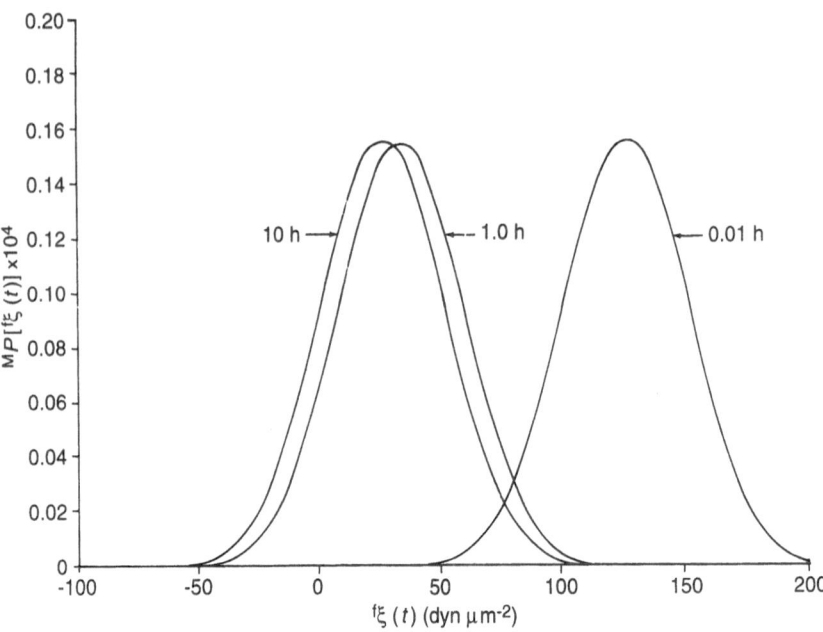

Fig. 10.15 Statistical distribution of microstress $^{\alpha}\xi(t)$.

2. All individual fibres of the fibrous system are assumed to be of the same class and identical in rheological response. This response is characterized by the parameters mentioned in the previous subsection concerning the viscoelastic response of a fibre segment and given in Table 10.3.
3. The statistical distribution of the parameter $^f\lambda$, equation (10.20b), is assumed to be 'Gaussian' with a mean value of 0.50 and a standard deviation of 0.05.

Following the assumptions above, the statistical distribution of the material operator $^{\alpha f}\Gamma(t)$ (equation (10.22)) of a structural element is determined. The time evolution of the latter statistical distribution is illustrated in Fig. 10.13. With reference to equation (10.27), an arbitrary statistical distribution of fibre microstress is assumed to exist internally within the intermediate domain at time $t = 0$. This distribution is 'Gaussian' with a mean value of 38.87 dyn μm^2. In accordance with equation (10.27), the time evolution of the statistical distribution of the microstrain is illustrated in Fig. 10.14. In order to illustrate the application of the model for the case of stress relaxation, an arbitrary statistical distribution of microstrain is assumed to exist locally in the microstructure of the intermediate domain at time $t = 0$. This distribution is assumed again for simplification to be 'Gaussian' with a mean value of 0.007 μm μm^{-1} and a standard deviation of 0.007 $\mu m \, \mu m^{-1}$. The resulting distribution of the fibre microstress is determined in accordance with equation (10.28). The time evolution of the latter distribution is given in Fig. 10.15.

REFERENCES

Arnott, S. and Wonacott, A. J. (1966) The refinement of the crystal and molecular structures of polymers using X-ray data and stereochemical constraints. *Polymers*, **7**, 157–66.

Axelrad, D. R. (1970) Stochastic analysis of the flow of two-phase media, in Proc. 5th Int. Congr. on Rheology (ed. S. Onagi), University of Tokyo Press, Tokyo, pp. 221–31.

Axelrad, D. R. (1971) Rheology of structured media. *Arch. Mech. Stosow.*, **23**(1), 131–40.

Axelrad, D. R. (1978) *Micromechanics of Solids*, Elsevier, New York.

Axelrad, D. R. (1984) *Foundations of the Probabilistic Mechanics of Discrete Media*, Pergamon, Oxford.

Axelrad, D. R. and Basu, S. (1974) Operational approach to the deformation of structured media, in *Stochastic Problems in Mechanics*, University of Waterloo Press, pp. 61–77.

Axelrad, D. R. and Jaeger, L. G. (1969a) Random theory of deformation in heterogeneous media, in Proc. Southampton 1969 Civil Engineering Materials Conf., Part I (ed. M. Te'eni), Wiley–Interscience, London, p. 571.

Axelrad, D. R. and Jaeger, L. G. (1969b) Local energy fluctuations in yielding in inhomogeneous media, in Proc. Southampton 1969 Civil Engineering Materials Conf., Part I (ed. M. Te'eni), Wiley–Interscience, London, p. 87.

Axelrad, D. R. and Kalousek, J. (1971) Stress holographic interferometry. *Micromechanics Lab. Rep. 71-7*, McGill University, Montreal.

Axelrad, D. R. and Provan, J. W. (1975) Microrheology of crystalline media. *Rheol. Acta*, **12**, 177–82.

Axelrad, D. R., Provan, J. W. and Basu, S. (1974) Analysis of the semi-group property and the constitutive relations of structured solids, in Proc. Conf. on Symmetry, Similarity and Group Theoretic Methods in Mechanics (eds P. G. Glockner and M. C. Singh), University of Calgary.

Bharucha-Reid, A. T. (1960) *Elements of the Theory of Markov Processes and its Applications*, McGraw-Hill, New York.

Battista, O. A. (ed.) (1964) *Synthetic fibers in Papermaking*, Interscience, New York.

Bikales, N. M. and Segal, L. (eds) (1971) *Cellulose and Cellulose Derivatives*, Wiley–Interscience, New York.

Bollmann, W. (1970) *Crystal Defects and Crystalline Interfaces*, Springer, New York.

Brezinski, J. P. (1956) The creep properties of paper. *Tappi*, **39**(2), 116–28.

Corte, H. (1966) Paper and board, in *Composite Materials* (ed. L. Holliday), Elsevier, New York, pp. 475–526.

Corte, H. and Kallmes, O. J. (1962) Statistical geometry of a fibrous network, in *The Formation and Structure of Paper*, Vol. 1 (ed. F. Bolam), Paper and Board Makers' Association, London, pp. 13–52.

Corte, H. and Schaschek, H. (1955) Physical nature of paper strength. *Papier*, **9**, 319–30.

Gardener, K. H. and Blackwell, J. (1974) The hydrogen bonding in native cellulose. *Biochim. Biophys. Acta*, **342**, 232–7.

Haddad, Y. M. (1975) Response behaviour of a two-dimensional fibrous network. PhD Thesis, McGill University, Montreal.

Haddad, Y. M. (1980) A theoretical approach to interfiber bonding of cellulose. *J. Colloid Interface Sci.*, **76**(2), 490–501.

Haddad, Y. M. (1984) A microstructural approach to the rheology of fibrous systems. *J. Colloid Interface Sci.*, **100**(1), 143–65.

Haddad, Y. M. (1985) A stochastic approach to the rheology of randomly structured fibrous networks. *Mater. Sci. Eng.*, **72**, 135–47.

Haddad, Y. M. (1986) A microstructural approach to the mechanical response of composite systems with randomly oriented, short fibres. *J. Mater. Sci.*, **21**, 3767–76.

Haddad, Y. M. (1987) Approche microstructurelle de la rhéologie des systèmes fibreux. I. Analyse théorique. *Res Mec.*, **22**, 243–65.

Haddad, Y. M. (1990) A microstructural approach to the mechanical response of polycrystalline systems: I. Theoretical analysis. *Res. Mec.*, **28**, 177–96.

Hill, R. L. (1967) The creep behaviour of individual pulp fibers under tensile loading. *Tappi*, **50**(8), 432–40.

Jacobson, R. A., Wunderlich, J. A. and Lipscomb, W. N. (1961) The crystal and molecular structure of cellulose. *Acta Crystallogr.*, **14**, 598–607.

Jayme, G. and Hunger, G. (1962) Electron microscope 2- and 3-dimensional classification of fibre bonding, in *The Formation and Structure of Paper*, Transaction of Oxford Symposium (1961), Vol. 1 (ed. F. Bolam), Technical Section of the British Paper and Board Makers' Association, London, pp. 135–70.

Kallmes, O. J. (1969) Technique for determining the fiber orientation distribution throughout the thickness of the sheet. *Tappi*, **52**(3), 482–5.

Kalousek, J. (1973) Experimental investigations of the deformation of structured media. PhD Thesis, McGill University, Montreal.

Lai, J. S. Y. and Findley, W. N. (1968) Stress relaxation of nonlinear viscoelastic materials under uniaxial strain. *Trans. Soc. Rheol.*, **12**, 259–80.

Manley, R. S. J. and Inoue, S. (1965) The fine structure of regenerated cellulose. *J. Polym. Sci. Polym. Lett. B*, **3**(3), 691–5.

Mark, R. E. (1967) *Cell Wall Mechanics of Tracheids*, Yale University Press, London.

McIntosh, D. C. and Leopold, B. (1962) Bonding strength of individual fibres, in transactions of the Oxford Symposium, Vol. 1, British Paper and Board Makers' Association, London, pp. 265–76.

Meredith, R. (ed.) (1956) *Mechanical Properties of Textile Fibres*, North-Holland, Amsterdam.

Meyer, K. H. and Misch, L. (1937) Positions des atoms dans le nouveau modèle spatial de la cellulose. *Helv. Chim. Acta*, **20**, 232–44.

Morse, P. M. (1929) Diatomic molecules according to the wave mechanics. II. Vibrational levels. *Phys. Rev.*, **34**, 57–64.

Nissan, A. H. (1959) The rheological behaviour of hydrogen bonded solids. *Trans. Faraday Soc.*, **55**, 2048–53.

Nissan, A. H. (1967) The significance of hydrogen bonding at the surface of cellulose network structure, in *Surfaces and Coatings Related to Paper and Wood*, Syracuse University Press, Syracuse, NY, pp. 221–65.

Nissan, A. H. and Sternstein, S. S. (1964) Cellulose fibre bonding. *Tappi*, **47**(1), 1–6.

Onogi, S. and Sasaguri, K. (1961) Elasticity of Paper and other fibrous sheets. *Tappi*, **44**(12), 874–80.

Page, D. H. and Tydeman, P. A. (1960) Fibre to fibre bonds. Part 2. A preliminary study of their properties in paper sheets. *Paper Technol.*, **1**(5), 519–30.

Page, D. H., Tydeman, P. A. and Hunt, M. (1962) A study of fibre to fibre bonding by direct observation, in Transactions of the Oxford Symposium, Vol. 1, British Paper and Board Makers' Association, London, pp. 171–93.

Paradowski, R. J. (1991) Pauling develops his theory of the chemical bond, in *Great Events from History II, Science and Technology Series*, Vol. 2 (1910–1931) (ed. F. N. Magill), Salem Press, Pasadena, CA, pp. 926–33.

Pauling, L. (1960) *The Nature of the Chemical Bond and Structure of Molecules and Crystals*, Cornell University Press, Ithaca, NY.

Rance, H. F. (1948) Some new studies in the strength properties of paper, in Proc. Great Britain and Ireland Paper Makers' Association, Vol. 29, pp. 449–76.

Rance, H. F. (1962) Introduction to symposium, in Transactions of the Oxford Symposium, Vol. 1, British Paper and Board Makers' Association, London, pp. 1–11.

Roelofsen, P. A. (1959) Plant cell wall, in *Encyclopedia of Plant Anatomy*, Gebrüder Bointrager, Berlin- Nicolassee.

Schulz, J. H. (1961) The effect of straining during drying on the mechanical and viscoelastic behaviour of paper. *Tappi*, **44**(10), 736–44.

Sokolov, N. D. (1959) On the quantum theory of the hydrogen bonding, in Symp. on Hydrogen Bonding, Ljubljana, 1957 (ed. D. Hadzi and W. H. Thompson), Pergamon, London, pp. 385–92.

Sternstein, S. S. and Nissan, A. H. (1962) A molecular theory of viscoelasticity of a three-dimensional hydrogen bonded network, in Transactions of the Oxford Symposium, Vol. 1, British Paper and Board Makers' Association, London, pp. 319–50.

Tonnesen, B. A. and Ellefsen, O. (1971) Investigations of the structure of cellulose and its derivatives, F. Submicrostructural investigations, in *Cellulose and Cellulose Derivatives*, Vol. V, Part IV (eds N. M. Bikales and L. Segal), Wiley, New York, pp. 265–304.

Van den Akker, J. A. (1962) Some theoretical considerations on the mechanical properties of fibrous structures, in *The Formation and Structure of Paper*, Transactions of the Oxford Symposium (1961), Vol. 1 (ed. F. Bolam), Technical Section, British Paper and Board Makers' Association, London, pp. 205–41.

Van den Akker, J. A. (1970) Structure and tensile characteristics of paper. *Tappi*, **53**(3), 388–400.

Van den Akker, J. A., Lathrop, A. L., Voelker, M. H. and Dearth, L. R. (1958) Importance of fibre strength to sheet strength. *Tappi*, **41**(8), 416–25.

FURTHER READING

Axelrad, D. R. and Basu, S. (1973) Operational approach to the deformation of structured media, in *Stochastic Problems in Mechanics* (eds S. T. Ariaratnam and H. H. E. Leipholz), Study No. 10, University of Waterloo, pp. 61–77.

Axelrad, D. R. and Basu, S. (1973) Mechanical relaxation of crystalline solids. *Adv. Mol. Relax. Process.*, **6**, 185–99.

Axelrad, D. R., Haddad, Y. M. and Atack, D. (1975) Stochastic deformation theory of a two-dimensional fibrous network, in Proc. 11th Annual Meet. (ed. G. J. Dvorak), Society of Engineering Science, Duke University, NC, pp. 166–7.

Axelrad, D. R., Basu, S. and Haddad, Y. M. (1976) Microrheology of cellulosic systems, in Proc. 7th Int. Congr. on Rheology, Gothenburg.

Basu, S. (1975) On a general deformation theory of structural solids. PhD Thesis, McGill University, Montreal.

Bernal, J. D. (1958) General introduction: structure arrangements of macromolecules. *Discuss. Faraday Soc.*, **25**, 7–18.

Blumenthal, R. C. and Getoor, R. K. (1963) *Markov Processes and Potential Theory*, Academic Press, New York.

Bochner, S. (1942) Stochastic processes. *Ann. Math.*, **48**, 1014.

Bourbaki, N. (1951) *Topologie Générale*, Hermann, Paris.

Doob, J. L. (1953) *Stochastic Processes*, Wiley, New York.

Dynkin, E. P. (1965) *Markov Processes*, Vols. 1 and 2, Academic Press, New York.

Ehrenfest, P. and Ehrenfest, T. (1959) *The Conceptual Foundations of the Statistical Approach in Mechanics*, Cornell University Press, Ithaca, NY.

Foguel, S. R. (1969) *The Ergodic Theory of Markov Processes*, Van Nostrand Reinhold, New York.

Gikhman, I. I. and Shorohod, A. V. (1969) *Introduction to the Theory of Random Processes*, Saunders, Philadelphia, PA.

Goffman, C. and Pedrick, G. (1965) *First Course in Functional Analysis*, Prentice-Hall, Englewood Cliffs, NJ.

Högfors, C. (1987) History-dependent systems. *Rheol. Acta*, **26**, 317–21.

Hopf, E. (1954) The general temporally discrete Markoff process. *J. Ration. Mech. Anal.*, **3**, 12–45.

Jongschaap, R. J. J. (1987) On the derivation of some fundamental expressions for the average stress tensor in systems of interacting particles. *Rheol. Acta*, **26**, 328–35.

Kakutani, S. (1940) Ergodic theorems and the Markoff processes with a stable distribution. *Proc. Imp. Acad. Jpn. Tokyo*, **16**, 49–54.

Kampe de Feriet, J. (1962) Statistical mechanics of continuous media, in Proc. 13th Symp. on Applied Mechanics, Hydrodynamic Instability, American Mathematical Society, pp. 165–98.

Kappos, D. A. (1969) *Probability Algebras and Stochastic Spaces*, Academic Press, New York.

Kendall, D. G. (1955) Some analytical properties of continuous stationary Markov transition functions. *Trans. Am. Math. Soc.*, **78**, 529–40.

Khinchine, A. I. (1949) *Mathematical Foundations of Statistical Mechanics*, Dover Publications, New York.

Kolmogorov, A. N. (1956) *Foundations of Probability Theory*, Chelsea, New York.

Moreau, J. J. (1971) Sur l'evolution d'un système elastoviscoplastique. *C.R. Acad. Sci., Ser. A*, **273**, 118–21.

Peterline, A. (1969) Bond rupture in highly oriented crystalline polymers. *J. Polym. Sci. A2*, **7**, 1151–63.

Peters, L. (1955) A note on nonlinear viscoelasticity. *Textile Res. J.*, **25** (March), 262–5.

Pinsker, M. S. (1964) *Information and Information Stability of Random Variables and Processes*, Holden-Day, San Francisco, CA.

Pugachev, V. S. (1965) *Theory of Random Functions*, Addison-Wesley, Reading, MA.

Renyi, A. (1970) *Foundations of Probability*, Holden-Day, San Francisco, CA.

Rosenblatt, M. (1971) *Markov Processes, Structure and Asymptotic Behaviour*, Springer, New York.

Schausberger, A., Knoglinger, H. and Janeschitz-Kriegl, H. (1987) The role of short chain molecules for the rheology of polystyrene melts. II. Linear viscoelastic properties. *Rheol. Acta*, **26**, 468–73.

Shimanouchi, T. and Mizushima, S. (1955) On the helical configuration of a polymer chain. *J. Chem. Phys.*, **23**(4), 707–11.

Simmons, G. F. (1963) *Introduction to Topology and Modern Analysis*, McGraw-Hill, New York.

Sneddon, I. N. (1971) In *Functional Analysis in Continuum Mechanics I* (ed. A. C. Eringen), Academic Press, New York.

Treves, F. (1967) *Locally Convex Spaces and Partial Differential Equations*, Springer, New York.

Van Kampen, N. G. (1962) Fundamental problems in statistical mechanics of irreversible processes, in *Fundamental Problems in Statistical Mechanics* (ed. E. D. G. Cohen), North-Holland, Amsterdam.

Yaglom, A. M. (1962) *Theory of Stationary Random Functions*, Prentice-Hall, New York.

Yosida, K. (1965) *Functional Analysis*, Springer, Berlin.

Appendix A

Introduction to tensors

In continuum mechanics we deal with physical events which are, in general, independent of both the position and the orientation of the observer. That is, if two coordinate systems, fixed at different locations in space and with different relative orientations, are used to observe and, hence, to formulate a physical law governing a particular physical event, the resulting physical law would hold valid in the two coordinate systems and in any coordinate system not moving with respect to the other two systems. In a proper description, such events are represented mathematically by tensorial quantities and the resulting physical laws are expressed in terms of tensorial equations. As a mathematical entity, a tensor has an existence independent of any coordinate system. Specifically, the components of a tensor in one coordinate system would determine the corresponding components of the same tensor in any other coordinate system not moving relative to the first one. Further, the tensorial equations governing specific physical events would be invariant under coordinate transformation. This invariance of tensor equations under coordinate transformation highlights the usefulness of tensor calculus in the study of continuum mechanics. Invariance of the form of a physical law referred to two frames of reference in general motion is, however, more difficult and would require the tools of general relative theory, although in some cases the postulated 'principle of material frame indifference' is used.

In dealing with general coordinate transformations between arbitrary curvilinear coordinate systems, the resulting tensors are conventionally referred to as general tensors. However, when transformation is carried out from one homogeneous coordinate system to another, the tensors are defined as Cartesian tensors. For simplicity, we limit ourselves in this appendix to tensors in three-dimensional Euclidean space.

Tensors may be classified by rank, or order, according to the particular form of the transformation law they obey. Such a classification determines the number of components a given tensor possesses in an N-dimensional space. In this context, a tensorial quantity is conventionally presented using indicial notation. In this notation, letter indices, either subscripts or superscripts, are appended to a general letter or symbol representing the quantity of interest.

A.1 INDICIAL NOTATION

In writing a set of N quantities $a_1, a_2, \cdots, a_N$ the notation a_i $(i = 1, 2, \cdots, N)$ is generally used. In this notation, the subscript i is referred to as the index and it implies that it may take on any integer value from the set $\{1, \cdots, N\}$. The number N is called the range of the index. Hence, the quantity a_i represents any element of the set $\{a_1, a_2, \cdots, a_N\}$ and simultaneously identifies the entire set. In a similar manner, one may use the notation x_i $(i = 1, 2, 3)$ to denote the coordinates of a point in a rectangular Cartesian coordinate system. Further, if we adopt the convention that a Latin index would have automatically the range of 3, then the coordinates of a point in a rectangular Cartesian coordinate system would simply be identified by x_i and it would be understood that x_i: x_1, x_2, x_3. On the other hand, if we are dealing with a two-dimensional space, we may use Greek indices with the understanding that a Greek index has a range of 2, e.g. x_α: x_1, x_2.

A.2 CLASSIFICATION OF TENSORIAL QUANTITIES

In tensorial notation, a system which depends on m indices is conventionally referred to as an mth-order system. As illustrated below, if we denote the range of the index by the letter N, then the number of components in the mth-order tensorial system would be given by N^m. A letter index may occur either once or twice in a given term. When an index appears unrepeated in a term, that index is known as 'free' index. The number and location of the free indices reveal directly the exact tensorial character of the quantity dealt with. For instance, the tensorial order of a given term is equal to the number of free indices appearing in that term. On the other hand, repeated indices are often referred to as 'dummy' indices as they can be replaced by any other letter not appearing as a free index without changing the meaning of the term in which they occur, i.e.

$$a_i b_i = a_j b_j = a_k b_k.$$

- A first-order tensor is denoted by a kernel letter bearing one free index only irrespective of the number of dummy indices that may be present. Examples of a first order tensor are

$$a_i, \quad a^i, \quad a_{ij}b_j, \quad c_{jkk}.$$

In a three-dimensional space, the number of components of a first-order system is 3 while, in a two-dimensional space, the number of components is 2:

$$a_{\alpha\beta}b_\beta = a_{\alpha 1}b_1 + a_{\alpha 2}b_2 = (a_{11}b_1 + a_{12}b_2, a_{21}b_1 + a_{22}b_2).$$

- A second-order tensor is identified by two free indices. Thus, a second-order tensor is specified by nine components in any coordinate system in a three-dimensional space, for example

$$a_{ij} = \begin{bmatrix} a_{11} & a_{12} & a_{13} \\ a_{21} & a_{22} & a_{23} \\ a_{31} & a_{32} & a_{33} \end{bmatrix}.$$

Other designations of a second-order system in a three-dimensional space may take the form a_{ij}, a^{ij}, $a^i_{\cdot j}$ where the dot indicates that the j is the second index.

Alternatively, a second-order system in a two-dimensional space would have four components, e.g.

$$a_{\alpha\beta} = \begin{bmatrix} a_{11} & a_{12} \\ a_{21} & a_{22} \end{bmatrix}.$$

- A third-order tensor has three free indices which correspond to either 27 components in a three-dimensional space or 8 components in a two-dimensional space. A third-order system may be constructed through a logical combination of the first-order and second-order tensors. Examples of a third-order tensor are

$$a_{ijk}, \quad a^{ijk}, \quad a^i_{\cdot jk}.$$

- A fourth-order tensor has four free indices which represent 81 components in a three-dimensional space and 16 components in a two-dimensional space. Examples of a fourth-order tensor are

$$a_{ijkm}, \quad a^{ijkm}, \quad a^{ij}_{km}, \quad a^i_{jkm}, \quad a_{\alpha\beta\gamma\delta}, \quad a^{\alpha\beta\gamma\delta}, \quad a^{\alpha\beta}_{\gamma\delta}, \quad a^{\alpha}_{\beta\gamma\delta}.$$

- A generalized system of the mth order, e.g. $a_{ijkl\cdots m}$ has m indices which concur with 3^m components in a three-dimensional space. Alternatively, in a generalized system of the χth order, e.g. $a_{\alpha\beta\gamma\cdots\chi}$, the number of indices corresponds to 2^χ components in a two-dimensional space.
- A tensorial system with no index, a scalar, is referred to as a zero-order system. It has one component only, for example a zero-order tensor a.

A.3 ADDITION AND MULTIPLICATION OF TENSORIAL SYSTEMS OF THE SAME ORDER AND RANGE

Addition of two systems $a_{ij\cdots k}$ and $b_{ij\cdots k}$ of the same order and magnitude has the following properties.

- Addition is commutative:

$$a_{ij\cdots k} + b_{ij\cdots k} = b_{ij\cdots k} + a_{ij\cdots k}.$$

- Addition is associative:

$$a_{ij\cdots k} + (b_{ij\cdots k} + c_{ij\cdots k}) = (a_{ij\cdots k} + b_{ij\cdots k}) + c_{ij\cdots k}.$$

- There exists a unique system, 0, such that

$$a_{ij\cdots k} + 0 = a_{ij\cdots k}.$$

- To every system $a_{ij\cdots k}$ there corresponds a unique system $-a_{ij\cdots k}$ such that

$$a_{ij\cdots k} + (-a_{ij\cdots k}) = 0.$$

A.4 MULTIPLICATION OF TENSORIAL SYSTEMS OF DIFFERENT ORDER AND OF THE SAME RANGE

An mth-order tensorial system may be multiplied with an nth-order system to produce an $(m + n)$th-order system with the following properties.

- Multiplication is commutative:

$$a_i b_{jk} = b_{jk} a_i.$$

- Multiplication is associative:

$$a_i(b_{jk} c_{ls}) = (a_i\, b_{jk})\, c_{ls}.$$

- Multiplication is distributive with respect to addition:

$$a_{ij}(b_k + c_k) = a_{ij}\, b_k + a_{ij}\, c_k.$$

- There exists a unique scalar, 1, called unity such that

$$1 \times a_{ij\cdots k} = a_{ij\cdots k}.$$

A.5 SUMMATION CONVENTION

If x_i is a set of N variables and b_i is a set of N constants, a linear form may be written as

$$\sum_{i=1}^{N} b_i x_i = b_1 x_1 + b_2 x_2 + \cdots + b_N x_N$$

and if b_{ij} is a set of N constants, a quadratic form may be expressed by

$$\sum_{i,j=1}^{N} b_{ij} x_i x_j = b_{11} x_1 x_1 + b_{12} x_1 x_2 + \cdots + b_{1N} x_1 x_N$$

$$+ b_{21} x_2 x_1 + \cdots$$

$$\vdots$$

$$+ b_{N1} x_N x_1 + \cdots + b_{NN} x_N x_N.$$

As shown above, the summation is carried out over repeated indices and, thus, the summation sign could be omitted. Hence, the following convention may be adopted: **the repetition of an index implies the summation over the range of that index in the absence of an explicit statement to the contrary.**

A.6 SYMMETRIC AND SKEW-SYMMETRIC SYSTEMS

A.6.1 Symmetric systems

A tensor is described as **symmetric** if its matrix of rectangular Cartesian components is symmetric. In a system of two or more indices, if the values of the elements do not change by interchanging two indices, the system is said to be symmetric with respect to these two indices. More generally, if the system is symmetric with respect to all indices, the system is referred to as completely symmetric. Examples of symmetric systems are as follows:

- symmetric in i and j,

$$a_{ijk} = a_{jik};$$

- completely symmetric,

$$a_{ij} = a_{ji},$$
$$a_{ijk} = a_{kij} = a_{jki} = a_{kji} = a_{ikj} = a_{jik}.$$

Symmetry is a tensor property in a real sense, i.e. if the matrix of a tensor is symmetric in one Cartesian coordinate system, it would be symmetric in all such systems. However, the product of two symmetric tensors might not be necessarily symmetric.

A.6.2 Skew-symmetric systems

The definitions of skew symmetry in tensors follow those for symmetry except that interchange of a pair of indices would change the sign of the tensor:

- skew symmetric in i and j,

$$b_{ijk} = -b_{jik};$$

- completely skew symmetric,

$$b_{ij} = -b_{ji},$$
$$b_{ijk} = b_{kij} = b_{jki} = -b_{kji} = -b_{ikj} = -b_{jik}.$$

In this situation, the following should be noted.

- If, in a component of a skew-symmetric system, any two indices are not distinct, the value of such a component must be zero. For example, if b_{ijk} is skew symmetric in j and k, then $b_{i11} = -b_{i11} = 0$.
- In the skew-symmetric system $b_{ij} = -b_{ji}$, the diagonal elements b_{11}, b_{22} and b_{33} are all zero.
- If the order of a completely skew-symmetric system is equal to the range of its indices, then the nonvanishing components of the system have only one distinct absolute numerical value. For example, if b_{ijk} is completely skew-symmetric, then the only distinct term is b_{123}.

A.6.3 An important property of all second-order tensors

Any second-order system may be represented as the sum of a symmetric and a skew-symmetric system. For instance, the second-order system a_{ij} may be expressed by

$$a_{ij} = a_{(ij)} + a_{[ij]}$$

in which the first term on the right-hand side is symmetric while the second is skew symmetric. These two terms are expressed respectively by

$$a_{(ij)} = \tfrac{1}{2}(a_{ij} + a_{ji}), \quad \text{symmetric}$$

and

$$a_{[ij]} = \tfrac{1}{2}(a_{ij} - a_{ji}), \quad \text{skew symmetric}.$$

Example A.1

If

$$a_{\alpha\beta} = \begin{bmatrix} \cos\theta & -\sin\theta \\ \sin\theta & \cos\theta \end{bmatrix}$$

then

$$a_{(\alpha\beta)} = \tfrac{1}{2}(a_{\alpha\beta} + a_{\beta\alpha}) = \begin{bmatrix} \cos\theta & 0 \\ 0 & \cos\theta \end{bmatrix}$$

and

$$a_{[\alpha\beta]} = \tfrac{1}{2}(a_{\alpha\beta} - a_{\beta\alpha}) = \begin{bmatrix} 0 & -\sin\theta \\ \sin\theta & 0 \end{bmatrix}$$

thus

$$a_{\alpha\beta} = a_{(\alpha\beta)} + a_{[\alpha\beta]}.$$

A.7 CARTESIAN TENSORS

In a rectangular Cartesian frame of reference, we choose as base vectors the unit vectors e_1, e_2, e_3 parallel to the coordinate axes. Thus, a vector $\mathbf{A}$ in terms of its components A_i is expressed by

$$\mathbf{A} = A_i\, e_i.$$

We may also consider a transformation of the unit vectors by choosing another rectangular Cartesian system with new unit vectors e_i' which would be related to the original (unprimed system) by

$$e_i' = a_{ji}\, e_j.$$

Thus, $\mathbf{A}$ may be expressed by

$$\mathbf{A} = A_i\, e_i = A_i'\, e_i'.$$

Taking dot products of both sides by e'_j and e_j respectively would result into these two equations:

$$A'_j = a_{ij} A_i;$$
$$A_j = a_{ji} A'_i. \tag{A.1}$$

That is, on a transformation of coordinates from one rectangular Cartesian system to another, the components of a vector A would obey the law given by (A.1). Thus, the following definition may be stated.

The components of any first-order system which obeys the law (A.1), on a coordinate transformation, are said to be the components of a Cartesian tensor of first order.

One notes that the first-order Cartesian tensor components are the components of a vector.

Consider now the system

$$A_i B_j = A_{ij}.$$

If A_i and B_j transform as vectors, one has

$$A'_{ij} = A'_i B'_j$$
$$= a_{ki} a_{mj} A_k B_m \tag{A.2}$$
$$= a_{ki} a_{mj} A_{km}.$$

If the components of a second-order system transform according to (A.2) then they are the components of a Cartesian tensor of second order.

In a similar manner, a third-order Cartesian tensor is a third-order system the components of which transform according to

$$A_{ijk} = a_{ip} a_{jm} a_{kn} A'_{pmn}. \tag{A.3}$$

The tensor transformation laws illustrated above may be generalized to Cartesian tensors of any order, i.e.

$$A'_{ijk\cdots} = a_{ip} a_{jm} a_{kn} \cdots A_{pmn\cdots}.$$

QUIZ

Define the Cartesian tensor of zeroth order.

A.8 SPECIAL TENSORS

The following tensors in rectangular Cartesian systems are used frequently.

A.8.1 The Kronecker delta

This is a second-order tensor defined as

$$\delta_{ij} = \begin{cases} 1 & \text{if } i = j \\ 0 & \text{if } i \neq j \end{cases} = \begin{bmatrix} 1 & 0 & 0 \\ 0 & 1 & 0 \\ 0 & 0 & 1 \end{bmatrix}.$$

Also

$$\delta_{\alpha\beta} = \begin{cases} 1 & \text{if } \alpha = \beta \\ 0 & \text{if } \alpha \neq \beta \end{cases} = \begin{bmatrix} 1 & 0 \\ 0 & 1 \end{bmatrix}.$$

QUIZ

Prove the substitution property of the Kronecker delta:

$$a_i \delta_{ij} = a_j, \quad a_{ij} = a_{ik}\delta_{kj} \quad \text{and} \quad a_\alpha \delta_{\alpha\beta} = a_\beta.$$

A.8.2 The alternating tensor

This exists for any completely skew-symmetric system in which the number of indices is equal to the range of indices. The most important are the second- and third-order alternating tensors defined respectively by

$$\varepsilon_{\alpha\beta} = \begin{cases} +1 & \text{if } \alpha, \beta \text{ is an even permutation of } 1, 2, \\ -1 & \text{if } \alpha, \beta \text{ is an odd permutation of } 1, 2, \\ 0 & \text{if } \alpha, \beta \text{ are indistinct,} \end{cases}$$

i.e.

$$\varepsilon_{\alpha\beta} = \begin{bmatrix} 0 & 1 \\ -1 & 0 \end{bmatrix}$$

and

$$\varepsilon_{ijk} = \begin{cases} +1 & \text{if } i, j, k \text{ is an even permutation of } 1, 2, 3, \\ -1 & \text{if } i, j, k \text{ is an odd permutation of } 1, 2, 3, \\ 0 & \text{if any two of } i, j, k \text{ are indistinct,} \end{cases}$$

i.e.

$$\varepsilon_{ijk} = \begin{bmatrix} 0 & 0 & 0 \\ 0 & 0 & 1 \\ 0 & -1 & 0 \end{bmatrix} \begin{bmatrix} 0 & 0 & -1 \\ 0 & 0 & 0 \\ 1 & 0 & 0 \end{bmatrix} \begin{bmatrix} 0 & 1 & 0 \\ -1 & 0 & 0 \\ 0 & 0 & 0 \end{bmatrix}$$

QUIZ

Show that the completely skew-symmetric systems $b_{\alpha\beta}$ and b_{ijk} can be written respectively as

$$b_{\alpha\beta} = \varepsilon_{\alpha\beta}\, b_{12} \quad \text{and} \quad b_{ijk} = \varepsilon_{ijk}\, b_{123}.$$

QUIZ

Prove the alternating tensor–Kronecker delta relationship (known as $\varepsilon - \delta$ relation):

$$\varepsilon_{\alpha\beta}\varepsilon_{\gamma\lambda} = \delta_{\alpha\gamma}\delta_{\beta\lambda} - \delta_{\alpha\lambda}\delta_{\beta\gamma}; \quad \varepsilon_{ijk}\varepsilon_{ipm} = \varepsilon_{jki}\varepsilon_{pmi} = \delta_{jp}\delta_{km} - \delta_{jm}\delta_{kp}.$$

A.9 DIVERGENCE AND GRADIENT OPERATIONS

A.9.1 Gradient of a scalar

A time-dependent scalar field $\phi = \phi(X_i, t)$, where X_i represent the components of a position vector, may be partially differentiated with respect to X_i. This partial differentiation is conventionally denoted by a comma, i.e.

$$\frac{\partial \phi}{\partial X_i} = \phi_{,i}$$

and the associated vector is represented by

$$\nabla \phi = \phi_{,i}\, \mathbf{e}_i.$$

The vector field $\nabla \phi$ is referred to as the gradient of the scalar field ϕ. In this context, the symbol ∇ may be regarded to be an operator such that

$$\nabla = _{,i}\, \mathbf{e}_i.$$

A.9.2 Gradient of a vector

For a vector $\mathbf{A} = A_i\, \mathbf{e}_i$ with components A_i being a time-dependent vector field $A_i = A_i(x_i, t)$, the partial derivative of each component with respect to the coordinates is expressed as

$$\frac{\partial A_i}{\partial x_j} = A_{i,j} = \begin{bmatrix} A_{1,1} & A_{1,2} & A_{1,3} \\ A_{2,1} & A_{2,2} & A_{2,3} \\ A_{3,1} & A_{3,2} & A_{3,3} \end{bmatrix}.$$

These nine functions are the components of the gradient of the vector $\mathbf{A}$.

A.9.3 Divergence of a vector

The quantity

$$A_{i,i} = A_{1,1} + A_{2,2} + A_{3,3}$$

is defined as the divergence of the vector **A** and is expressed by

$$\text{div } \mathbf{A} = A_{i,i}.$$

QUIZ

Show that

$$\text{div } \mathbf{A} = \mathbf{V} \cdot \mathbf{A} = A_{i,i}.$$

A.9.4 Laplacian operator

Following the definitions of the operations above, one may consider the divergence of the gradient of a scalar:

$$\mathbf{V} \cdot \mathbf{V}\phi = \phi_{,ii}$$
$$= \nabla^2 \phi.$$

The operator ∇^2 is usually referred to as the Laplacian operator.

A.9.5 Curl of a vector

The vector $\mathbf{V} \times \mathbf{A}$ is referred to as the curl of a vector and may be shown to be given by

$$\mathbf{V} \times \mathbf{A} = {}_{,i}\mathbf{e}_i \times A_j \mathbf{e}_j$$
$$= A_{j,i}\mathbf{e}_i \times \mathbf{e}_j$$
$$= A_{j,i}\varepsilon_{ijk}\mathbf{e}_k$$
$$= \varepsilon_{ijk}A_{j,i}\mathbf{e}_k$$

A.10 TENSOR FIELDS

If the components of a tensor quantity depend in some manner on the coordinates of a point, for example

$$A_{ij\cdots k} = A_{ij\cdots k}(X_1, X_2, X_3),$$

these components are said to be point functions and the tensor quantity is referred to as tensor field. Further, a tensor field could be time dependent if its components, in addition to being dependent on the coordinates are also time dependent, for instance the time-dependent tensor field

$$A_{ij\cdots k} = A_{ij\cdots k}(x_1, x_2, x_3, t).$$

An important property of tensor fields is that if all components of a tensor field vanish in one coordinate system, they vanish likewise in all coordinate systems which can be obtained by admissible transformations.

In continuum mechanics, one is usually concerned with those tensor fields the components of which, together with all partial derivatives with respect to both the coordinates and time, are continuous functions.

PROBLEMS

A.1 If $a_i = (1, 4, 8)$ and $b_i = (6, -4, -3)$ find the components of $c_i = a_i + b_i$.

A.2

1. If $a_i = (1, 3, 6)$ and

$$b_{ij} = \begin{bmatrix} 1 & 3 & 6 \\ 4 & 0 & 0 \\ 3 & 1 & 2 \end{bmatrix}$$

find the components
   (a) $C_{ijk} = a_i b_{jk}$ and
   (b) C_{iij}.
2. Is $C_{iij} = C_{jii}$?

A.3 Let δ_{ij} denote the Kronecker delta and ε_{ijk} the alternating tensor.

1. Show that
   (a) $\delta_{ij} \delta_{ij} = 3$,
   (b) $\varepsilon_{ijk} \varepsilon_{jki} = 6$ and
   (c) $\delta_{ij} = \delta_{ik}$
2. If

$$\Delta(u) = \begin{bmatrix} u_{11} & u_{12} & u_{13} \\ u_{21} & u_{22} & u_{23} \\ u_{31} & u_{32} & u_{33} \end{bmatrix}$$

show that

$$\varepsilon_{ijk}\, \Delta(u) = \varepsilon_{pmn}\, u_{ip}\, u_{jm}\, u_{kn}.$$

and

$$6\Delta(u) = \varepsilon_{ijk}\, \varepsilon_{pmn}\, u_{ip}\, u_{jm}\, u_{kn}.$$

3. Using tensor notation, show that

$$\mathbf{A} \times (\mathbf{B} \times \mathbf{C}) = (\mathbf{A} \cdot \mathbf{C})\mathbf{B} - (\mathbf{A} \cdot \mathbf{B})\mathbf{C}$$

where $\mathbf{A}$, $\mathbf{B}$ and $\mathbf{C}$ are vectors.

A.4

1. If a_{ij} is symmetric and b_{ij} is skew symmetric show that $a_{ij}\,b_{ij} = 0$.
2. Furthermore, if d_{ijk} is skew symmetric in the indices i and j, show that

$$a_{ij}\,d_{ijk} = 0.$$

A.5 If $\alpha b_{ij} + \beta b_{ji} = 0$ show that either

1. $\alpha + \beta = 0$ and b_{ij} is symmetric or
2. $\alpha - \beta = 0$ and b_{ij} is skew symmetric.

A.6 Show that $a_{ij}\,\delta_{ik}\,\delta_{jl} = a_{kl}$.
A.7 Prove that $\delta_{ij}\,\delta_{ik}\,\delta_{jk} = 3$.
A.8 Verify that $\varepsilon_{\alpha\beta}\,\varepsilon_{\delta\gamma} = \delta_{\alpha\delta}\,\delta_{\beta\gamma} - \delta_{\alpha\gamma}\,\delta_{\beta\delta}$.
A.9 Show that

1. $\varepsilon_{\alpha\beta}\,\varepsilon_{\gamma\beta} = \varepsilon_{\alpha\gamma}$ and
2. $\varepsilon_{\alpha\beta}\,\varepsilon_{\alpha\beta} = 2$.

A.10

1. Using the indicial notation, prove the following vector identities:
   (a) $\mathbf{V} \times \mathbf{V}\phi = \mathbf{0}$;
   (b) $\mathbf{V}\cdot\mathbf{V} \times \mathbf{a} = 0$.
2. If A_{lm} is a second-order tensor, show that its derivative with respect to x_n, i.e. $\partial A_{lm}/\partial x_n$, is a Cartesian tensor of the third order.
3. For arbitrary tensors $\mathbf{P}$ and $\mathbf{Q}$, both of order unity, show that

$$\mu = (\mathbf{P} \times \mathbf{Q}) \cdot (\mathbf{P} \times \mathbf{Q}) + (\mathbf{P} \cdot \mathbf{Q})^2 = P^2 Q^2.$$

4. If A_{ij} is a symmetric tensor and B_{ij} a skew-symmetric tensor, show that

$$A_{ij}\,B_{ij} = 0.$$

A.11

1. If $b = \det b_{ij}$, verify that $b = \varepsilon_{ijk}\,b_{1i}\,b_{2j}\,b_{3k}$.
2. Show that $\varepsilon_{ijk}\,b_{ri}\,b_{sj}\,b_{tk}$ is skew symmetric in any pair of the indices r, s, t. Hence demonstrate that

$$\varepsilon_{rst}\,\varepsilon_{ijk}\,b_{ri}\,b_{sj}\,b_{tk} = 6b.$$

3. Show that

$$\text{div } \mathbf{T} = \mathbf{V}\cdot\mathbf{T} = T_{i,i}.$$

A.12 Show that

1. $\varepsilon_{ijm}\,\varepsilon_{ijk} = 2\,\delta_{mk}$ and
2. $\varepsilon_{ijm}\,\varepsilon_{ijm} = 6$.

A.13 If

$$a_{ij} = \begin{bmatrix} \frac{1}{2} & -\frac{1}{2} & 0 \\ \frac{1}{2} & \frac{1}{2} & 0 \\ 0 & 0 & 1 \end{bmatrix}$$

and the components of the vector **A** in the unprimed coordinate system are

$$A_i = (1, 2, 3),$$

find A_i'.

A.14 Show that $\varepsilon_{ijk} \, a_j b_k \, c_i = \varepsilon_{ijk} \, a_i b_j c_k$.

A.15 Show, using Cartesian tensor methods, that

1. $\nabla \cdot \nabla \times \mathbf{v} = 0$,
2. $\nabla \times \nabla Q = 0$,
3. $\nabla \cdot (Q\mathbf{v}) = \nabla Q \cdot \mathbf{v} + Q\nabla \cdot \mathbf{v}$,
4. $\nabla \times (Q\mathbf{v}) = \nabla Q \times \mathbf{v} + Q\nabla \times \mathbf{v}$ and
5. $\nabla \cdot (\mathbf{u} \times \mathbf{v}) = \nabla \times (\mathbf{u} \cdot \mathbf{v})$.

FURTHER READING

Aris, R. (1962) *Vectors, Tensors and the Basic Equations of Fluid Mechanics*, Prentice-Hall, Englewood Cliffs, NJ.

Bishop, R. L. and Goldberg, S. I. (1968) *Tensor Analysis and Manifolds*, Dover Publications, New York.

Borg, S. F. (1963) *Matrix–Tensor Methods in Continuum Mechanics*, Van Nostrand, New York.

Brillouin, L. (1946) *Les Tenseurs en Mécanique et en Élasticité*, Dover Publications, New York.

Coburn, N. (1955) *Vector and Tensor Analysis*, Macmillan, New York.

Ericksen, J. L. (1960) Tensor fields, in *Encyclopedia of Physics*, Vol. 3/1 (ed. S. Flügge), Springer, Berlin, pp. 794–859.

Hay, G. E. (1953) *Vector and Tensor Analysis*, Dover Publications, New York.

Jefferys, H. (1931) *Cartesian Tensors*, Cambridge University Press, Cambridge.

Landau, L. and Lifshitz, E. (1951) *The Classical Theory of Fields* (translated from Russian by H. Hammermesh), Addison-Wesley, Reading, MA.

Lass, H. (1950) *Vector and Tensor Analysis*, McGraw-Hill, New York.

Levi-Civita, T. (1927) *The Absolute Differential Calculus* (translated from Italian by M. Long), Blackie, London.

Lichnerowicz, A. (1958) *Eléments de Calcul Tensoriel*, Armand Colin.

McConnell, A. J. (1946) *Applications of the Absolute Differential Calculus*, Blackie, London.

Michal, A. D. (1947) *Matrix and Tensor Calculus with Applications to Mechanics, Elasticity and Aeronautics*, Wiley, New York.

Ricci, G. and Levi-Civita, T. (1901) Methodes du calcul différentiel absolu et leurs applications. *Math. Ann.*, **54**, 125–201.

Sokolnikoff, I. (1964) *Tensor Analysis*, Wiley, New York.

Spain, B. (1953) *Tensor Calculus*, Interscience, New York.

Spiegel, M. R. (1959) *Theory and Problems of Vector Analysis and an Introduction to Tensor Analysis*, Schaum, New York.

Synge, J. and Schild, A. (1949) *Tensor Calculus*, University of Toronto Press, Toronto.

Truesdell, C. A. and Toupin, R. A. (1954) The classical field theories, in *Encyclopedia of Physics*, Vol. 3, Part 1 (ed. F. Flügge), Springer, Berlin.

Wills, A. P. (1938) *Vector Analysis with an Introduction to Tensor Analysis*, Prentice-Hall, Englewood Cliffs, NJ.

Appendix B

Delta and step functions

B.1 THE DELTA FUNCTION $\delta(t)$

The delta function is defined by

$$\delta(t) = \lim_{a \to 0} \delta(t; a) \tag{B.1}$$

where a is a parameter of arbitrary positive value and the function $\delta(t; a)$ is given by

$$\delta(t; a) = \frac{1}{\pi} \int_0^{\pi} \exp(-apt) \cos (pt) \, dp$$

$$= \frac{a}{\pi(a^2 + t^2)} . \tag{B.2}$$

Thus, by combining (B.1) and (B.2), the delta function may be defined as

$$\delta(t) = \frac{1}{\pi} \lim \frac{a}{a^2 + t^2}$$

$$= \lim_{a \to 0} \frac{1}{\pi} \int_0^{\infty} \exp(-ap) \cos (pt) \, dp. \tag{B.3}$$

Following the above, it can be shown that the delta function $\delta(t)$ has the properties

$$\delta(t) = \begin{cases} \infty, & t = 0, \\ 0, & t \neq 0, \end{cases} \tag{B.4}$$

and

$$\int_{-\infty}^{\infty} \delta(t) \, dt = 1. \tag{B.5}$$

The dimension of the delta function is the reciprocal of the dimension of its argument. The delta function is considered as an even function, i.e.

$$\delta(-t) = \delta(t). \tag{B.6}$$

Further, an important property of the delta function is the so-called 'shifting property', that is

$$\int_{-\infty}^{\infty} f(t)\delta(t) \, dt = f(0), \tag{B.7}$$

i.e. the operation of integrating over $f(t)\delta(t)$ shifts the function $f(t)$ to $f(0)$. Equation (B.7) would remain valid if one changed the interval of integration from $-\infty \le t \le \infty$ to $0 \le u \le t$. That is,

$$\int_0^t f(u)\delta(u) \, du = f(0). \tag{B.8}$$

Further, through a change of variable in (B.5), it follows that

$$\delta(t/c) = c\delta(t) \tag{B.9}$$

with the dimension c/t. The function $c\delta(t)$ appearing in equation (B.9) is often referred to as an impulse of strength c.

The definition of the delta function can be extended to include the argument $(t - t')$. Thus, with reference to (B.4), one can write

$$\delta(t - t') = \begin{cases} \infty, & t = t', \\ 0, & t \ne t', \end{cases} \tag{B.10}$$

with, in view of (B.5),

$$\int_{-\infty}^{\infty} \delta(t - t') \, dt = 1. \tag{B.11}$$

The function $\delta(t - t')$ is kown as the 'shifted' delta function. In view of equation (B.6), the function $\delta(t - t')$ is treated as an even function. That is

$$\delta(t - t') = \delta(t' - t).$$

The shifting property (B.7) can be also applied to $\delta(t - t')$, i.e.

$$\int_{-\infty}^{\infty} f(t)\delta(t - t') \, dt = \int_{-\infty}^{\infty} f(t - t')\delta(t) \, dt = f(t'). \tag{B.12}$$

Further, in analogy to (B.8), one can write

$$\int_0^t f(u)\delta(t - u) \, du = \int_0^t f(t - u)\delta(u) \, du = f(t). \tag{B.13}$$

B.2 THE STEP 'HEAVISIDE' FUNCTION $H(t)$

Integrating the function $\delta(t; a)$ of (B.2) leads to a function $H(\infty; a) = 1$ such that

$$H(t; a) = \frac{1}{2} + \frac{1}{\pi} \arctan \frac{t}{a}. \tag{B.14}$$

In (B.14), as the value of the parameter a decreases, the value of the function $H(t; a)$ approaches a straight line at $H(t; a) = 1$ for $0 \leq t \leq \infty$. The resulting function at the limit as $a \to 0$ is known as the 'unit step function' and is given the notation $H(t)$. The latter is often referred to in the literature as the 'Heaviside' unit step function (the name is taken after the British physicist Heaviside (1850–1925).

Combining (B.3) and (B.14), it can be shown that

$$H(t) = \frac{1}{2} + \frac{1}{\pi} \lim_{a \to 0} \int_0^\infty \exp(-ap) \frac{\sin pt}{p} \, dp \tag{B.15}$$

from which it is apparent that

$$\delta(t) = \frac{dH(t)}{dt}, \tag{B.16}$$

i.e. the delta function $\delta(t)$ is the derivative of the unit step function $H(t)$. Thus, the latter is defined as

$$H(t) = \begin{cases} 0, & t < 0, \\ 1, & t > 0. \end{cases} \tag{B.17}$$

For $t = 0$, the unit step function $H(t)$ is undefined unless one distinguishes between $t = 0^-$ and $t = 0^+$ as the last point of negative time and the first point of positive time, respectively (Flügge, 1975).

One can also demonstrate that the Heaviside unit step function $H(t)$ is an odd function of its argument, i.e.

$$H(t) = -H(-t) \tag{B.18}$$

By its definition (B.17), the unit step function $H(t)$ can be used as a restrictive device to limit the values of a given function to its values for a particular range of the argument t.

The definition of the Heaviside function may be extended to include the argument $t - t'$ such that

$$H(t - t') = \begin{cases} 0, & t < t', \\ 1, & t > t'. \end{cases} \tag{B.19}$$

REFERENCE

Flügge, W. (1975) *Viscoelasticity*, Springer, New York.

FURTHER READING

Carslaw, H. S. and Jaeger, J. C. (1941) *Operational Methods in Applied Mathematics*, Oxford University Press, London.

Churchill, R. V. (1958) *Operational Methods*, McGraw-Hill, New York.

Goldman, S. (1949) *Transformation Calculus and Electrical Transients*, Constable, London.

Tschoegl, N. W. (1989) *The Phenomenological Theory of Linear Viscoelastic Behaviour. An Introduction*, Springer, New York.

Appendix C

Laplace transformation

C.1 INTEGRAL TRANSFORMS

In this appendix we introduce briefly the basic concepts and properties concerning the operation of integral transformation of a function $F(t)$ in t space into another function in s space. If we denote the integration operator by I, then the integral transform is expressed by

$$I[F(t)] = \int_a^b \Gamma(t, s)F(t)\ \mathrm{d}t = f(s) \tag{C.1}$$

where $\Gamma(t, s)$ denotes some prescribed function of the variable t and a parameter s. The function $f(s)$ is called the integral transform of $F(t)$. The class of function $F(t)$ and the range of the parameter s are to be specified in a manner such that the integral (C.1) exists. Hence, the transformation $I[F(t)]$ applies to all integrable functions whereby the function $f(s)$ may be interpreted as the image of the original function $F(t)$ under this transformation.

In the above example, an inverse transformation exists in the sense that, when the image function $f(s)$ is given, a function $F(t)$ exists which has this image. The inverse transformation of (C.1) is expressed as

$$F(t) = I^{-1}[f(s)]. \tag{C.2}$$

An integral transformation $I[F(t)]$ is described as linear if, for every pair of functions $F_1(t)$ and $F_2(t)$ and for each pair of constants a and b, the following relation is satisfied:

$$I[aF_1(t) + bF_2(t)] = aI[F_1(t)] + bI[F_2(t)]. \tag{C.3}$$

That is, in the case of a linear transformation, the integral transform of a linear combination of two functions is the same linear combination of the transforms of these functions.

Integral transforms have many physical applications. Linear integral transformations are particularly useful in solving problems in differential equations. In this context, with certain kernels $\Gamma(t, s)$, the transformation (C.1) when applied to prescribed linear differential forms in $F(t)$ changes those forms into algebraic

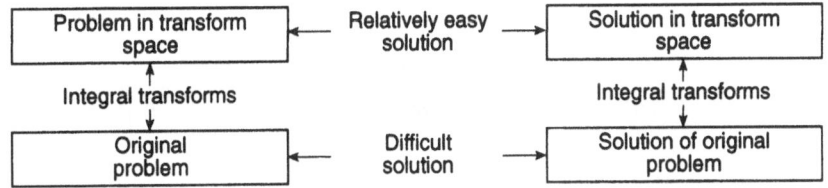

Fig. C.1 Use of integral transforms to solve differential equations.

expressions in $f(s)$ that would involve certain bounding values of the object function $F(t)$. Hence, classes of problems in ordinary differential equations transform into much simpler algebraic problems to solve. Accordingly, if an inverse transformation is possible, the solution of the original problem can be determined (Fig. C.1).

Within the realm of the theory of linear integral transformation, two special classes of integral transforms are of particular importance, i.e. those of the operational mathematics of Laplace and Fourier transformations. We deal in the present appendix with the definition and basic properties of Laplace transforms whilst in Appendix D we introduce those pertaining to Fourier transformation.

C.2 DEFINITION OF LAPLACE TRANSFORM

Laplace transformation is a form of operational mathematics that is of significant importance in the treatment of problems concerning differential equations. Laplace (1749–1827) and Cauchy (1789–1857) were two of the earlier contributors to the development of Laplace transform and the pertaining operational calculus. For a comprehensive review of the subject matter, reference is made to Churchill (1958) and Doetsch (1974), among others.

If a function $F(t)$, defined for all positive values of the variable t, is multiplied by a kernel function $\exp(-st)$ of the variable t and a parameter s and integrated with respect to t, an image function $\bar{f}(s)$ is expressed as

$$\int_0^\infty \exp(-st)F(t)\,dt = \bar{f}(s). \tag{C.4}$$

We note in (C.4) that the integration is carried out over the infinite interval from $t = 0$ to $t = \infty$.

Expression (C.4) is known as the Laplace transformation of the original function $F(t)$ into the image function $\bar{f}(s)$. This transformation is given here the notation $\mathscr{L}[F(t)]$. Hence, with reference to (C.4)

$$\mathscr{L}[F(t)] = \int_0^\infty \exp(-st)F(t)\,dt = \bar{f}(s)$$

The image function $\bar{f}(s)$ is referred to as the Laplace transform of the object function $F(t)$. Although the transform operator may be considered to be real, it generally assumes complex values.

Example C.1

Determine the Laplace transform of $F(t) = 1$ for $t > 0$.

$$\mathscr{L}[F(t)] = \mathscr{L}(1) = \int_0^\infty \exp(-st) \, dt = \left[-\frac{1}{s} \exp(-st) \right]_0^\infty.$$

Thus, for $s > 0$,

$$\mathscr{L}(1) = \frac{1}{s}.$$

Example C.2

Determine the Laplace transform of $F(t) = \exp(bt)$ for $t > 0$, where b is a constant.

$$\mathscr{L}[F(t)] = \mathscr{L}[\exp(bt)] = \int_0^\infty \exp(bt) \exp(-st) \, dt = \left[\frac{1}{b-s} \exp[-(s-b)t] \right]_0^\infty.$$

Thus, for $s > b$,

$$\mathscr{L}[\exp(bt)] = \frac{1}{s-b}.$$

Example C.3

Consider the Laplace transform of the unit step function:

$$H(t_1, t) \begin{cases} = 0 & \text{for } 0 < t < t_1, \\ = 1 & \text{for } t > t_1. \end{cases}$$

$$\mathscr{L}[H(t_1, t)] = \int_t^\infty \exp(-st) \, dt = \left[-\frac{1}{s} \exp(-st) \right]_0^\infty.$$

Thus, for $s > 0$,

$$\mathscr{L}[H(t_1, t)] = \frac{\exp(-t_1 s)}{s}.$$

Example C.4

The Laplace transforms of many other functions can be determined. For instance,

$$\mathscr{L}(t^2) = \frac{2}{s^3},$$

$$\mathscr{L}(\sin at) = \frac{a}{s^2 + a^2}$$

and

$$\mathcal{L}(\cos bt) = \frac{s}{s^2 + b^2}.$$

C.3 EXISTENCE OF LAPLACE TRANSFORMATION

Two conditions need to be satisfied in order for the Laplace transform $\bar{f}(s)$ of $F(t)$ to exist, that is $F(t)$ be sectionally continuous in every finite interval for the variable $t \geq 0$ and $F(t)$ be of exponential order as t tends to ∞. These conditions for the existence of the Laplace transform of a function are sufficient rather than necessary conditions. However, they are convenient for the majority of applications (e.g. Churchill, 1958).

A function $F(t)$ is considered to be sectionally continuous on a finite interval $0 \leq t \leq t_1$ if it is such that the interval $(0, t_1)$ can be subdivided into a finite number of intervals, in each of which $F(t)$ is continuous and has finite limits as t approaches either such limit of the subinterval from inside. The integral of every function of this class over the interval $(0, t_1)$ exists; it is the sum of the integrals of the continuous functions over the subintervals.

An example of a sectionally continuous function is the unit step function $H(\tau, t)$, where

$$H(\tau, t) \begin{cases} = 0 & \text{for } 0 < t < \tau, \\ = 1 & \text{for } t > \tau. \end{cases}$$

It is a sectionally continuous function in the interval $0 \leq t \leq t_1$ for every positive number t_1.

The second condition for the existance of Laplace transform $\bar{f}(s)$ for a function $F(t)$ is that $F(t)$ be of exponential order as the variable t tends to infinity. In other words, $F(t)$ must not grow at a greater rate than that of exponential as $t \to \infty$. This can be expressed as

$$\lim_{t \to \infty} \exp(-\alpha t)F(t) = 0 \qquad (C.5)$$

under the condition that a constant α exists.

An example of a function of exponential order is the function $\exp(2t)$. It is of the order of $\exp(\alpha t)$ as $t \to \infty$ for $\alpha \geq 2$. The unit step function $H(t_1, t)$, mentioned above, as well as the function t^n are also of the order $\exp(\alpha t)$ as $t \to \infty$ for any positive α. On the other hand, the function $\exp(t^2)$ is not of exponential order. It is further emphasized that not every function of s is a transform. The class of functions $\bar{f}(s)$ that are transforms is limited by several conditions concerning the continuity of $\bar{f}(s)$. Under these conditions, $\bar{f}(s)$ is continuous when $s > \alpha$ and $\bar{f}(s)$ vanishes as s tends to infinity (Churchill, 1958).

C.4 TRANSFORMS OF DERIVATIVES

A fundamental property of Laplace transformation is that concerned with the Laplace transformation of derivatives. This property enables us to replace the operation of differentiation of order n by a single algebraic operation on the transform. As a first step, we consider the Laplace transform of the first derivative of the original function $F(t)$. Let $F(t)$ be continuous with a sectionally continuous derivative $F'(t)$ in every finite interval $0 \leq t \leq t_1$. The function $F(t)$ is, further, assumed to be of exponential order as $t \to \infty$. Then,

$$\mathscr{L}[F'(t)] = \int_0^\infty \exp(-st)F'(t)\,dt = \left[\exp(-st)F(t)\right]_0^\infty + s\int_0^\infty \exp(-st)F(t)\,dt$$

$$= [\exp(-st)F(t)]_0^\infty + s\bar{f}(s).$$

Since $F(t)$ is of exponential order $\exp(\alpha t)$ then, for s greater than α, the first derivative on the right-hand side of the above expression becomes $-F(0)$ and accordingly

$$\mathscr{L}[F'(t)] = s\bar{f}(s) - F(0). \tag{C.6}$$

To obtain the transformation of the derivative of the second order, consider $F'(t)$ be continuous and $F''(t)$ be sectionally continuous in each finite interval. Also, let $F(t)$ and $F'(t)$ be of exponential order as t tends to ∞. Thus, for $s > \alpha$, it can be shown that

$$\mathscr{L}[F''(t)] = s^2\bar{f}(s) - sF(0) - F'(0). \tag{C.7}$$

The same procedure above can be applied to obtain the Laplace transformation of the nth derivative of $F(t)$. Let the function $F(t)$ and its first $n - 1$ derivatives be continuous. Also, consider $F^n(t)$ to be sectionally continuous in every finite interval $0 \leq t \leq t_1$ and $F(t)$, $F^1(t)$, $\cdots$, $F^{n-1}(t)$ to be of exponential order $\exp(\alpha t)$ as the variable t tends to ∞. Accordingly, it can be shown, by mathematical induction, that the Laplace transform of the nth derivative of $F(t)$, for $s > \alpha$, is given by

$$\mathscr{L}[F^n(t)] = s^n f(s) - s^{n-1}F^1(0) - \cdots - sF^{n-2}(0) - F^{n-1}(0) \tag{C.8}$$

where

$$F^k(0) = \left.\frac{d^k F(t)}{dt^k}\right|_{t=0} \tag{C.8b}$$

Example C.5

Consider $\mathscr{L}(t)$. The function $F(t) = t$ and its derivative $F'(t) = 1$ are continuous and of exponential order $\exp(\alpha t)$ for any positive α. Hence, with reference to (C.6),

$$\mathscr{L}[F'(t)] = s\mathscr{L}[F(t)] - F(0)$$

$$\mathscr{L}(1) = s\mathscr{L}(t).$$

However, $\mathcal{L}(1) = 1/s$ (Example C.1); then

$$\mathcal{L}(t) = 1/s^2.$$

Example C.6

Determine $\mathcal{L}(\sin at)$. The function $F(t) = \sin(at)$ and its derivatives are all continuous and of exponential order $\exp(\alpha t)$ for $\alpha > 0$. Thus, in view of (C.7),

$$\mathcal{L}[F''(t)] = s^2\mathcal{L}[F(t)] - sF(0) - F'(0) \quad (s > 0).$$

On substitution for $F(t) = \sin(at)$ in the above expression, it follows that

$$-a^2\mathcal{L}(\sin at) = s^2\mathcal{L}(\sin at) - s,$$

i.e.

$$\mathcal{L}(\sin at) = \frac{a}{s^2 + a^2} \quad (s > 0).$$

C.5 INITIAL VALUE THEOREM

If $F(t)$ is Laplace transformable, then the behaviour of $F(t)$ in the neighbourhood of $t = 0$ corresponds to the behaviour of $s\bar{f}(s)$ in the neighbourhood of $s = 0$.

The initial value $F(0)$ of $F(t)$ can be obtained from the transform $f(s)$ through the relation

$$\lim_{t \to 0} F(t) = \lim_{s \to 0} s\bar{f}(s). \tag{C.9}$$

This result is of particular interest as it may be generalized to obtain $F(t)$ for small values of the variable t; if $\bar{f}(s)$ can be expanded in a power series of terms involving $(1/s)^n$, $n \geq 1$, then, in view of the linearity property (C.3), a term-by-term inversion could be applied.

C.6 THE INVERSE TRANSFORM

If

$$\mathcal{L}[F(t)] = \bar{f}(s)$$

then the inverse Laplace transform is

$$F(t) = \mathcal{L}^{-1}[\bar{f}(s)]. \tag{C.10}$$

That is, $F(t)$ is the inverse Laplace transform of $\bar{f}(s)$.

In the strict sense of the concept of uniqueness of functions, the inverse Laplace transform is not unique. A theorem due to Lerch (Carslaw and Jaeger, 1941) concerning the uniqueness of the inverse transform is of interest here. It states that if two functions $F_1(t)$ and $F_2(t)$ have the same Laplace transform $\bar{f}(s)$, then

$$F_2(t) = F_1(t) + \phi(t) \tag{C.11}$$

where $\phi(t)$ is a null function. The latter is expressed by

$$\int_0^{t_1} \phi(t) \, dt = 0 \qquad \text{(C.12)}$$

for every positive t_1.

Hence, a given transform function $\bar{f}(s)$ cannot have more than one inverse transform $F(t)$ that is continuous for each positive t. On the other hand, it is possible that a function $\bar{f}(s)$ would not have a continuous inverse transform.

C.7 SHIFTING THEOREM

If the inverse transformation of $\bar{f}(s)$ is $F(t)$, then the inverse transformation of $\exp(-as)\,\bar{f}(s)$ is given by

$$\mathscr{L}^{-1}[\exp(-as)\bar{f}(s)] = F(t-a)H(t-a) \qquad \text{(C.13)}$$

where $H(t-a)$ is the unit step function and a is a constant.

C.8 BASIC PROPERTIES OF LAPLACE TRANSFORM

C.8.1 Linearity

An important property of Laplace transform and its inverse is linearity. The latter follows from the definition of the transform. Thus, recalling (C.3)

$$\mathscr{L}[aF_1(t) + bF_2(t)] = a\mathscr{L}[F_1(t)] + b\mathscr{L}[F_2(t)] = a\bar{f}_1(s) + b\bar{f}_2(s) \qquad \text{(C.14)}$$

where a and b are constants. That is, the Laplace transform of a linear combination of two functions is the linear combination of the transforms of these functions. It can also be demonstrated that the linearity property can be extended to a linear combination of more than two functions. Such functions again must be sectionally continuous and each be of exponential order for their Laplace transforms to exist.

The linearity property applies also to the inverse of the transform. Thus, the inverse of (C.14) can be written as

$$\mathscr{L}^{-1}[a\bar{f}_1(s) + b\bar{f}_2(s)] = aF_1(t) + bF_2(t) = a\mathscr{L}^{-1}[\bar{f}_1(s)] + b\mathscr{L}^{-1}[\bar{f}_2(s)] \qquad \text{(C.15)}$$

C.8.2 Substitution (or shift of origin)

Let the object function $F(t)$, of exponential order $\exp(\alpha t)$, be such that its Laplace integral converges when $s > \alpha$. Recalling the Laplace transform expression (C.4), one may replace the argument of the transform $\bar{f}(s)$ by $s - a$ where a is a constant; then

$$\bar{f}(s-a) = \int_0^\infty \exp[-(s-a)t]F(t) \, dt = \int_0^\infty \exp(-st)\exp(at)F(t) \, dt.$$

Then, for $s - a > \alpha$, one has

$$\bar{f}(s - a) = \mathscr{L}[\exp(at)F(t)]. \tag{C.16}$$

That is, the substitution of $s - a$, where a is a constant, for the parameter s of $\bar{f}(s)$ would translate into multiplying the original function $F(t)$ by the function $\exp(at)$ as expressed in (C.16).

Example C.7

Consider the Laplace transform

$$\mathscr{L}(t^m) = \frac{m!}{s^{m+1}} \qquad (m = 1, 2, \cdots; s > 0).$$

Replacing the argument s by $s - a$, leads to

$$\frac{m!}{(s - a)^{m+1}} = \mathscr{L}[t^m \exp(at) \cos (bt)] \qquad (s > 0).$$

Example C.8

Consider the following Laplace transform:

$$\mathscr{L}(\cos bt) = \frac{s}{s^2 + b^2} \qquad (s > 0).$$

Replacing the parameter s by $s + a$ gives

$$\frac{s + a}{(s + a)^2 + b^2} = \mathscr{L}[\exp(-at) \cos (bt)] \qquad (s > -a).$$

C.8.3 Change of scale

One may also replace the argument of the object function $F(t)$ from t to at where a is a real positive constant; then

$$\mathscr{L}[F'(at)] = \frac{1}{a}\bar{f}\left(\frac{s}{a}\right). \tag{C.17}$$

C.8.4 Translation

Consider translation of the argument of the object function $F(t)$ from t to $t - \tau$ where both t and τ are variables and $F(t) = 0$ for $t < 0$; then

$$\mathscr{L}[F(t - \tau)] = \exp(-s\tau)\bar{f}(s). \tag{C.18}$$

C.8.5 Differentiation of transforms

It can be shown that

$$\frac{d^n}{dt^n}\bar{f}(s) = \mathcal{L}[(-1)^n t^n F(t)].\tag{C.19}$$

C.8.6 Integration of transforms

The following property can be proved:

$$\int_0^\infty \bar{f}(s)\, ds = \int_0^\infty \frac{F(t)}{t}\, dt.\tag{C.20}$$

C.8.7 Transform of integral

Consider the integral

$$\int_0^t F(\tau)\, d\tau.$$

Then

$$\mathcal{L}\left[\int_0^t F(\tau)\, d\tau\right] = \frac{1}{s}\bar{f}(s)$$

and

$$\mathcal{L}\left[\underbrace{\int_0^t d\tau \int_0^t d\tau \cdots \int_0^t F(\tau)\, d\tau}_{n}\right] = \frac{1}{s^n}\bar{f}(s).\tag{C.21}$$

C.8.8 Transform of convolution integral

Consider two functions $F(t)$ and $G(t)$, both sectionally continuous and of exponential order. The convolution of $F(t)$ and $G(t)$ is defined by

$$\int_0^t F(t)G(t - \tau)\, dt$$

and is conventionally denoted by $F(t) * G(t)$. The following properties of the convolution of functions $F(t)$, $G(t)$ and $J(t)$ defined over $-\infty < t < \infty$ can be verified:

- commutativity,

$$F(t) * G(t) = G(t) * F(t);$$

- associativity,

$$F(t) * [G(t) * J(t)] = [F(t) * G(t)] * J(t) = F(t) * G(t) * J(t);$$

- distributivity,

$$F(t) * [G(t) + J(t)] = F(t) * G(t) + F(t) * J(t);$$

- Titchmarsh theorem,

$$F(t) * G(t) = 0$$

implies that $F(t) = 0$ or $G(t) = 0$.

The Laplace transform of a convolution integral is expressed as

$$\mathscr{L}[F(t) * G(t)] = \bar{f}(s)\bar{g}(s) \tag{C.22}$$

with the property

$$\bar{f}(s)\bar{g}(s) = \bar{g}(s)\bar{f}(s).$$

That is, the Laplace transform of the convolution integral is also commutative.

Some operations for Laplace transformations are given in Table C.1. Meantime, examples of Laplace transforms are shown in Table C.2.

Table C.1 Operations for Laplace transformation

$F(t)$	$\bar{f}(s)$
$F(t)$	$\int_0^\infty \exp(-st)F(t)\, dt$
$aF(t) + bJ(t)$	$a\bar{f}(s) + b\bar{J}(s)$
$F'(t)$	$s\bar{f}(s) - F(+0)$
$\int_0^t F(\tau)\, d\tau$	$\dfrac{1}{s}\bar{f}(s)$
$\int_0^t F(t - \tau)G(\tau)d\tau = f * G$	$\bar{f}(s)\bar{g}(s)$
$tF(t)$	$-\bar{f}'(s)$
$t^n F(t)$	$(-1)^n \bar{f}^{(n)}(s)$
$\exp(at)F(t)$	$\bar{f}(s - a)$
$F(t - a)$, where $F(t) = 0$ for $t < 0$	$\exp(-as)\bar{f}(s)$
$\dfrac{1}{a} F\left(\dfrac{t}{a}\right)$ $(a > 0)$	$\bar{f}(as)$
$F(t)$, where $F(t + a) = F(t)$	$\dfrac{\int_0^a \exp(-st)F(t)\, dt}{1 - \exp(-as)}$
$F(t)$, where $F(t + a) = -F(t)$	$\dfrac{\int_0^a \exp(-st)F(t)\, dt}{1 + \exp(-as)}$

Table C.2 Examples of Laplace transforms

$F(t)$	$\bar{f}(s)$
1	$\dfrac{1}{s}$
t	$\dfrac{1}{s^2}$
$\exp(at)$	$\dfrac{1}{s-a}$
$t^n \ (n = 1, 2, \ldots)$	$\dfrac{n!}{s^{n+1}}$
$\dfrac{1}{t^{1/2}}$	$\left(\dfrac{\pi}{s}\right)^{1/2}$
$\exp(at) - \exp(bt) \quad (a > b)$	$\dfrac{a-b}{(s-a)(s-b)}$
$t\exp(at)$	$\dfrac{1}{(s-a)^2}$
$\dfrac{1}{(n-1)!}\,t^{n-1}\exp(at)$	$\dfrac{1}{(s-a)^n} \quad (n = 1, 2, \ldots)$
$\dfrac{1}{a-b}[\exp(at) - \exp(bt)]$	$\dfrac{1}{(s-a)(s-b)}$
$\dfrac{1}{a-b}[a\exp(at) - b\exp(bt)]$	$\dfrac{s}{(s-a)(s-b)}$
$\sin at$	$\dfrac{a}{s^2 + a^2}$
$\cos at$	$\dfrac{s}{s^2 + a^2}$
$\dfrac{1}{a}\sin at$	$\dfrac{1}{s^2 + a^2}$
$t^n \exp(at) \quad (n = 1, 2, \ldots)$	$\dfrac{n!}{(s-a)^{n+1}}$
$\exp(-at)\sin bt$	$\dfrac{b}{(s+a)^2 + b^2}$
$\dfrac{1}{a^2}(1 - \cos at)$	$\dfrac{1}{s(s^2 + a^2)}$

Table C.2 *(continued)*

$F(t)$	$\bar{f}(s)$
$\dfrac{1}{a^3}(at - \sin at)$	$\dfrac{1}{s^2(s^2 + a^2)}$
$\dfrac{t}{2a}\sin at$	$\dfrac{s}{(s^2 + a^2)^2}$
$t \cos at$	$\dfrac{s^2 - a^2}{(s^2 + a^2)^2}$
$\dfrac{1}{b}\exp(at)\sin bt$	$\dfrac{1}{(s - a)^2 + b^2}$
$\exp(at)\cos bt$	$\dfrac{s - a}{(s - a)^2 + b^2}$
$\dfrac{1}{a}\sin at - \dfrac{1}{b}\sin bt$	$\dfrac{b^2 - a^2}{(s^2 + a^2)(s^2 + b^2)}$
$\cos at - \cos bt$	$\dfrac{(b^2 - a^2)s}{(s^2 + a^2)(s^2 + b^2)}$
$\dfrac{\cos at - \cos bt}{b^2 - a^2}$	$\dfrac{s}{(s^2 + a^2)(s^2 + b^2)} \quad (a^2 \neq b^2)$
$\dfrac{1}{2a^3}(\sin at - at \cos at)$	$\dfrac{1}{(s^2 + a^2)^2}$
$\dfrac{1}{2a}(\sin at + at \cos at)$	$\dfrac{s^2}{(s^2 + a^2)^2}$
$\sinh at$	$\dfrac{a}{s^2 - a^2}$
$\cosh at$	$\dfrac{s}{s^2 - a^2}$
$(1 + a^2t^2)\sin at - at \cos at$	$\dfrac{8a^3s^2}{(s^2 + a^2)^3}$
$\dfrac{1}{2a^3}\sin at \sinh at$	$\dfrac{s}{s^4 + 4a^4}$
$\dfrac{1}{2a^2}(\cosh at - \cos at)$	$\dfrac{s}{s^4 - a^4}$
$\sin at \cosh at - \cos at \sinh at$	$\dfrac{4a^3}{s^4 + 4a^4}$

Table C.2 *(continued)*

$F(t)$	$\bar{f}(s)$
$\dfrac{1}{(\pi t)^{1/2}} \cos 2(at)^{1/2}$	$\dfrac{1}{s^{1/2}} \exp(-a/s)$
$\dfrac{1}{(\pi t)^{1/2}} \cosh 2(at)^{1/2}$	$\dfrac{1}{s^{1/2}} \exp(a/s)$
$\dfrac{1}{t} [\exp(bt) - \exp(at)]$	$\log \dfrac{s-a}{s-b}$
$\dfrac{1}{t} \sin at$	$\arctan \dfrac{a}{s}$
$-\dfrac{(b-c)\exp(at) + (c-a)\exp(bt) + (a-b)\exp(ct)}{(a-b)(b-c)(c-a)}$ where a, b and c are distinct constants	$\dfrac{1}{(s-a)(s-b)(s-c)}$
$\dfrac{t^n}{n!} \quad (n = 0, 1, 2, \ldots)$	$\dfrac{1}{s^{n+1}}$
$\dfrac{1}{4a^3} (\sin at \cosh at - \cos at \sinh at)$	$\dfrac{1}{s^4 + 4a^4}$
$\cos at \cosh at$	$\dfrac{s^3}{s^4 + 4a^4}$
$\dfrac{1}{2a^3} (\sinh at - \sin at)$	$\dfrac{1}{s^4 - a^4}$
$\dfrac{1}{2a} (\sinh at + \sin at)$	$\dfrac{s^2}{s^4 - a^4}$
$\dfrac{1}{2} (\cosh at + \cos at)$	$\dfrac{s^3}{s^4 - a^4}$

PROBLEMS

C.1 Show the validity of the following transformations, where a, b, and c are constants:

1. $\mathscr{L}(c + bt) = \dfrac{as + b}{s^2}$;

2. $\mathscr{L}(\sinh ct) = \dfrac{c}{s^2 - c^2}$;

3. $\mathscr{L}[\exp(at)] = \dfrac{1}{s - a}$;

4. $\mathscr{L}(t^n) \qquad (n = 1, 2, \cdots) = \dfrac{n!}{s^{n+1}}$;

5. $\mathscr{L}([\exp(at) - \exp(bt)] \qquad (a > b) = \dfrac{a - b}{(s - a)(s - b)}$;

6. $\mathscr{L}\left(\dfrac{1}{a} \sin at - \dfrac{1}{b} \sin bt\right) = \dfrac{b^2 - a^2}{(s^2 + a^2)(s^2 + b^2)}$;

7. $\mathscr{L}(\cos at - \cos bt) = \dfrac{(b^2 - b^2)s}{(s^2 + a^2)(s^2 + b^2)}$.

C.2 Find the Laplace transform of each of the following functions:

1. $\sin t + 2 \cos t$;
2. $\cos^2 t$;
3. $\sin t \cos t$.

C.3 Find

$$\mathscr{L}^{-1}\left(\frac{s + 1}{s^2 + 2s}\right)$$

and

$$\mathscr{L}^{-1}\left[\frac{a^2}{s(s + a)^2}\right].$$

C.4 Obtain the following inverse transforms where a and b are constants:

$$\mathscr{L}^{-1}\left[\frac{a^2}{s(s^2 + a^2)}\right];$$

$$\mathscr{L}^{-1}\left[\frac{b}{s(s + b)}\right];$$

$$\mathscr{L}^{-1}\left[\frac{b^3}{s(s + b)^3}\right].$$

REFERENCES

Carslaw, H. S. and Jaeger, J. C. (1941) *Operational Methods in Applied Mathematics*, Oxford University Press, London.

Churchill, R. V. (1958) *Operational Mathematics*, McGraw-Hill, New York.

Doetsch, G. (1974) *Introduction to the Theory and Application of the Laplace Transformation*, Springer, Berlin.

FURTHER READING

Abramowitz, M. and Stegun, I. A. (eds) (1965) *Handbook of Mathematical Functions*, Dover Publications, New York.

Cost, T. L. (1964) Approximate Laplace transform inversion in viscoelastic stress analysis. *AIAA* Journal, **2**, 2157–66.

Doetsch, G. (1950, 1955, 1956) *Handbuch der Laplace-Transformation*, Vols 1, 2 and 3, Birkhauser, Basel.

Erdeli, A., Magnus, W., Oberhettinger, F. and Tricomi, F. (1954) *Tables of Integral Transforms*, Vols 1 and 2, McGraw-Hill, New York.

Kreyszig, E. (1972) *Advanced Engineering Mathematics*, 3rd edn, Wiley, New York, pp. 147–87.

Schapery, R. A. (1962) Approximate methods of transform inversion for viscoelastic stress analysis, in Proc. 4th US Natl Congr. of Applied Mechanics, pp. 1075–85.

Tranter, C. J. (1956) *Integral Transforms in Mathematical Physics*, 2nd edn., Methuen, London.

Widder, D. V. (1941) *The Laplace Transform*, Princeton University Press, Princeton, NJ.

Appendix D

Fourier transformation

D.1 DEFINITION OF FOURIER TRANSFORM

Similar to Laplace transformations (Appendix C), Fourier transforms are linear integral transformations with operational properties under which differential functions are converted into algebraic forms involving boundary values.

The Fourier transformation of a sectionally continuous function $F(t)$, defined for all positive values of the variable t, is denoted here by $S[F(t)]$ and expressed by

$$S[F(t)] = \int_{-\infty}^{\infty} \exp(-i\omega t)F(t) \, dt = f(\omega). \tag{D.1}$$

With reference to (D.1), the Fourier transform is based on the kernel function $\exp(-i\omega t)$ and, hence, on this kernel's real and imaginary parts, i.e. $\cos \omega t$ and $\sin \omega t$ where ω is a constant parameter. Since such a kernel's functions are often used to describe the propagation of waves in different media, Fourier transforms are used extensively in such studies for extraction of information from different phases of waves, particularly when phase information is involved.

In equation (D.1), it is understood that $f(\omega)$ is the Fourier transform, or the image, of its object function $F(t)$ and that $F(t)$ is the inverse transform of $f(\omega)$. If t represents the time parameter, for instance, then equation (D.1) and its inverse imply that $F(t)$ can be analysed into an integral sum of harmonic oscillations over a continuous range of frequencies. If $F(t)$ exists only for the range $t > 0$, then a special simplification may be possible in terms of the finite Fourier sine transform $f_s(\omega)$ and the finite Fourier cosine transform $f_c(\omega)$. These forms possess a complete symmetry and could be used in a situation where the variable t represents the time and $F(t)$ implies some stimulus applied to a particular system from zero time onwards. We shall introduce the Fourier sine and cosine transforms later in this appendix.

D.2 RELATIONS BETWEEN FOURIER PAIRS

D.2.1 Linearity

If $f_1(\omega)$ and $f_2(\omega)$ are the Fourier transforms of $F_1(t)$ and $F_2(t)$, respectively, and a and b are two arbitrary constants, then

$$S[aF_1(t) + bF_2(t)] = aS[F_1(t)] + bS[F_2(t)] = af_1(\omega) + bf_2(\omega). \tag{D.2}$$

D.2.2 Scaling

If a is a real constant, then

$$S[F(at)] = \frac{1}{|a|} f\left(\frac{\omega}{a}\right). \tag{D.3}$$

D.2.3 Symmetry

If $F(t)$ is an even function, then its Fourier transform $f(\omega)$ is also even. Also, if the object function $F(t)$ is an odd function, then $f(\omega)$ is also odd.

D.2.4 Shifting

If the function $F(t)$ is shifted by a constant a, then its Fourier spectrum remains the same but its phase angle is adjusted by a linear term $a\omega$:

$$S[F(t \pm a)] = f(\omega)\, \exp(\pm i\omega t) = A(\omega)\, \exp\{i[\phi(\omega) \pm a\omega]\}. \tag{D.4}$$

With ω_0 as a real constant, the Fourier integral of $\exp(i\omega t)$ is obtained by shifting $f(\omega)$ by ω_0. That is,

$$S[\exp(i\omega_0 t)F(t)] = f(\omega - \omega_0). \tag{D.5}$$

D.2.5 Differentiation

If $S[F(t)] = f(\omega)$, then

$$S\left[\frac{d^n F(t)}{dt^n}\right] = (i\omega)^n f(\omega) \tag{D.6}$$

and

$$S[(-it)^n F(t)] = \frac{d^n f(\omega)}{d\omega^n}. \tag{D.7}$$

D.3 FINITE FOURIER SINE TRANSFORMS

Consider a function $F(x)$ that is sectionally continuous and defined over the interval between $x = 0$ and $x = \pi$. The Fourier sine transformation of $F(x)$ on that interval is denoted here by $S_n[F(x)]$ and is expressed as

$$S_n[F(x)] = \int_0^\pi F(x) \sin nx \, dx = f_s(n) \qquad (n = 1, 2, \cdots) \qquad (D.8)$$

where $f_s(n)$ is the finite sine transform. This transformation sets up a correspondence between functions $F(x)$ defined within the interval $0 < x < \pi$ and se-
quences of numbers $f_s(n)$ $(n = 1, 2, \cdots)$.

For example, the function $F(x) = 1$ has the transform

$$f_s(n) = \int_0^\pi \sin(nx) \, dx = \frac{1 - (-1)^n}{n} \qquad (n = 1, 2, \cdots).$$

Also, the function $F(x)$ $(0 < x < \pi)$ has the transform

$$f_s(n) = \int_0^\pi x \sin(nx) \, dx = \pi \frac{(-1)^{n+1}}{n} \qquad (n = 1, 2, \cdots).$$

In order to obtain an inversion formula for the transformation (D.8), consider
both the object function $F(x)$ and its first derivative $F'(x)$ to be sectionally continuous
functions. Let, also, $F(x)$ be defined at each point x_0 of discontinuity, $0 < x_0 < \pi$, by
its mean value:

$$F(x_0) = \tfrac{1}{2}[F(x_0 + 0) + F(x_0 - 0)] \qquad (0 < x_0 < \pi). \qquad (D.9)$$

It follows, according to the classical theory of Fourier series, that the Fourier sine
series for $F(x)$ converges to the function

$$F(x) = \frac{2}{\pi} \sum_{n=1}^\infty \sin(nx) \int_0^\pi F(\xi) \sin(n\xi) \, d\xi \qquad (0 < x < \pi). \qquad (D.10)$$

Thus, with reference to (D.8),

$$F(x) = \frac{2}{\pi} \sum_{n=1}^\infty f_s(n) \sin(nx) \qquad (0 < x < \pi) \qquad (D.11)$$

which is the inversion formula for the Fourier sine transformation (D.8).

In the class of sectionally continuous functions with sectionally continuous
derivatives of the first order, there is only one function with a given transform as
demonstrated by (D.11). In other words, the inverse transformation is unique. It is
apparent that both the transformation $S_n[F(x)]$ and its inverse are linear transformations.

Recalling (D.8), the Fourier sine transform of a function on an interval $0 < x < c$ is
expressed in terms of the transform on the standard interval $(0 < x < \pi)$ by the
substitution $x' = \pi x/c$ in (D.8), i.e.

$$\int_0^c F(x) \sin\left(\frac{n\pi x}{c}\right) dx = \frac{c}{\pi} \int_0^\pi F\left(\frac{cx'}{\pi}\right) \sin(nx') \, dx' = \frac{c}{\pi} S_n\left[F\left(\frac{cx}{\pi}\right)\right]. \qquad (D.12)$$

As an example, the function $F(x)$ $(0 < x < c)$ has the sine transform

$$S_n(x)\big|_c = \frac{c}{\pi} S_n\left(\frac{cx}{\pi}\right) = \frac{c^2}{\pi} \frac{(-1)^{n+1}}{n} \qquad (n = 1, 2, \cdots).$$

An important property of $S_n[F(x)]$ is as follows. If $F(x)$ and $F'(x)$ are continuous and $F''(x)$ is sectionally continuous, then

$$\int_0^\pi F''(x) \sin nx \, dx = [F'(x) \sin nx \,]_0^\pi - n \int_0^\pi F'(x) \cos nx \, dx$$

$$= [-n \cos nx F(x)]_0^\pi - n^2 \int_0^\pi F(x) \sin nx \, dx$$

which can be written as

$$S_n[F''(x)] = -n^2 S_n[F(x)] + n[F(0) - (-1)^n F(\pi)]. \tag{D.13}$$

That is, the finite Fourier sine transformation resolves the differential form $F''(x)$ into a linear algebraic form in the transform $f_s(n)$ and the boundary values $F(0)$ and $F(\pi)$ as expressed by (D.13). Formulae for the transforms of other derivatives $F^{(2n)}(x)$ of even order may be determined in the same manner. This property is employed in the construction of tables of Fourier transforms

Example D.1

Consider $F(x) = x^2$; then, $F''(x) = 2$ and

$$S_n(2) = -n^2 S_n(x^2) - n(-1)^n \pi^2.$$

Also,

$$S_n(2) = 2S_n(1) = 2[1 - (-1)^n]/n.$$

Thus,

$$S_n(x^2) = \frac{\pi^2}{n} (-1)^{n-1} - \frac{2}{n^3} [1 - (-1)^n].$$

D.4 FINITE FOURIER COSINE TRANSFORMS

The finite Fourier cosine transformation of a function $F(x)$, $0 < x < \pi$, is denoted here by $C_n[F(x)]$ and expressed by

$$C_n[F(x)] = \int_0^\pi F(x) \cos nx \, dx = f_c(n) \qquad (n = 0, 1, 2, \cdots) \tag{D.14}$$

where $f_c(n)$ is the resulting finite Fourier cosine transform. Consider both the object function $F(x)$ and its first derivative to be sectionally continuous; also, $F(x)$ is defined, at each point of discontinuity within the interval, by its mean value (D.9). Thus, the inverse transformation $F(x)$ is given by the Fourier cosine series

$$F(x) = \frac{1}{\pi} f_c(0) + \frac{2}{\pi} \sum_{n=1}^{\infty} f_c(n) \cos nx \qquad (0 < x < \pi). \tag{D.15}$$

As in the case of the Fourier sine transform, the cosine transform of each sectionally continuous function $F(x)$ exists and it is unique.

In a corresponding manner to the sine transform, the cosine transformation resolves the differential form $F''(x)$ into an algebraic form in $f_c(n)$ and the boundary values $F'(0)$ and $F'(\pi)$, i.e.

$$C_n[F''(x)] = -n^2 f_c(n) - F'(0) - (-1)^n F'(\pi) \qquad (n = 0, 1, 2, \cdots,). \tag{D.16}$$

This is again under the condition that $F(x)$ and $F'(x)$ are continuous and $F''(x)$ is sectionally continuous.

D.5 JOINT PROPERTIES OF SINE AND COSINE TRANSFORMS

Under the conditions that $F(x)$ is continuous and $F'(x)$ is sectionally continuous, it can be shown that

$$S_n[F'(x)] = -nC_n[F(x)] \qquad (n = 1, 2, \cdots) \tag{D.17}$$

and

$$C_n[F'(x)] = nS_n[F(x)] - F(0) + (-1)^n F(\pi) \qquad (n = 0, 1, 2, \cdots). \tag{D.18}$$

Alternative formulations of the joint properties of S_n and C_n are

$$S_n[G(x)] = -nC_n\left[\int_0^x G(x')\,dx'\right] \qquad (n = 1, 2, \cdots) \tag{D.19}$$

and

$$C_n\left[G(x) - \frac{1}{\pi} g_c(0)\right] = nS_n\left[\int_0^x G(x')\,dx' - \frac{x}{\pi} g_c(0)\right] \qquad (n = 0, 1, \cdots) \tag{D.20}$$

where $G(x)$ is any sectionally continuous function and $g(n)$ is its Fourier transform.

D.6 FOURIER SINE AND COSINE TRANSFORMS OVER UNBOUNDED INTERVALS

Consider $F(x)$ to be a function defined on a specified unbounded interval, sectionally continuous on each finite subinterval, and that the integral of $F(x)$ over the unbounded interval exists. If k denotes a real parameter, the Fourier sine transformation of $F(x)$ is expressed as

$$S_k[F(x)] = \int_0^\infty F(x) \sin(kx)\,dx = f_s(k) \qquad (x \geq 0, k \geq 0). \tag{D.21}$$

At the same time, the Fourier cosine transformation is defined by

$$C_k[F(x)] = \int_0^\infty F(x) \cos kx \, dx = f_c(k) \qquad (x \geq 0, \, k \geq 0). \qquad (D.22)$$

When $F'(x)$ is also sectionally continuous on each finite subinterval $0 \leq x \leq x_1$, then $F(x)$ may be represented by either the Fourier sine or cosine integral formula. That is,

$$F(x) = \frac{2}{\pi} \int_0^\infty f_s(k) \sin kx \, dk = \frac{2}{\pi} S_x[f_s(k)] \qquad (D.23)$$

and

$$F(x) = \frac{2}{\pi} \int_0^\infty f_c(k) \cos kx \, dk = \frac{2}{\pi} C_x[f_c(k)] \qquad (D.24)$$

i.e. the inverse transforms are given by the transforms themselves.

When $F(x)$ is continuous and $F'(x)$ is sectionally continuous and $F(\infty) = 0$, it can be shown that the Fourier sine and cosine transforms are interconnected by the following two relations:

$$S_k[F'(x)] = -kC_k[F(x)]; \qquad (D.25)$$

$$C_k[F'(x)] = kS_k[F(x)] - F(0). \qquad (D.26)$$

Further, when the function $F(x)$ is replaced by its first derivative $F'(x)$, the following relations can be written:

$$S_k[F''(x)] = -k^2 f_s(k) + kF(0) \qquad (D.27)$$

and

$$C_k[F''(x)] = -k^2 f_c(k) - F'(0). \qquad (D.28)$$

In the derivation of (D.27) and (D.28), it is assumed that $F(x)$ and $F'(x)$ are both continuous and integrable, that $F''(x)$ is sectionally continuous and that $F(\infty) = F'(\infty) = 0$. The same procedure can be used for obtaining the transform of $F^{(2n)}(x)$.

PROBLEMS

D.1 Show that

$$S_n^{-1}\left[\frac{1 - (-1)^n}{n^3}\right] = \frac{x}{2}(\pi - x)$$

and

$$S_n^{-1}\left[\frac{1}{n^3}(-1)^{n+1}\right] = \frac{x(\pi^2 - x^2)}{6\pi}.$$

D.2 If $F(x)$ and $F'(x)$ are continuous except that $F'(x)$ has a jump b at $x = c$, where $0 < c < \pi$, and if $F''(x)$ is sectionally continuous, show that

$$S_n[F''(x)] = -n^2 f_s(n) + n[F(0) - (-1)^n F(\pi)] - b \sin nc.$$

D.3 Where a is a constant, prove the following:

1. $f_c(n + a) = C_n[F(x) \cos ax] - S_n[F(x) \sin ax]$;
2. $f_s(n + a) = S_n[F(x) \cos ax] + C_n[F(x) \sin ax]$;
3. $2C_n[F(x) \sin ax] = f_s(n + a) - f_s(n - a)$;
4. $2S_n[F(x) \cos ax] = f_s(n - a) + f_s(n + a)$;

5. $S_r[\exp(ax)] = \dfrac{r}{r^2 + a^2}$.

D.4 Prove the following relations where a is a constant:

1. $2S_r[F(x) \cos ax] = f_s(r + a) + f_s(r - a)$;

2. $C_r[\exp(-ax)] = \dfrac{a}{r^2 + a^2}$;

3. $C_r[\exp(-ax^2)] = \dfrac{1}{2}\left(\dfrac{\pi}{a}\right)^{1/2} \exp\left(-\dfrac{r^2}{4a}\right)$;

4. $C_r\left(\dfrac{a}{x^2 + a^2}\right) = \dfrac{\pi}{2} \exp(-ar)$.

FURTHER READING

Bochner, S. and Chandrasekhoran, L. (1947) *Fourier Transforms*, Princeton University Press, Princeton, NJ.

Churchill, R. V. (1941) *Fourier Series and Boundary Value Problems*, McGraw-Hill, New York.

Churchill, R. V. (1958) *Operational Mathematics*, 2nd edn, McGraw-Hill, New York.

Donoghue, W. F. (1969) *Distributions and Fourier Transforms*, Academic Press, New York.

Paley, R. and Wiener, N. (1934) *Fourier Transforms in the Complex Domain*, American Mathematical Society, Providence, RI.

Sneddon, I. N. (1951) *Fourier Transforms*, McGraw-Hill, New York.

Titchmarch, E. C. (1937) *Introduction to the Theory of Fourier Integrals*, Clarendon, Oxford.

Tranter, C. J. (1956) *Integral Transforms in Mathematical Physics*, 2nd edn, Methuen, London.

Wiener, N. (1933) *The Fourier Integral*, Cambridge University Press, London.

Author index

Subject index